Lecture Not

MW00814781

The Editorial Policy for Proceedings

The series Lecture Notes in Physics reports new developments in physical research and teaching – quickly, informally, and at a high level. The proceedings to be considered for publication in this series should be limited to only a few areas of research, and these should be closely related to each other. The contributions should be of a high standard and should avoid lengthy redraftings of papers already published or about to be published elsewhere. As a whole, the proceedings should aim for a balanced presentation of the theme of the conference including a description of the techniques used and enough motivation for a broad readership. It should not be assumed that the published proceedings must reflect the conference in its entirety. (A listing or abstracts of papers presented at the meeting but not included in the proceedings could be added as an appendix.)
When applying for publication in the series Lecture Notes in Physics the volume's editor(s) should submit sufficient material to enable the series editors and their referees to make a fairly accurate evaluation (e.g. a complete list of speakers and titles of papers to be presented and abstracts). If, based on this information, the proceedings are (tentatively) accepted, the volume's editor(s), whose name(s) will appear on the title pages, should select the papers suitable for publication and have them refereed (as for a journal) when appropriate. As a rule discussions will not be accepted. The series editors and Springer-Verlag will normally not interfere with the detailed editing except in fairly obvious cases or on technical matters.
Final acceptance is expressed by the series editor in charge, in consultation with Springer-Verlag only after receiving the complete manuscript. It might help to send a copy of the authors' manuscripts in advance to the editor in charge to discuss possible revisions with him. As a general rule, the series editor will confirm his tentative acceptance if the final manuscript corresponds to the original concept discussed, if the quality of the contribution meets the requirements of the series, and if the final size of the manuscript does not greatly exceed the number of pages originally agreed upon.
The manuscript should be forwarded to Springer-Verlag shortly after the meeting. In cases of extreme delay (more than six months after the conference) the series editors will check once more the timeliness of the papers. Therefore, the volume's editor(s) should establish strict deadlines, or collect the articles during the conference and have them revised on the spot. If a delay is unavoidable, one should encourage the authors to update their contributions if appropriate. The editors of proceedings are strongly advised to inform contributors about these points at an early stage.
The final manuscript should contain a table of contents and an informative introduction accessible also to readers not particularly familiar with the topic of the conference. The contributions should be in English. The volume's editor(s) should check the contributions for the correct use of language. At Springer-Verlag only the prefaces will be checked by a copy-editor for language and style. Grave linguistic or technical shortcomings may lead to the rejection of contributions by the series editors.
A conference report should not exceed a total of 500 pages. Keeping the size within this bound should be achieved by a stricter selection of articles and not by imposing an upper limit to the length of the individual papers.
Editors receive jointly 30 complimentary copies of their book. They are entitled to purchase further copies of their book at a reduced rate. As a rule no reprints of individual contributions can be supplied. No royalty is paid on Lecture Notes in Physics volumes. Commitment to publish is made by letter of interest rather than by signing a formal contract. Springer-Verlag secures the copyright for each volume.

The Production Process

The books are hardbound, and quality paper appropriate to the needs of the authors is used. Publication time is about ten weeks. More than twenty years of experience guarantee authors the best possible service. To reach the goal of rapid publication at a low price the technique of photographic reproduction from a camera-ready manuscript was chosen. This process shifts the main responsibility for the technical quality considerably from the publisher to the authors. We therefore urge all authors and editors of proceedings to observe very carefully the essentials for the preparation of camera-ready manuscripts, which we will supply on request. This applies especially to the quality of figures and halftones submitted for publication. In addition, it might be useful to look at some of the volumes already published.
As a special service, we offer free of charge LATEX and TEX macro packages to format the text according to Springer-Verlag's quality requirements. We strongly recommend that you make use of this offer, since the result will be a book of considerably improved technical quality.
To avoid mistakes and time-consuming correspondence during the production period the conference editors should request special instructions from the publisher well before the beginning of the conference. Manuscripts not meeting the technical standard of the series will have to be returned for improvement.

For further information please contact Springer-Verlag, Physics Editorial Department V, Tiergartenstrasse 17, W-6900 Heidelberg, FRG

Jean-Daniel Fournier Pierre-Louis Sulem (Eds.)

Large Scale Structures in Nonlinear Physics

Proceedings of a Workshop
Held in Villefranche-sur-Mer, France
13-18 January 1991

Springer-Verlag Berlin Heidelberg GmbH

Editors

Jean-Daniel Fournier
Pierre-Louis Sulem
CNRS, URA 1362
Observatoire de la Côte d'Azur
B. P. 139, F-06003 Nice Cédex, France

ISBN 978-3-662-13836-6 ISBN 978-3-540-46469-3 (eBook)
DOI 10.1007/978-3-540-46469-3

Originally published by Springer-Verlag Berlin Heidelberg New York in 1991
Softcover reprint of the hardcover 1st edition 1991

Typesetting: Camera ready by author

58/3140-543210 - Printed on acid-free paper

Preface

This volume contains the proceedings of a workshop initiated by a suggestion of Herbert M. Fried to bring together scientists whose research focuses on large-scale structures in various fields of non-linear physics. The meeting, organized by Jean-Daniel Fournier, Herbert M. Fried and Pierre-Louis Sulem, was held at the "Citadelle" in Villefranche sur Mer (France), 13 – 18 January 1991 and was attended by 45 participants.

Coherent states, convective and turbulent patterns, inverse cascades, interfaces and cooperative phenomena in fluids and plasmas were discussed, together with the implementation of concepts of statistical mechanics to particle physics and nuclear matter. Special attention was devoted to phenomena such as mixing, fast dynamo and predictability, which display macroscopic features, even though generated by small-scale dynamical processes. In this context, homoclinic structure, the KAM theorem, Lyapunov stability and singularities were addressed. A lecture was delivered on a new perturbative technique for non-linear classical and quantum fields. Finally, new results concerning the analysis of hierarchically organized objects were presented.

In collaboration with Springer-Verlag, a special effort was made in order that these proceedings be attractive to a large audience. Authors were asked to put their contribution in perspective. Moreover, about one third of the articles are extended versions of conference talks, intended to provide an introductory background to the reader. To all the authors and to the Managing Editor of **Lecture Notes in Physics**, we wish to express our gratitude. We also thank all the lecturers and participants for the quality of their presentations and the interest of their comments. Finally, we are grateful to Tran Thanh Van for valuable advice. The meeting benefitted from partial support from the CNRS (astrophysics and mechanical engineering departments), the ACCES program of the Ministère de la Recherche et de la Technologie and the City of Nice through the Lépine Committee.

J.-D. Fournier and P.-L. Sulem
Nice, July 1991

Contents

THE PHASE-DIFFUSION EQUATION AND ITS REGULARIZATION FOR NATURAL CONVECTIVE PATTERNS

T. Passot and A.C. Newell
Department of Mathematics, University of Arizona, Tucson AZ, 85721

Abstract

We first present the phase diffusion and mean drift equation which describe convective patterns in large aspect ratio containers and for arbitrary Rayleigh and Prandtl numbers. Some applications are presented such as the prediction of the selected wavenumber or the instability of foci. We propose in a second step a regularized form of the phase diffusion equation able to reproduce the formation and dynamics of defects.

I. Introduction

Many interesting behaviors of pattern-forming systems described by non-linear partial differential equations can be captured by standard perturbative techniques when the stress parameter is small and/or there exists a spectral gap between the microscopic and the macroscopic scales. We are mainly interested here in the latter case, and more particularly in the case of Rayleigh-Bénard convection where only rolls are formed.

A physical system close to a bifurcation can be described by a low-dimensional dynamics if the number of modes whose eigenvalue cross the imaginary axis at the bifurcation threshold (the order parameters) is small. Although these order parameters, which determine the macroscopic order of the system, still interact with the damped modes, the "slaving principle", (rigorously demonstrated for ODE's), states indeed that these decaying modes can be eliminated and explicitly expressed by the order parameters. In the class of systems leading to pattern formation, this may be compromised due to the spatial extension of the system (an infinite number of modes are squeezed in the immediate vicinity of the critical point). When one direction can be privileged (either because the system is not rotationaly invariant as it is the case for liquid crystals, or because one chooses to look at a particular direction only), it is possible to derive amplitude equations (such as the Ginzburg-Landau or the Newell-Whitehead-Segel equations [1]) which describe the behavior of the order parameters on a slow time scale. These equations contain slow spatial derivatives to account for the mode coupling in the bandwidth around each normal mode and thus are called envelope equations. They can be reduced to a generic form depending only on the type of bifurcation considered and the symmetries of the problem. Another approach can be adopted to treat the fully two-dimensional case [2], where equations for the order parameters are derived which contain the fully spatio–temporal dynamics on a fast varying scale (such as in the Swift-Hohenberg equation [3]). It will be observed in the conclusion that the approach we propose in this paper is close to this one but it applies even far from the bifurcation point.

Usually an order parameter is a complex field $\mathcal{A}$, corresponding to an amplitude A and a phase ϕ. It often happens that, due to some additional gauge invariances of the amplitude equations ($\mathcal{A} \to \mathcal{A}\exp(i\phi)$), corresponding to physical invariances of the system (translational invariance), the dynamics of $\mathcal{A}$, close to a particular solution $\mathcal{A}_0$, can be reduced to the evolution of the phase variable ϕ only. This reduction is possible if the amplitude A remains slaved to phase gradients on the so-called phase branch, graph of the phase eigenvalue $\lambda(q)$ in terms of the wavenumber q of the phase perturbation. Homogeneous phase perturbations of a given solution $\mathcal{A}_0$ are marginal so that $\lambda(0) = 0$. The phase branch determines the linear terms of the phase equations; the nonlinear terms are determined through the elimination of the damped modes. It is however now well-known [4] that the phase description may break down when phase instabilities are present. For example the phase equation corresponding to the complex Ginzburg-Landau equation is the Kuramoto-Sivashinski equation which is known to develop a form of weak turbulence whereas the long term development of the phase instability in the complex Ginzburg-Landau equation always leads to the appearance of defects in the system, due to the coupling of phase and amplitude modes.

When the stress parameter is well above its critical value leading to the onset of the spatial pattern under investigation, it is still possible (and in fact it is only possible) to derive a large–scale description of the dynamics in terms of phase variations. The undisturbed pattern is indeed translationally invariant, and thus the phase mode is still a marginal mode. The amplitude of the nonlinear waves building the pattern is however always slaved to phase gradients, except possibly near defects. An exact derivation of the phase equation for rolls with no privileged direction has been achieved recently in the case of Rayleigh-Bénard convection by Cross and Newell [5] and Newell, Passot and Souli [6-7]. The technique was already known in other fields: for modulated nondissipative nonlinear train waves as developed by Whitham [8] or for modulated traveling waves in reaction-diffusion equations as exposed by Howard and Kopell [9] (see also Kuramoto [10]). In all these examples, use is made of the inverse aspect ratio ϵ as an expansion parameter, and not of the stress parameter as previously. This relies on the observation that almost everywhere in the convective pattern, a local wavevector $\mathbf{k}$ can be defined which changes slowly throughout the container. Therefore, ignoring the dependence in the vertical direction, the field variables can be represented by locally periodic functions $f(\theta; R_{(i)})$ where f is 2π periodic in θ and $R_{(i)}$ represents the collection of stress parameters of the system. The field variable f varies over distances of the order of the roll size whereas the amplitude A (the norm of f e.g.) or the wavevector (the gradient of θ) vary over distances of the size of the box. This fact allows for the definition of a large–scale phase $\Theta(X, Y, T) = \epsilon\theta$ where $X = \epsilon x$, $Y = \epsilon y$ and $T = \epsilon^2 t$ are the scales of the long wavelength disturbances. The phase θ refers here to the total phase of the pattern whereas ϕ defined above corresponds to the deviation about the phase of the basic pattern $\theta_0 = k_0 x$. Note also at this point that in contrast to the case of the Newell-Whitehead-Segel equations, direction Y is here scaled in the same way as direction X

since we want to preserve rotational invariance. To leading order, and when no mean flow effects are present, the phase obeys a universal quasi-linear diffusion equation

$$\tau(k)\Theta_T + \nabla\cdot\mathbf{k}B(k) = 0 \tag{1}$$

where $\tau(k) > 0$ and $B(k)$ are calculable functions of the wavenumber $k = |\mathbf{k}|$. To this order of approximation, the amplitude A of the rolls is slaved to the wavenumber k through the relation

$$A^2 = \mu^2(k) \tag{2}$$

Equation (1) generalizes the fixed orientation equation of Pomeau and Manneville [11]. At finite Prandtl numbers a large scale horizontal mean flow, generated by the curvature of the rolls, advects the phase contours. Its presence is due to the existence of another marginal mode, namely a large-scale varying pressure field. The next section is devoted to a derivation of these coupled phase-mean drift equations and Section 3 is concerned with some predictions made on the behavior of convective patterns using the previous equations. In particular it is shown that all previous theories are contained in the present formalism and that moreover some new instabilities can be analyzed, such as the one of circular target patterns.

The difficulty with equation (1) is that it is ill-posed for some values of the wavenumber, outside the nonlinear stability region. Section 4 is devoted to the definition of a regularizing scheme for that equation, illustrated on a microscopic pattern forming model, the Swift-Hohenberg equation. Section 5 exposes some numerical experiments on the regularized equation and Section 6 is the conclusion.

II. The phase diffusion–mean drift equations

The starting point of the calculation is the determination of stable fully nonlinear straight parallel roll solutions of the following Oberbeck-Boussinesq equations:

$$\sigma(\partial_t\mathbf{u} + \mathbf{u}\cdot\nabla\mathbf{u}) = -\nabla p + T\hat{\mathbf{z}} + \nabla^2\mathbf{u} \tag{3a}$$

$$\partial_t T + \mathbf{u}\nabla T = Rw + \nabla^2 T \tag{3b}$$

$$\nabla\cdot\mathbf{u} = 0 \tag{3c}$$

where the parameter R represents the Rayleigh number and σ is the inverse Prandtl number. The temperature is T, the pressure p and the velocity $\mathbf{u}$ has components (u, v, w). Rigid-rigid boundary conditions are considered which state that $\mathbf{u} = T = 0$ at the top and bottom boundaries: $z = \pm\frac{1}{2}$. The existence of such a 2π periodic solution $\mathbf{v} = \mathbf{f}(\theta, z, k)$ (where $\mathbf{v} = (u, v, w, T, p)$ and $\theta = kx$) is assured by the previous work of Busse and Colleagues [12-19]; $\mathbf{f}$ is calculated, using a Galerkin technique as in [12].

Now we must look for modulated solutions and solve the linear equations obtained after inserting in (3a-3c) the following expansions in powers of the inverse aspect ratio ϵ:

$$\partial_z \to \partial_z$$

$$\partial_{\mathbf{x}} \to \mathbf{k}\partial_\theta + \epsilon\nabla_{\mathbf{X}}$$

$$\partial_t \to \epsilon\Theta_T\partial_\theta + \epsilon^2\partial_T$$

$$\mathbf{v} = \mathbf{v_0} + \epsilon\mathbf{v_1} + \epsilon^2\mathbf{v_2} + \cdots$$

Cross and along the roll velocities are also introduced: $\tilde{u} = \hat{\mathbf{k}} \cdot \mathbf{u}$ and $\tilde{v} = (\hat{\mathbf{k}} \times \mathbf{u}) \cdot \hat{\mathbf{z}}$ where $\hat{\mathbf{k}} = \frac{\mathbf{k}}{k}$. We end up at order ϵ and ϵ^2, with linear systems of the form $M\mathbf{v_j} = \mathbf{g_j}$ $(j = 1, 2)$, where M is the operator obtained by linearizing (3a-3c) about $\mathbf{v} = \mathbf{f}$. This operator is singular since due to translational invariance, $\frac{\partial f}{\partial \theta}$ is element of its kernel. We remark that $\frac{\partial f}{\partial p}$ is also a marginal mode corresponding to the fact that the pressure is at order 0 defined up to a large–scale varying constant p_s. When solving for $\mathbf{v_1}$ (resp. $\mathbf{v_2}$), solvability conditions have thus to be applied to $\mathbf{g_1}$ (resp. $\mathbf{g_2}$), leading respectively to the phase and mean drift equations. The compatibility conditions state that the right-hand sides of the linear systems must be orthogonal to the kernel of the adjoint operator $M^\dagger$. Since M is not self-adjoint, we must solve for the homogeneous adjoint boundary value problem separately. One of its solution, corresponding to the pressure mode is trivial; it reads $\mathbf{v}^\dagger = (0, 0, 0, 0, 1)$. The corresponding solvability condition is satisfied for $j = 1$ but becomes nontrivial for $j = 2$: it states that the "large–scale" divergence of the field $< \mathbf{u_1} >$ ($<>$ denotes averaging over θ and z) must be zero, but we know that $< \mathbf{u_1} >$ will contain non divergence-free terms due to the contributions of the slow horizontal Reynolds stresses: $\sigma \partial_X u_0^2 + \sigma \partial_Y (u_0 v_0)$ and $\sigma \partial_X (u_0 v_0) + \sigma \partial_Y v_0^2$. It appears clearly now why me must include the slowly varying pressure p_s at order 0, whose gradients contribute at order ϵ in the resolution of the velocity field: the pressure p_s is precisely determined by requiring that $\mathbf{u_1}$ be solenoidal and this equation is just another form of the mean drift equation that we usually prefer to write in terms of a stream function ψ after eliminating p_s ($< \mathbf{u} >= (\hat{\mathbf{k}} \times \nabla\psi) \cdot \hat{\mathbf{z}}$). We also see that at infinite Prandtl number no Reynolds stress is present to drive the mean flow.

The system is closed but the difficulty is that we must solve for the fluctuating parts of $\mathbf{v_1}$; no approximation for the vertical structure of the mean flow will suffice in order to get a reasonable agreement with experiment, especially at low Prandtl numbers. We have to use an extremely robust inversion method to solve these singular equations and we use a technique based on a singular value decomposition of the matrix obtained after projecting the operator M on the appropriate basis. At this order of approximation, the equations read:

$$\Omega(A, k, R, \sigma) = 0, \tag{4a}$$

$$\Theta_T + \rho(k)\mathbf{V} \cdot \nabla\Theta + \frac{1}{\tau(k)} \nabla \cdot \mathbf{k}B(k) = 0, \tag{4b}$$

$$\hat{\mathbf{z}} \cdot \nabla \times \hat{\mathbf{k}}\alpha(k)(\hat{\mathbf{k}} \times \nabla\psi) \cdot \hat{\mathbf{z}} - \nabla \cdot \hat{\mathbf{k}}\beta(k)(\hat{\mathbf{k}} \cdot \nabla\psi)$$
$$= \hat{\mathbf{z}}\nabla \times (\sigma\mathbf{k}\nabla \cdot \mathbf{k}A^2 - \frac{\hat{\mathbf{k}}}{\tau_\alpha(k)} \nabla \cdot \mathbf{k}B_\alpha(k)) - \nabla \cdot \hat{\mathbf{k}}(\nabla \times \mathbf{k}B_\beta(k)) \cdot \hat{\mathbf{z}} \tag{4c}$$

where $\mathbf{k} = \nabla\Theta$, $\mathbf{V} = \nabla \times \psi\hat{\mathbf{z}}$, the quantities $\rho(k)$, $B(k)$, $\tau(k)$, $\alpha(k)$, $\beta(k)$, $B_\alpha(k)$, $\tau_\alpha(k)$, $B_\beta(k)$ are all functions of k which are explicitly calculated. The first equation, otained by fixing the periodicity of the basic roll solution, gives the amplitude A as a function of the wavenumber k, given the Rayeigh and Prandtl numbers. The vertical structure of the mean drift velocity is different for the "along the roll" or the "across the roll" components and this is why the left–hand side of equation (4c) is not simply a Laplacian. The first term of the right–hand side comes from the horizontal Reynolds stress whereas the second and third terms essentially come from the vertical Reynolds stress. The appropriate boundary conditions are that ψ is a streamline and that the roll axis is perpendicular to the boundary: $\hat{\mathbf{k}} \cdot \hat{\mathbf{n}} = 0$. For the special case of circular target patterns however, the correct boundary condition is rather: $\hat{\mathbf{k}} \times \hat{\mathbf{n}} = 0$. These equations are translationaly and rotationaly invariant and also Galilean invariant, even though the original system is not. In the next section we will show that the preceding equations can predict the nonlinear stability region of straight parallel rolls (usually called the Busse balloon) and that moreover they can also predict the stability of circular target patterns.

III. Some results

Linearizing equations (4a-4c) about straight parallel rolls ($\Theta_0 = k_0 X$, $\psi_0 = 0$), it is easy to find the loci (in the (R,k) plane, for a given σ) of the Eckhaus (for a purely X dependent perturbation), zig–zag (for a purely Y dependent perturbation) or skewed-varicose border, limits of the nonlinear stability domain. The skewed-varicose instability occurs when the maximum of the growth rate corresponds to a perturbation containing both X and Y dependences. It is the modification of the Eckhaus instability in presence of mean drift and generally leads to formation of pairs of dislocations. Figure 1 displays some of these results where we can check that our theory reproduces the long-wavelength instabilities of the Busse balloon to within a very good accuracy.

However, in natural patterns, the influence of sidewall boundaries is such that the rolls tend to form circular patches. It is then natural to investigate the stability of exactly circular target patterns. Moreover there is a possibility to check our calculations and conjectures with laboratory experiments performed by Steinberg, Ahlers and Cannell [21]. These authors reproduce this model, using a lateral forcing on the sidewalls to initiate the circular pattern. For an exactly circular pattern (4b) becomes:

$$\Theta_T + \frac{1}{\tau r}\frac{\partial}{\partial r}(rkB) = 0 \tag{5}$$

where r is the radius in polar coordinate. It is easy to see that equation (5) leads to stationary solutions with $k = k_B$ such that $B(k_B) = 0$, the focus acting as a source of new rolls if $k < k_B$ and as a sink if $k > k_B$. Therefore the curvature is acting as a wavenumber selection mechanism, first observed by Pomeau and Manneville [22], even in natural patterns where fractions of foci singularities are present (especially in the corners of the containers). Using a functional which is shown to decrease almost every time (except perhaps when dislocations play a significant role) it can be argued [23] that on the time scale ϵ^{-2}, the horizontal diffusion time, patches form which satisfy the boundary condition $\mathbf{k} \cdot \mathbf{n} = 0$ along portions of the boundary, and in which $k \to k_B$ almost everywhere. On figure 1a (resp. 1b), we displayed the locus of k_B and the selected wavenumber for different Rayleigh numbers as found in experiments done respectively by Steinberg Ahlers and Cannell [21] and Heutmaker and Gollub [24]. We find a good agreement between the observed dominant wavenumber and k_B.

When non–axisymmetric perturbations are superimposed on the circular target $\Theta_0 = k_B r$, a non–zero mean–drift field will be generated which tends to increase the deformation and opposes to the action of the diffusive terms. Above a certain value of

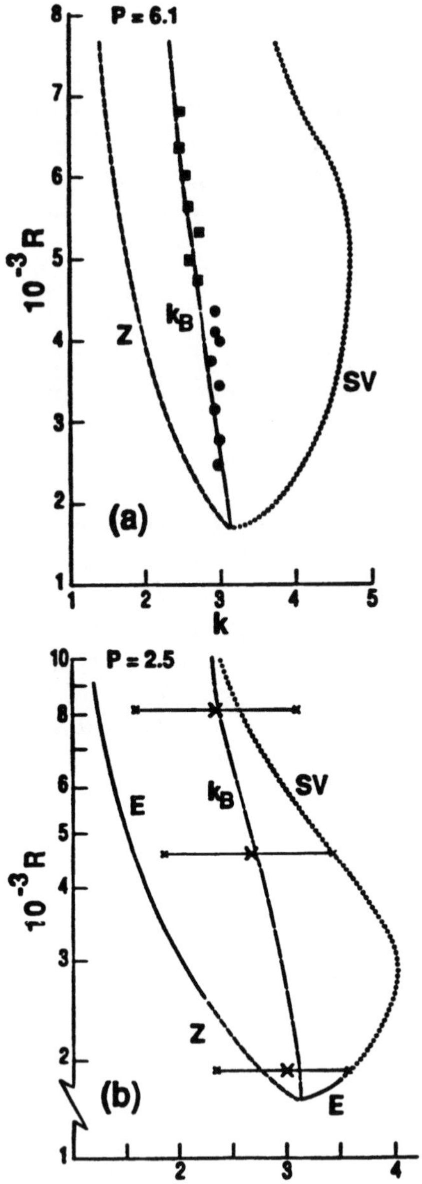

Figure 1 The long–wave borders of the Busse balloon marked E (Eckhaus), Z (zig–zag) and SV (skewed varicose), and the zeros k_B of $B(k)$ for two different Prandtl numbers: (a) 6.1 and (b) 2.5. In (a) the wavenumber of the $3\frac{1}{2}$ and 3 roll equilibrium states is displayed (circle and squares respectively) as the Rayleigh number is increased (taken from Steinberg Ahlers and Cannel [21]). In (b) the maxima (x) and the support of the wavenumber distribution are superimposed (taken from Heutmaker and Gollub [24]).

the Rayleigh number (for a given Prandtl number), the circular patch will be linearly unstable. Nonlinear saturation is experimentally observed to occur which leads to an off-centered target as displayed in figure 2. This "focus instability" is important in initiating time dependence (even in natural patterns), as observed when the selected wavenumber lies well inside the Busse balloon. At the point where k_B crosses the skewed-varicose boundary we rather suggest that dislocations would be nucleated all over the container and that the spatial coherence will be lost. The values of the critical Rayleigh number at which the focus instability appears, as calculated by our theory, is consistent with experiments [21], but corrections to account for the finiteness of the box are certainly necessary. Moreover it appears that in the center of the patch the amplitude of the rolls is no longer slaved [7] and this is important to calculate a precise value of the instability threshold. The study of this latter point requires the determination of coupled PDE's governing the behavior of both the phase and amplitude variables. This point is now addressed in the following Section.

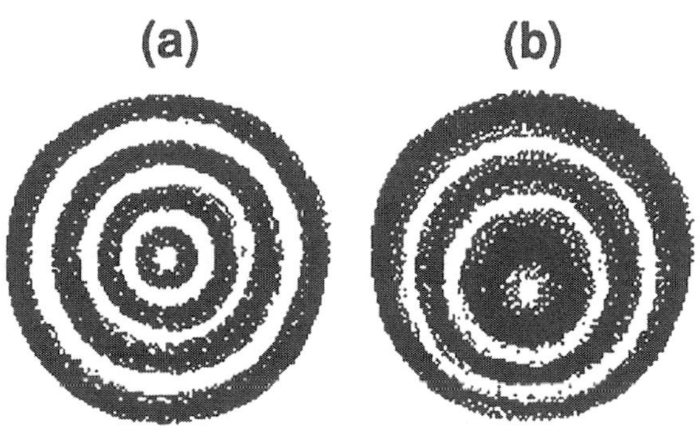

Figure 2 Equilibrium patterns taken from Steinberg, Ahlers and Cannel [21] at $P = 6.1$, aspect ratio 6 and for two different Rayleigh numbers: (a) a stable state three roll state at $R = 3R_C$ and (b) its distortion at $R = 4R_C$.

IV. Regularization of the phase equation

We shall now start with general considerations and show the limitation of the preceding approach, trying to bring up some new features in order to cure the problems. Let us introduce, for the sake of general discussions, a typical pattern forming equation of the form:

$$\partial_t w + F(\frac{\partial}{\partial x}, R, w) = 0, \tag{6}$$

invariant under space and time translations and parity symmetry. As an example we will use the real and complex Swift-Hohenberg equations because they are simpler, but all the ideas will go through for the full Oberbeck–Boussinesq equations even with mean drift effects present. We assume that (6) admits stable steady periodic solutions $w_0(\theta; R, k)$ invariant under parity symmetry $\theta \to \theta$. The period in θ is 2π. This defines the wavenumber k, or equivalently an amplitude A, so that w_0 can also be considered as a function of A. Since the pattern breaks the space translational symmetry it can undergo phase perturbations. We will assume that in the parameter ranges chosen, no other symmetries are broken, (i.e. there are no secondary bifurcations) and then the phase mode is the only marginal mode, except, as we shall discuss, at the marginal stability boundary.

The phase diffusion equation (1) can also be obtained for (6); it was previously derived as a solvability condition of the equation which governs the correction w' to the almost periodic field w_0. This solvability condition is made necessary by the existence of a periodic null solution $v_1 = \frac{\partial w_0}{\partial \theta}$ of the operator obtained by linearizing the original microscopic equations about the field w_0. The null solution v_1 exists because of the translational invariance. There is also a second null solution, $v_2 = \frac{\partial w_0}{\partial A} + \frac{\theta}{2k^2}\frac{\partial w_0}{\partial \theta}\frac{\partial k^2}{\partial A}$, (i.e. $v_2 = \frac{dw_0}{dA} = \frac{\partial w_0}{\partial A} + \frac{\partial w_0}{\partial \theta}\frac{\partial \theta}{\partial k^2}\frac{\partial k^2}{\partial A}$ and $\frac{\partial \theta}{\partial k^2} = \frac{\theta}{2k^2}$), but, because of the presence of the second term, v_2 is not periodic and thus in general does not lead to an additional solvability condition. It is however periodic at the boundaries of the marginal stability curve given by $A^2 = \mu^2(R,k) = 0$ so that for values of k at which μ^2 is zero, there is indeed a second solvability condition which turns out to involve a correction to the amplitude equation (2). We shall see shortly why this correction is necessary.

The difficulty with (1) is that it is ill-posed in the sense that

$$\nabla \cdot \mathbf{k}B = (B + \frac{\theta_x^2}{k}\frac{dB}{dk})\theta_{xx} + 2\frac{\theta_x\theta_y}{k}\frac{dB}{dk}\theta_{xy} + (B + \frac{\theta_y^2}{k}\frac{dB}{dk})\theta_{yy}$$

($\mathbf{k} = \nabla\theta = (\theta_x, \theta_y)$), is only negative definite and therefore (1) is only stable when both $B(k)$ and $\frac{d}{dk}kB$ are negative. This occurs for a band of wavenumbers $k_Z(R) < k < k_E(R)$ ($B(k_Z) = 0$, $\frac{d}{dk}kB|_{k=k_E} = 0$) which is the nonlinear stability domain or "Busse balloon". As soon as the wavenumber k wanders outside the Busse balloon, there is a rapid growth of the short scales and unphysical instabilities. We must therefore regularize the phase diffusion equation.

The nature of the instabilities when $k < k_Z$ or $k > k_E$ are different [25]: At the zig–zag border the instability is a supercritical–type bifurcation and it saturates when the rolls develop a zig–zag pattern. The regularization of (1) can thus simply be achieved by going to the next orders of the expansion, which consists in adding to the phase diffusion equation terms proportional to $\epsilon^2\nabla^2\nabla\cdot\mathbf{k}$. There are other nonlinear terms but this one is the most important, capable of achieving a balance with $\nabla \mathbf{k} B$ after introducing the scaling $\tilde{Y} = \epsilon^{\frac{1}{2}} y$ in the "along the roll direction". On the other hand, at the Eckhaus border, the instability leads to a readjustment of the roll wavelength (or for the skewed–varicose instability when mean flow effects are present, it leads to formation of dislocations), during which process the amplitude approaches zero locally and the wavenumber wanders way outside the right border of the marginal curve. It appears clearly now that even though the preceding type of regularization is sufficient at the beginning of the instability process, it is not at all sufficient to describe the whole story.

It will be necessary to supplement the phase diffusion equation by a Newell–Whitehead–type equation when k approaches k_r, the right border of the marginal stability curve. This is possible precisely because at k_r the second solvability condition leads to a correction to the amplitude equation. However when this equation is derived not at the minimum of the marginal stability curve but at some point (k_0, R_0) for $R_0 > R_c$, the critical Rayleigh number, additional terms are present compared to the Newell-Whitehead equation which lead to an unbounded transfer of energy to any $k < k_0$ and one must write an equation for $W = A\exp(i\theta)$, where θ is the total phase, to remedy this problem.

In order to match properly the "generalized Ginzburg–Landau equation" derived at k_r and the phase diffusion equation with the bilaplacian term, valid close to k_E, one must find coupled PDE's for the phase and an amplitude mode valid in the range $k_E < k < k_r$. This system does not need to contain exact higher order terms because the process which leads k to approach and exceed k_r is very rapid. The solution appears to be attracted to a robust singular solution of the reverse nonlinear heat equation in which the high k modes are slaved to the low k ones. This is analogous to the regularization of strictly hyperbolic equations: None of the large scale dynamics (shock strength and speed) are affected by the choice of the artificial viscosity. However it is not clear how to derive an equation for the amplitude A in this range of wavenumbers using the usual asymptotic expansions discussed in the preceding Section because there is no second solvability condition.

Before introducing another approach, let us mention a possible extension of this formalism in some simple cases, showing the existence of a coupling to an amplitude mode even though this one is not marginal. For more details on that procedure see [26] and [27]. The solution of the linear equation for the corrections w_i to the basic field w_0 can be computed (at least in principle), using the method of variations of parameters starting from the n solutions of the homogeneous linear problem (n is the degree of the system). Of these n solutions, only the one which corresponds to the space translation symmetry $\frac{\partial w_0}{\partial \theta}$ is periodic of period 2π. The companion solution, $\frac{d}{dk^2} w_0 = k\partial_k w_0 + \theta\partial_\theta w_0$ will only be periodic when the amplitudes of w_0 is very small. The general solution for w_i will have n arbitrary constants, one of which is determined by the solvability condition and is just the phase diffusion equation or its correction, and the rest are determined by insisting w_i is 2π periodic. In the cases where any of the remaining $(n-1)$ solutions is almost periodic, the corresponding constant becomes unbounded. For example, for the case of the complex Swift-Hohenberg equation:

$$\partial_t w + (1+\Delta)^2 w - Rw + w^2 w^* = 0, \tag{7}$$

the linearized operator is $Lv = \partial_t v + ((1+k^2\partial_\theta^2)^2 - R + 2\mu^2)v + w_0^2 v^*$, $w_0 = A\exp(i\theta)$, the solvability condition demands that the imaginary part of the coefficient of $\exp(i\theta)$ on the R.H.S. is zero and $v = \frac{e^{i\theta}}{A^2}\Re(R.H.S.)$ which becomes unbounded as $A \to 0$ (see [5]). In this case the asymptotic expansion becomes nonuniform and must be modified. In fact, this is an indicator that a new order parameter must be introduced, in this case, an active amplitude. However, other possibilities for unbounded constants exist. If for example, in the real Swift-Hohenberg situation, the third harmonic $3k$ of the fundamental mode with wavenumber k lies sufficiently inside the marginal curve so that it becomes competitive with the original periodic solution, then a new phase θ_2 with local period $\frac{2\pi}{3}$ must be introduced, together with its companion amplitude. This would first occur (and we checked it does) at a value of the Rayleigh number R for which k given by $(k^2-1)^2 = R$ lies in the left marginal stability boundary and $3k$ given by $(9k^2-1)^2 = R$ lies on the right marginal stability boundary (i.e. $R = 0.64$ and $k = 1/\sqrt{5}$). In such circumstances, we would be forced to introduce a more general quasiperiodic solution $w(\theta_1, \theta_2; k_1, k_2)$ and seek two phase diffusion equations. In this paper, we are going to avoid the latter situation which we might call mode resonance and concentrate on the former where the amplitude of the chosen mode becomes an active order parameter.

How do we modify the amplitude equation which to this point is algebraic, which means the amplitude is slaved? In the case of the complex Swift–Hohenberg equation, the solution is obvious because we have the exact functional form of $w_0 = A\exp(i\theta)$ which when substituted in (7) gives exactly:

$$A\Theta_T + \mathcal{L}_1^1 A + \epsilon^2 \mathcal{L}_3 A = 0 \tag{8a}$$

$$R - A^2 - (1-k^2)^2 = \frac{\epsilon^2}{A}(A_T + \mathcal{L}_2^1 A + \epsilon^2 \mathcal{L}_4 A). \tag{8b}$$

where

$$\begin{aligned}
\mathcal{L}_0 &= (1+k^2\partial_\theta^2)^2 \\
\mathcal{L}_1 &= (1+k^2\partial_\theta^2)D_1 + D_1(1+k^2\partial_\theta^2) \\
\mathcal{L}_2 &= (1+k^2\partial_\theta^2)D_2 + D_2(1+k^2\partial_\theta^2) + D_1^2\partial_\theta^2 \\
\mathcal{L}_3 &= D_1 D_2 + D_2 D_1 \\
\mathcal{L}_4 &= D_2^2
\end{aligned}$$

and

$$\begin{aligned}
D_1 &= 2\mathbf{k}\cdot\nabla + \nabla\cdot\mathbf{k} \\
D_2 &= \nabla^2.
\end{aligned}$$

$\mathcal{L}_j^p$ means that we replace $\frac{\partial^2}{\partial\theta^2}$ in $\mathcal{L}_j$ by $-p^2$.

A simple "trick" can also be worked out in simple instances, as for example for obtaining the nonlinear Schrödinger equation for slowly varying wave trains starting from Whitham's theory [26]. It consists in adding a free parameter to the right-hand side of the equations and use this extra degree of freedom to impose another constraint on the constants of integration such as to remove from the solution this term, which diverges when the amplitude is small. In the case of a conservative nonlinear wave equation the extra free parameter is obtained by adding explicit higher order corrections to the dispersion relation. It can also be obtained, for dissipative systems, by expanding the stress parameter and the form that the extra condition will take is also a correction of the dispersion or eikonal relation.

In general however we will not have the luxury of such exact solutions and we will now discuss an alternative approach to obtaining (8a) and (8b) as asymptotic expansions directly from the equation (6). The first step is to use the scale separation (expressed as $\partial_{\mathbf{x}} \to \mathbf{k}\partial_\theta + \epsilon\nabla_{\mathbf{X}}$) to expand F as $\Sigma\epsilon^i F_i$. We could express w as a Fourier series in θ: $w = \Sigma_{n=-\infty}^{n=+\infty}\alpha_n e^{in\theta}$ where the Fourier coefficients α_n are slowly varying functions of space and time. This representation would however give rise to an infinite number of coupled equations for the coefficients and it is not easy to see how to close the system. The more natural choice of basis is the set of eigenfunctions of the operator obtained by linearizing F about the periodic solution w_0. Let us consider w_0 of parity S (symmetric) to be the solution of $F_0(w_0) = 0$ and denote by L the linear operator obtained by linearizing $F_0(w)$ about w_0: $L = \frac{\delta F_0}{\delta w}|_{w_0}$. A complete stability analysis about the periodic solution w_0 should lead to the full basis of the Bloch functions and their associated Floquet exponents. Since here we are interested only in large scale instabilities, we will only consider those Bloch eigenmodes which have the same periodicity as the basic nonlinear solution. This means in particular, we work in parameter ranges so as to exclude the secondary instabilities spoken about in the last paragraph and which, for the real Swift–Hohenberg equation, involve modes which are superharmonic of the basic pattern. We are thus led to consider a discrete set of eigenmodes ξ_i (and $\xi_i^\dagger$ for the adjoint operator $L^\dagger$) corresponding to the eigenvalue λ_i, where $\lambda_{i+1} \geq \lambda_i$. We have $\lambda_0 = 0$ and $\xi_0 \propto \partial_\theta w_0$. Let us normalize the eigenmodes by imposing $< \xi_i^\dagger|\xi_j >= \delta_{ij}$ and $< \xi_i|\xi_i >= 1$. Due to the assumption about the parity of the basic solution the eigenmodes are either antisymmetric (ξ_{iA}) corresponding the phase translation or the introduction of new phases, or symmetric (ξ_{iS}) corresponding to the companion amplitudes. The marginal mode ξ_0 is of parity A, the least damped mode in a neighborhood of k_r is of parity S and it corresponds to the amplitude mode. For the wavenumbers considered, all the eigenvalues λ_i, $i \geq 1$ are positive.

Let us write the perturbation w' as $w_s + w_p$ where w_s is the dominant symmetric correction that we can write $\alpha_S\xi_{1S}$ and where w_p represents the passive or slaved modes. Substituting this expression into (6), we then project the equation onto the adjoint modes $\xi_0^\dagger$, $\xi_{1S}^\dagger$. The first equation obtained is the usual phase diffusion equation containing the dominant correction terms and the second one is a correction to the eikonal equation. We get

$$< \xi_0^\dagger|\partial_\theta(w_0 + w_s) > \Theta_T + < \xi_0^\dagger|F_1(w_0 + w_s) > +\epsilon^2 < \xi_0^\dagger|F_3(w_0 + w_s) >= 0 \quad (9a)$$

$$\epsilon^2 < \xi_{1S}^\dagger|\partial_T(w_0 + w_s) > + < \xi_{1S}^\dagger|F_0(w_0 + w_s) > +\epsilon^2 < \xi_{1S}^\dagger|F_2(w_0 + w_s) >= 0 \quad (9b)$$

Away from the borders of the neutral curve, λ_{1S} is not small and (9b) can be solved algebraically for α_S (expand $< \xi_{1S}^\dagger|F_0(w_0 + w_s) >$ as $< \xi_{1S}^\dagger|F_0(w_0) > + < \xi_{1S}^\dagger|\frac{\delta F_0}{\delta w} \cdot w_s >= 0 + \lambda_{1S}\alpha_S$). In this case we are left with a single equation for the phase Θ.

In the low amplitude limit, w_0 can be approximated by $\mu\xi_{1S}$ and then $w_0 + w_s$ is $(\mu + \alpha_{1S})\xi_{1S} = A\xi_{1S}$. Then $< \xi_{1S}^{\dagger}|F_0(w_0 + w_S) > = < \xi_{1S}^{\dagger}|F_0(A\xi_{1S}) >$ which can be calculated directly. However this latter term takes on a generic form at low amplitude and for a supercritical–type bifurcation, independently of the nonlinearity involved in F_0. It reads: $g(k)A(A^2 - \mu^2)$, where $g(k)$, which corresponds to the damping rate of an amplitude perturbation to the roll solution, is simply $\frac{\lambda_{1S}}{2\mu^2}$. In the low amplitude limit, one finally obtains that (9a), (9b) are

$$< \xi_0^{\dagger}|\xi_{1S} > A\Theta_T + < \xi_0^{\dagger}|F_1(A\xi_{1S}) > + \epsilon^2 < \xi_0^{\dagger}|F_3(A\xi_{1S}) > = 0, \tag{10a}$$

$$\epsilon^2 \partial_T A + g(k)A(A^2 - \mu^2) + \epsilon^2 < \xi_{1S}^{\dagger}|F_2(A\xi_{1S}) > = 0. \tag{10b}$$

The structure of F_1 suggests strongly that after multiplying (11a) by A and approximating the $O(\epsilon^2)$ term in (10a) by its more relevant term, we shall obtain the following equations:

$$\Theta_T + \frac{1}{\tilde{\tau}(k)A^2}\nabla \cdot (\mathbf{k}A^2\tilde{B}(k)) + \epsilon^2\eta\nabla^4\Theta = 0 \tag{11a}$$

$$\epsilon^2 A_T + g(k)A(A^2 - \mu^2(k)) - \epsilon^2\zeta(\nabla^2 A, (\mathbf{k}\nabla)^2 A, \cdots) = 0 \tag{11b}$$

which are the usual phase and algebraic amplitude equations when $O(\epsilon^2)$ terms are neglected. This in turn defines $\tilde{\tau}$ and $\tilde{B}$ by the relations: $\mu^2\tilde{\tau}(k) = \tau(k)$ and $\mu^2\tilde{B}(k) = B(k)$.

For the real Swift-Hohenberg equation at small amplitudes, where we can approximate w by a one Galerkin mode expansion for the basic roll solution (i.e. $w = \sqrt{2}A\cos(\theta)$), we have $\mu^2(k) = \frac{2}{3}(R-(1-k^2)^2)$, $\tilde{\tau}(k) = 1$, $\tilde{B}(k) = 2(1-k^2)$ and $g(k) = \frac{3}{2}$.

The function ζ entering at the $O(\epsilon^2)$ level in (11b) is in general complicated. However we shall not need to know its particular form since it contains terms which are irrelevant as long as A is not small, in which case it would mean that the wavenumber is close to the marginal stability border where a separate analysis is performed as shown below. The coefficient of the bilaplacian term in (11a) is calculated by looking at the most unstable wavelength of the zig–zag instability and in this case only the "along the roll" term (or Θ_{YYYY} if the basic pattern is $\Theta_0 = k_0X$) is relevant. In the other case where the basic pattern undergoes an Eckhaus instability, the same parameter η multiplies also the "across the roll" (or Θ_{XXXX}) term and now serves exactly as an artificial viscosity whose exact value will not influence the final resulting pattern.

We now carry out an analysis close to the marginal stability border and try to extend the region of validity of equation (11a-b) for values of the wavenumber greater than the marginal value k_r. This will in particular allow for a description of defects.

One might first try to match (11a), (11b) with the Newell–Whitehead–Segel equation derived for wavenumber k_r and values of the Rayleigh R well away from R_c. Considering eq. (6), we will denote by $\lambda(R, k)$ the eigenvalue of the operator $L_0 = \frac{\delta F}{\delta w}|_{w=0}$ for a mode whose wavenumber is slightly inside the marginal stability curve ($\lambda(R, k)$

and $-\lambda_{1S}(k)$ are equivalent in the limit k goes to k_r). It is easy to see that the complex amplitude equation for the marginal mode takes the form:

$$\partial_\tau \mathcal{A} = -\frac{\partial\lambda}{\partial R}|_0 (R-R_0)\mathcal{A} - \frac{\partial\lambda}{\partial(ik_j)}|_0 \nabla_j \mathcal{A} - \frac{\epsilon}{2}\frac{\partial^2\lambda}{\partial(ik_j)\partial(ik_l)}|_0 \nabla_j\nabla_l \mathcal{A}$$
$$- g_0|\mathcal{A}|^2\mathcal{A} + O(\epsilon|\mathcal{A}|^4 A) + O(\epsilon^2), \qquad (12)$$

where ∂_τ stands for $\partial_{T_1} + \epsilon\partial_{T_2}$, $T_j = \epsilon^j t$. The index 0 means that the expression has to be evaluated at k_r (possibly by taking a limit). One has of course $\lambda(R, k_r) = 0$ and g_0 identifies here with the Landau constant which is also equal to $g(k_r)$, the limit as k goes to k_r of $-\frac{\lambda(R,k)}{2\mu^2(k)}$. The term containing first order derivatives in $\mathcal{A}$ is at the origin of an amplitude instability of the $\mathcal{A}$ = constant solution, leading to a decrease of the wavenumber towards values closer to the band center. It is thus natural to look for an equation governing the evolution of $W = \mathcal{A}e^{ik_r x}$; it is clear that its linear part reads in Fourier space:

$$\partial_t \hat{W} + \lambda(R,k)\hat{W} = 0$$

where $\mathbf{k} = \mathbf{k_r} + \nabla\phi$. We will now look for an approximation to this equation by expanding $\lambda(R,k)$ near k_r as a quadratic spline polynomial in k^2 (λ is indeed a function of k^2 due to rotational invariance.) Setting:

$$\lambda(R,k^2) = \alpha_0 - \beta_0 k^2 + \gamma_0 k^4, \qquad (13)$$

the equation for W can now be written:

$$\epsilon^2\partial_T W + (\alpha_0 + \epsilon^2\beta_0\nabla^2 + \epsilon^4\gamma_0\nabla^4)W + g_0|W|^2 W = 0. \qquad (14)$$

If it happens that the coefficient γ_0 is negative, a higher order polynomial should be used to fit the function $\lambda(R,k)$ in such a way that the highest order Laplacian of equation (14) is a damping term. The above approximation is made necessary especially in situations where the growth rate contains a denominator which is also a function of k^2. This happens for example when the time derivative term of w in (6) contains also spatial derivatives (c.f. model I in [5]). We can derive the first–order linear phase equation for (14) in order to relate its coefficients to the one of the phase equation (1) linearized about $\Theta = k_r X$. Identification of the coefficients leads to:

$$\lambda' = -\frac{B}{\tau}|_0 \qquad (15)$$

$$\lambda'' = -\frac{\mu^2}{2\tau}(\frac{B}{\mu^2})'|_0 \qquad (16)$$

where the prime denotes differentiaiton with respect to k^2. The equation (14) is only valid around k_r and in order to match with the phase diffusion–amplitude equations valid inside the marginal stability curve, it will be convenient to have the coefficients α_0, β_0, and γ_0 become functions of k and calculated from $\tau(k)$, $B(k)$ and $g(k)$ by demanding that to leading order the phase diffusion (1) and amplitude equations (2) (parts of (11a) and (11b)) should be recovered from (20). We find

$$\gamma(k) = -\frac{\mu^2}{2\tau}\frac{d}{dk^2}(\frac{B}{\mu^2}) \tag{17}$$

$$\beta(k) = \frac{B}{\tau} + 2k^2\gamma \tag{18}$$

$$\alpha(k) = k^2\beta - k^4\gamma - g\mu^2. \tag{19}$$

It is easy to find, using (15) and (16), that the functions $\alpha(k)$, $\beta(k)$ and $\gamma(k)$ have indeed the right limits α_0, β_0 and γ_0 as k approaches k_r.

The equation

$$\epsilon^2\partial_T W + (\alpha(k) + \epsilon^2\beta(k)\nabla^2 + \epsilon^4\gamma(k)\nabla^4)W + g(k)|W|^2W = 0 \tag{20}$$

is therefore the canonical and universal equation for describing the behavior of the complex order parameter $W = Ae^{i\frac{\Theta}{\epsilon}}$ constructed from the amplitude and phase (and not to be confused with the exact solution except in the special case of the complex Swift-Hohenberg equation itself) near the right-hand border of the marginal stability curve. For the complex Swift-Hohenberg equation where $g(k) = 1$, $\tau = A^2$, $B = A^2\frac{dA^2}{dk^2} = -2(k^2-1)(R-(k^2-1)^2)$ and $\mu^2 = R-(k^2-1)^2$, we have $\gamma = 1$, $\beta = 2$ and $\alpha = 1 - R$ so that (20) is indeed the complex Swift-Hohenberg equation.

In the regime well inside the marginal stability curve where the amplitude is certainly slaved and where a one-Galerkin mode approximation for w_0 ceases to be a good approximation, it is appropriate to write (11a) and (11b) as

$$W_t - g(k)W(\mu^2 - |W|^2) + i\frac{W}{|W|^2}(\frac{1}{\tilde{\tau}(k)}\nabla\cdot\mathbf{k}|W|^2\tilde{B}(k) + \eta\nabla^2\nabla\cdot\mathbf{k}|W|^2) = 0. \tag{21}$$

so that it will be easier to match with eq. (20). The time and spatial derivatives are performed with respect to the small scales $(\mathbf{x}, t)$. The bilaplacian term has been slightly altered since it is more convenient to introduce the combination $\mathbf{k}|W|^2$ which is well defined throughout the whole field (even at defects). This formulation is valid as long as the modulus of the damping rate of the "least damped" mode is large enough so that $\mu^2 - |W|^2$ stays always negligible. If this condition is not met, the pattern is close to a secondary bifurcation or a small-scale instability and in both cases it is certainly not possible to describe the dynamics by the present formalism. We exclude this possibility.

The matching between (20) and (21) is now easy to perform by introducing a function $\delta(k)$ which is zero near k_{Mr} and smoothly goes to unity when k approaches the center of the band. The form of this function does not have any influence on the precision of the formulation. We obtain:

$$\begin{aligned} &W_t + (\alpha(k) + \beta(k)\nabla^2 + \gamma(k)\nabla^4)W + g(k)|W|^2W \\ &+ i\delta(k)\frac{W}{|W|^2}(\frac{1}{\tilde{\tau}(k)}\nabla\cdot\mathbf{k}|W|^2\tilde{B}(k) + \eta\nabla^2\nabla\cdot\mathbf{k}|W|^2) = 0. \end{aligned} \tag{22}$$

It must be noted here that the coefficients α, β and γ are different from those given in (17)-(18)-(19) and in particular now depend on the function δ.

This equation contains exact regularizing terms only at the border k_r; for values of k between k_E and k_r these terms are taken simpler for convenience. We conjecture and verify numerically in the next Section that it does not affect the overall dynamics. All evidence suggests that there is no growth of spurious instabilities and it would appear that all high k modes which are generated are simply slaved. When defects form, the dynamics accelerate and the shape of the wavenumber around the center of the defect seems to be very robust with respect to modification of the higher order terms. It would appear that the solution is attracted to a "singular" solution of the nonlinear heat equation. This point will be illustrated in the next Section where some numerical experiments are made using eq. (22).

V. Numerical simulations of the regularized phase diffusion equation

We choose to illustrate the capabilities of eq. (22) by displaying the basic behavior of solutions initiated with simple patterns whose wavenumber is close to the Eckhaus and the zig–zag border. The coefficients of (22) are calculated taking the real Swift–Hohenberg as a microscopic model. For more details the reader can refer to [28].

We will consider initial conditions for which the wavenumber is initially not too far from to the borders of the nonlinear stability domain which for $R = 2$ is delimited by $k = 1.$ at the zig–zag border and $k = 1.37$ at the Eckhaus border. When we are not too far from the zig–zag boundary, the zig–zag instability is saturated by the regularization of the phase-diffusion equation and the coefficients α, β, γ play no role. Notice for k in this region, $\delta = 1$ and we are in effect solving (11a-b). On the other hand when we start outside the Eckhaus boundary, defects are nucleated and near defects, wavenumbers near both k_r and k_l are present. In this case we need to use the full equation (22).

The regularized equation is amazingly robust. It gives results in close agreement with the original Swift-Hohenberg model even for the choices of stress parameter and initial conditions for which secondary instabilities are potentially present and for which the saturation of the zig–zag instability occurs at significantly finite amplitudes.

The integration of (22) is done with a standard pseudo-spectral method using a temporal scheme which mixes an Adams-Bashforth for the nonlinear terms and an exact integration of the linear terms. Each coefficient α, β, and γ is split into its constant limiting value and a remainder. We use periodic boundary condition on a spatial grid containing 32 collocation points and the time step is 0.1. Due to this spatial resolution we are limited to a rather large value of ϵ and of η. We chose $\epsilon = .25$. We take for all runs $\eta = 1$. We verified that a change of these values do not affect the dynamics except for determining the wavelength at which the zig–zag instability saturates.

We display here two calculations. The first one concerns the evolution of an Eckhaus unstable pattern at $k = 1.45$ with a slight modulation transverse to the rolls tending to squeeze them in the middle of the container and a perturbation superimposed in the "along the roll" direction which leads to the formation of two pairs of defects, one in the middle of the box and the other one on the edge. Figures 3a and 3c display the contours of the real and imaginary part of the field W and show the pinching of the roll and the defects. A cut of the wavenumber is shown in figure 3b. The defects move (they climb) and an equilibrium is obtained where quasi–circular patches are formed on the edges (Fig. 3d). This state seems to be stationary at least for as long as we

(a)

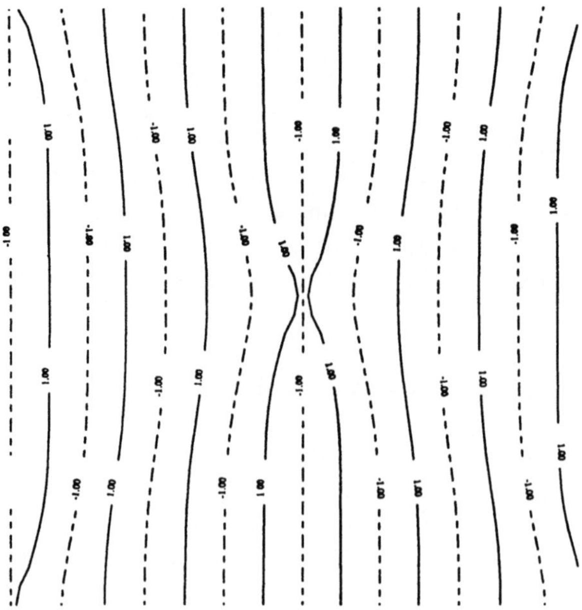

(b)

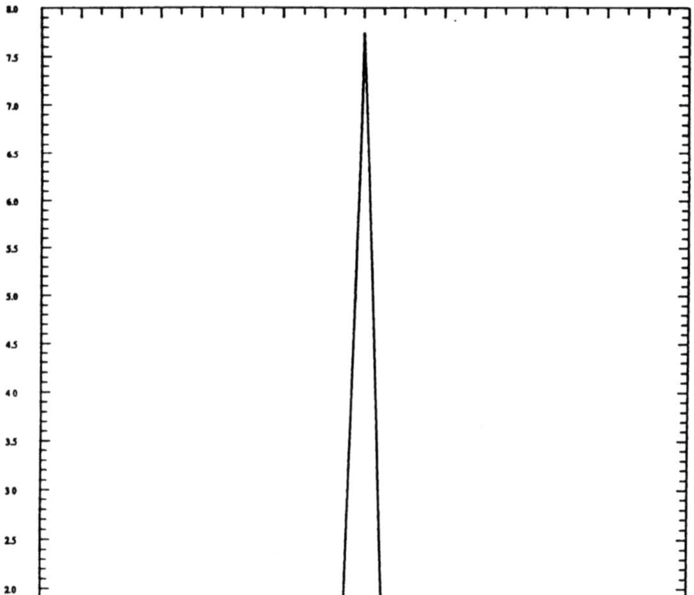
8.0
7.5
7.0
6.5
6.0
5.5
5.0
4.5
4.0
3.5
3.0
2.5
2.0
1.5
1.0
0
2
4
6
8
10
12
14
16
18
20
22
24
26
28
30
32

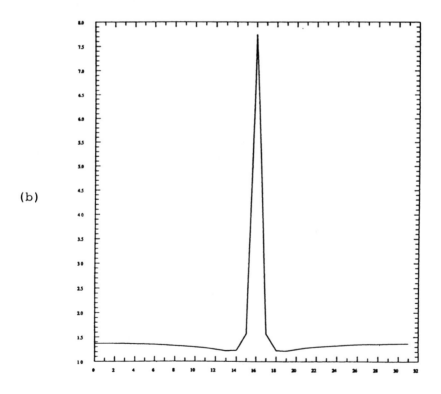

(c)

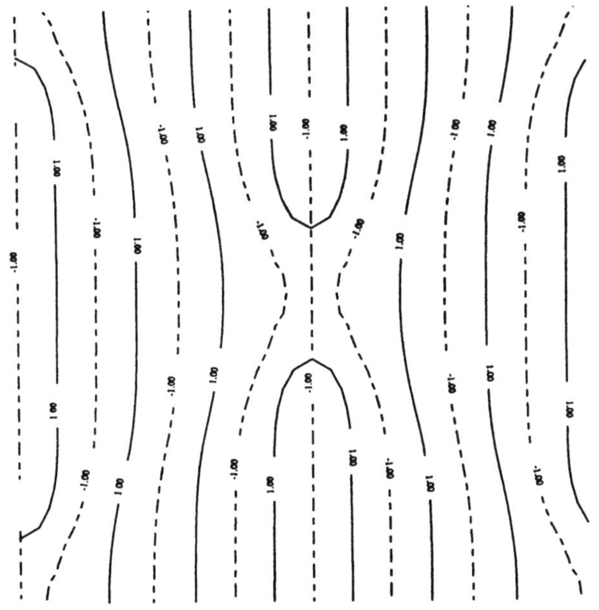

(d)

Figure 3 Contour lines of $\Re(w) = 0$ (value $+1$) and $\Im(w) = 0$ (value -1). This sequence corresponds to the temporal evolution of an Eckhaus unstable pattern with $w(t=0) = A\exp(ik_0(x + \alpha \sin x \cos y))$. We have $\eta = 1.$, $k_0 = 1.45$ and $\alpha = 0.01$.
(a) Pinching of the rolls;
(b) cut of the wavenumber at the middle of the box at the same time;
(c) two pairs of defects;
(d) final state;

integrated ($t = 80$). The symmetry of the pattern has not been broken in this case. In another experiment with $k = 1.4$ (not shown) the symmetry finally breaks and after gliding the defects annihilate each other. The final state is then a straight parallel roll pattern containing one less roll.

In a second experiment we investigated the nonlinear evolution of a zig–zag instability. Starting with a pattern at $k \doteq 0.7$ with a small perturbation along the roll we end up with a saturated state as shown in figure 4 where the wavenumber of the pattern stabilizes around 1. The number of wavelengths in the direction of the perturbation depends on the value of η. It is interesting to note that the perturbation is not periodic in the direction along the roll. This is also observed when integrating the real Swift-Hohenberg equation. For this value of k as an initial condition, the value of the destabilizing diffusion coefficient is quite large. It is interesting to note that we get a good agreement for the final state with the original equation. We must remark however that the early development of the zig–zag instability is different in the two cases. For the basic equation, the instability begins at small scale, concentrated in a certain region of the pattern. The equation (22) leads however to a large scale development uniformly throughout the direction of the perturbation.

Figure 4 Final state of a pattern initially zig–zag unstable with $k_0 = 0.7$, $\eta = 1.$, $w(t = 0) = A\exp(ik_0(x + \alpha \sin y))$ and $\alpha = 0.01$. This state is almost stationnary and $k \sim 1.06$.

VI. Conclusion

As a first concluding remark we would like to mention that the equation (22) we derived in Section 4, or its extension taking into account the mean drift term (in progress), is certainly quite general and bridges the gap which existed between qualitative and quantitative models. The first ones, such as the Swift-hohenberg [3] (or its generalizations [29],[30]), give a good overall picture of convective patterns but do not allow for exact comparisons with experiments, and cannot extend to large Rayleigh numbers. The second ones were up to now restricted to the case of simple patterns, close to onset (e.g. the Newell–Whitehead–Segel equation), or to the case of "defect-free" patterns (the phase diffusion equation as presented in [5] and [7]).

The equation we present here must include a small scale behavior of the order parameters A and Θ since it can reproduce the formation and dynamics of defects. But it does not need to capture all the small–scale features of the underlying pattern, such as boundary layers. In this sence it is much easier to handle than the microscopic model!

Proper boundary conditions should also be derived for (22) since it contains fourth order derivatives. This work is currently in progress.

Acknowledgments: Research support from the Arizona Center for Mathematical Sciences, sponsored by AFOSR Contract FQ8671-900589 (AFOSR-90-0021) with the University Research Initiative Program at the University of Arizona, is gratefully acknowledged.

References

1 A.C. Newell and J.A. Whitehead, "Finite bandwidth, finite amplitude convection", J. Fluid Mech. **38** (1969), 279. and L.A. Segel, J. Fluid Mech. **38** (1969) 203.
2 M. Bestehorn and H. Haken, "Traveling waves and pulses in a two–dimensional large–aspect–ratio system", Phys. Rev. A **42** (1990) 7195.
3 J. Swift and P.C. Hohenberg, "Hydrodynamic fluctuations at the convective instability", Phys. Rev. A **15**, (1977) 319.
4 P. Coullet, L. Gil and J. Lega, "Defect mediated turbulence", Phys. Rev. Lett. **62** (1989) 1619.
5 M.C. Cross and A.C. Newell, "Convection Patterns in Large Aspect Ratio Systems", Physica **10D** (1984) 299.
6 A.C. Newell, T. Passot and M. Souli, "Convection at Finite Rayleigh Numbers in Large-Aspect-Ratio Containers", Phys. Rev. Lett. **64** (1990) 2378.
7 A.C. Newell, T. Passot and M. Souli, "The phase diffusion and mean drift equations for convection at finite Rayleigh numbers in large containers", J. Fluid. Mech **220**, (1990) 187-252.
8 G.B. Whitham, "Linear and nonlinear waves", Wiley-Interscience, New-York (1974).

9 L.N. Howard and N. Kopell, "Slowly Varying Waves and Shocks Structures in Reaction–Diffusion Equations", S. Appl. Math. **56** (1977) 95.
10 Y. Kuramoto, "Chemical Oscillations, Waves and Turbulence", Springer Series in Synergetics **19**, (1984) H. Haken Ed., Springer–Verlag.
11 Y. Pomeau and P. Manneville, "Stability and fluctuations of a spatially periodic convective flow", Journal de Physique- Lettres **23** (1979), L609.
12 F.H. Busse, "On the stability of two–dimensional convection in a layer heated from below", J. Math. Phys. **46** (1967) 149.
13 F.H. Busse and J.A. Whitehead, "Instabilities of convection rolls in a high Prandtl number fluid", J. Fluid Mech., **47** (1971) 305.
14 F.H. Busse and J.A. Whitehead, "Oscillatory and collective instabilities in large Prandtl number convection", J. Fluid Mech., **66** (1974) 67.
15 F.H. Busse, "Nonlinear properties of convection", Rep. Prog. Phys., **41** (1978) 1929.
16 F.H. Busse, "Transition to turbulence in Rayleigh–Bénard convection", Hydrodynamic Instabilities and the Transition to Turbulence, edited by H.L. Swinney and J.P. Gollub (1981) 97, (Berlin: Springer Verlag).
17 R.M. Clever and F.H. Busse, "Transition to time dependent convection", J. Fluid Mech., **65** (1974) 625.
18 R.M. Clever and F.H. Busse, "Instabilities of convection rolls in a fluid of moderate Prandtl number", J. Fluid Mech., **91** (1979) 319.
19 R.M. Clever and F.H. Busse, "Large wavelength convection rolls in low Prandtl number fluid", J. Appl. Math. Phys. Z. angew. math. Phys. **29** (1978) 711.
20 Y. Pomeau and P. Manneville, "Wavelength selection in axisymmetric cellular structures", J. Physique Lett. **42** (1981) 1067.
21 V. Steinberg, G. Ahlers and D.S. Cannell, "Pattern formation and wavenumber selection by Rayleigh-Bénard convection in a cylindrical container", Physica Scripta, **T13** (1985) 135.
22 Y. Pomeau and P. Manneville, "Wavelength selection in axisymmetric cellular structures", J. Physique Lett., (1981), **42**, 1067.
23 A.C. Newell, "Two Dimensional Convection Patterns in Large Aspect Ratio Systems", Lecture Notes in Num. Appl. Anal. **5** (1982), 205.
24 M.S. Heutmaker and J.P. Gollub, "Wave–vector field of convective flow patterns", Phys. Rev. A **35** (1987) 242.
25 S. Fauve, "Large Scale Instabilities of Cellular Flows", in "Instabilities and Nonequilibrium Structures", (1987) 63-88, Eds. E. Tirapegui and D. Villaroel, D. Reidel Publishing Company.
26 A.C. Newell, "Solitons in Mathematics and Physics", CBMS-NSF vol. **48** (1985), SIAM.
27 J.C. Luke, "A perturbation Method for Nonlinear Dispersive Wave Problems", Proc. Roy. Soc. London Ser. A **292** (1966), 403.
28 T. Passot and A.C. Newell, "The regularization of the phase diffusion equation for natural convective patterns", submitted to J. Physique.
29 P. Manneville, "A two-dimensional model for three-dimensional convective patterns in wide containers", J. Physique **44** (1983) 759.
30 H.S. Greenside, M.C. Cross and W.M. Coughran, Jr., "Mean Flows and the Onset of Chaos in Large-Cell Convection", Phys. Rev. Lett. **60** (1988), 2269.

Asymptotic time behavior of nonlinear classical field equations

YVES POMEAU

Laboratoire de Physique Statistique,
24 Rue Lhomond
75231, Paris Cedex 05
France

Abstract*

Hamiltonian systems with many degrees of freedom, like large assemblies of interacting particles in a box are described by the Gibbs-Boltzmann statistics, as far as their average properties are concerned. This does not hold for the long time behavior of classical nonlinear field equations, as noticed already by Jeans, because of the infinite heat capacity of this field. Thus nonlinear (and nonintegrable) classical fields cannot relax for long times toward an ill defined thermal equilibrium. Here I consider an example of this relaxation problem, that is the long time evolution of solutions of the nonlinear Schrödinger equation, in the defocussing case. I show, under some assumptions that for long times there is a cascade toward smaller and smaller scales, introducing a kind of dissipation in a system that is formally reversible, and I give the scaling laws for this.

A famous piece of history in physics is the discovery of quantum mechanics through the realization that the Gibbs-Boltzmann statistics does not hold for the blackbody radiation. A classical field, as the electromagnetic field in a cavity has infinetely many degress of freedom, and by the Ehrenfest principle each of these degrees should get the energy $\frac{k_B T}{2}$ (k_B=Boltzmann constant, T=absolute temperature), showing that the heat capacity of such a system is infinite. In the present note I shall not consider such grand questions, but merely the following point: given a nonlinear classical field equation (equation 1 below), what is the asymptotic behavior of a smooth solution of it ? This is connected with the blackbody problem because, for a classical field, one cannot give a meaning to the Gibbs-Boltzmann weight $\exp(-\frac{H}{k_B T})$, where H is the energy. Again because of the infinite heat capacity one cannot relate the temperature to the energy of the system, as a Lagrange multiplier. Similar difficulties would appear when trying to define a microcanonical ensemble: then the heat capacity of the

* This summarizes part of a talk presented at the workshop on "Large-scale structures in nonlinear physics"held at Villefranche-sur-Mer, France, January 14-18, 1991.

infinitely many degrees of freedom forbids one to share a finite energy equally among those infinitely many degrees of freedom.

This kind of question has received recently a good deal of attention in the particular case of the long time dynamics of solutions of the Euler equations for inviscid fluids in 2dimension of space (2D) [1-4]. There, by introducing the Young measures [5], one can treat consistently the infinite number of degrees of freedom related to the small scale fluctuations, and then explain, for instance, why do those flows form large coherent structures together with messy small scale fluctuations. Those results are remarkable in the sense that they introduce both a kind of equilibrium in a non obvious sense and a kind of irreversible behaviour, for equations that are formally reversible.

I already pointed out that this kind of question has more than a theoretical interest, as it leads to a better understanding of some of the observed features in real fluid turbulence. Below, I examine another problem, that is the long time evolution of the nonlinear Schrödinger equation, in the defocussing case. This equation is a possible dynamical equivalent of the Landau-Ginzburg equation for superfluids and superconductors. However, the matter is rather complicated [6] and so I do not claim to have here something that applies directly to any kind of real superfluid or superconducting material. Instead it should help to understand that irreversibility may be more subtle than the mere formal introduction of damping terms in the equations of motion, a concept that could have some relevance outside of pure theory.

After convenient dimensionalization to get rid of irrelevant coefficients, the equation I am going to consider can be written:

$$i\frac{\partial\Psi}{\partial t} = \Delta\Psi + \Psi - |\Psi|^2\Psi \qquad (1).$$

This is the socalled nonlinear Schrödinger equation (NLS), where Δ is the ordinary Laplacian, and Ψ a complex valued smooth function of the vector **x** representing the position in space and of the time t. I shall not consider any specific geometry, but assume , for instance that either (1) is posed with periodic boundary conditions or that Ψ satisfies the Dirichlet conditions on some boundaries.

This equation has been shown to have [7] a smooth behaviour for smooth initial data in a rather wide range of conditions. In space dimensions 2 and higher, the <u>focussing</u> case [that is the same equation as (1), but with a plus-instead of minus-sign for the cubic term], this is known to lead to singularities after a finite time and for smooth initial data [8]. In dimension 1, the focussing case is integrable [9]. I shall consider now the question of the asymptotic behaviour of solutions of (1). The dynamical system defined by equation (1) is Hamiltonian, because it can be written as:

$$i\frac{\partial\Psi}{\partial t} = -\frac{\delta H}{\delta\Psi^*} \qquad (2),$$

where the energy $H\{\Psi\}$ is the functional:

$$H\{\Psi\} = \int dx\, (|\nabla\Psi|^2 - |\Psi|^2 + \frac{1}{2}|\Psi|^4) \qquad (3),$$

and $\frac{\delta}{\delta\Psi^*}$ is a Fréchet derivative, Ψ^* being the complex conjugate of Ψ.

Indeed the system has the usual Noether invariants (constants of the motion in the course of time): the momenta (if the boundary conditions allow it) and the energy H, plus the one related to invariance of H{.} under a global phase change of Ψ, this is the "number of particles":

$$N\{\Psi\} = \int dx \ |\Psi|^2 \qquad (4).$$

Now I shall develop the following idea: given N, there exists a solution $\Psi_0 e^{i\mu t}$ of (1) that minimizes the energy $H\{\Psi\}$. In a loose sense, one would expect that, given N again, the system will tend to relax to this state of lowest energy. It cannot do that formally in NLS, because of the energy conservation. However, the very existence of infinitely many degrees of freedom will make the system to diffuse in phase space toward these degrees of freedom. In other words fluctuations at smaller and smaller scales will appear, resulting from the nonlinear interaction between fluctuations of larger wavelength. Small scale fluctuations contribute to the energy mostly through the "kinetic energy" term $|\nabla\Psi|^2$ in the integrand of H{.}, and thus can yield a contribution to H{.} that is finite although the corresponding contribution to N, not involving the space derivative becomes negligible. In other terms, the amplitude of the small scale fluctuations becomes negligible as time goes on, but not their gradient. We shall make this more definite and give estimates of the asymptotic behavior of the various quantities involved.

Given N, the minimum of H{.} is reached for the function Ψ_0 that is the solution of:

$$\mu\Psi_0 + \Delta\Psi_0 + \Psi_0 - |\Psi_0|^2\Psi_0 = 0 \qquad (5),$$

where μ is a Lagrange mutiplier allowing to specify N. Equation (5) can be solved in 1D, with Dirichlet b.c. at x=0 and L. This makes appear elliptic functions. Two interesting limits can be investigated: N small and N large.

At small N, one expects Ψ_0 small, and thus one may neglect the cubic term in (5). This means that $-(\mu + 1)$ must be close to the largest eigenvalue, say $-(\mu_f + 1)$ of the Laplace operator in the domain of definition of Ψ_0. Then a standard perturbative method allows to relate the difference $\mu - \mu_f$ to N, the final result being:

$$\mu - \mu_f = - \frac{\int dx \ |\Psi'|^4}{\int dx \ |\Psi'|^2} \sim N,$$

where $\Psi'(x)$ is proportional to the eigenfunction of the Laplace operator with eigenvalue $-(\mu_f$ $+1)$.

At large N, one expects most of the domain to be filled with an uniform Ψ_0, because this minimizes the kinetic energy term. From (5) this constant value is such that:

$|\Psi_0|^2 \approx |\mu - \mu_f|^{1/2}$, and this is $\rho = N/\Omega$, Ω= volume of the support of Ψ. The crossing between the two regimes (small and large N) occurs when a typical length scale of this domain is of the same order as $\rho^{-1/2}$.

From this, the minimum energy $H_0(N)$ is a well defined concept. Indeed a solution of (1) with smooth initial data, if it remains smooth in the course of time (as assumed), will conserve this energy, and so cannot relax toward a "wavefunction" realizing this energy minimum. However, the things are not that simple because, as already said, the difference between H, initial (and forever) value of the energy and $H_0(N)$ can be taken care of by small scale fluctuations with a negligible contribution to N. Below, we shall develop a coherent schema for this based upon the following principles. First, we assume that, for long times the solution of (1) tends pointwise toward the function $\Psi_0(\mathbf{x})\, e^{i\mu t}$ minimizing the energy for N given (see in the remarks at the end a less strong statement for that). Let $\delta\Psi$ be the difference at time t between the actual value of $\Psi(\mathbf{x},t)$ and $\Psi_0(\mathbf{x})\, e^{i\mu t}$. The typical length scale for $\delta\Psi$ remains of the same order of magnitude as the small amplitude of $\delta\Psi$ itself: this keeps constant the "kinetic energy" density $|\nabla\Psi|^2$, and ensures the conservation of energy. This gives a first relation. Then the second relation is obtained by assuming that $\delta\Psi$ is given by the weak turbulence equations [10]: this is because the small scales are fed by the nonlinear interactions between modes and also because one can stop the Poincaré expansion in the strength of the nonlinearity at the first nontrivial order, where two fluctuations of wavenumber $\mathbf{k}_1$ and $\mathbf{k}_2$ can feed resonantly a fluctuation with the wavenumber $\mathbf{k}= \mathbf{k}_1+\mathbf{k}_2$.

The weak turbulence equations cannot be solved explicitely. However, we assume that they have a self similar behavior for long times. This yields another relationship between the exponents for the decay of $\delta\Psi$ and of the associated length scale. This is also consistent with the assumption of weak turbulence, as the higher order terms can be shown then to be truly negligible.

Weak turbulence theory is an extension of the Poincaré expansion to conservative nonlinear fields[10]. Let us detail some crucial steps in the derivation of the equations of weak turbulence for the present case. Formally this is a perturbation theory for solutions of (1) close to a ground state $\Psi_0\, e^{i\mu t}$. The function Ψ_0 depends on the position in a nontrivial way, as it has to be a solution of (5). We shall be concerned here with a very large system, that is with a system with a "macroscopic" size, much larger than the "microscopic" unit length. As it costs (kinetic) energy for Ψ_0 to be nonuniform in space (because of the $|\nabla\Psi|^2$ term in H) , the ground state Ψ_0 will be as uniform as possible in space, but for boundary layers of "microscopic" thickness of order 1. As the perturbations to Ψ_0 we wish to study have a wavelength much less than the size of the system, we may take Ψ_0 as uniform in their study. From (5), the constant value of $|\Psi_0|^2$ is $(1+\mu)$. If we limit ourselves then to linear perturbations, they have to satisfy the coupled equations:

$$i\frac{\partial\delta\Psi}{\partial t} = \Delta\delta\Psi + \delta\Psi - 2|\Psi_0|^2\delta\Psi - \Psi_0^2\, e^{2i\mu t}\,\delta\Psi^* \quad (6.a),$$

$$-i\frac{\partial\delta\Psi^*}{\partial t} = \Delta\delta\Psi^* + \delta\Psi^* - 2|\Psi_0|^2\delta\Psi^* - \Psi_0^{*2}\, e^{-2i\mu t}\,\delta\Psi \quad (6.b).$$

It is more convenient to introduce two auxiliary functions, $\delta(\mathbf{x},t)$ and its complex conjugate, related to $\delta\Psi$ and $\delta\Psi^*$ as:

$\delta\Psi = i\Psi_0 e^{i\mu t}\delta$ and $\delta\Psi^* = -i\Psi_0^* e^{-i\mu t}\delta^*$. Those functions satisfy the equations(recall that Ψ_0 is almost independent of **x**):

$$i\frac{\partial\delta}{\partial t} = (1+\mu+\Delta)\delta - (1+\mu)\delta^* \quad (7.a)$$

and

$$-i\frac{\partial\delta^*}{\partial t} = (1+\mu+\Delta)\delta^* - (1+\mu)\delta \quad (7.b)$$

For perturbations uniform in space (Δ=0), the solution is:

$\delta = 2(1+\mu)C_0 t + C_1 + iC_0$,where C_0 and C_1 are arbitrary real constants. This twofold degeneracy (a constant and a secular terms instead of two exponentials in the generic case) of the eigenproblem is linked to the two conservation relations (N and H). At nonzero wavenumber **k**, that is by replacing practically Δ by $-k^2$ in (7), one gets eigenfunctions depending on time as exp[iω(**k**)t], with ω(**k**) root of:

$$\omega^2(\mathbf{k}) = k^4 + 2(1+\mu)k^2. \quad (8)$$

This dispersion relation shows that there are two different regimes in the "short" wavelength (/large wavenumber) limit. If the wavenumber k is much bigger than the inverse size of the system ($\sim L^{-1} \sim \Omega^{-1/D}$, D dimension of space), but much smaller than $(1+\mu)^{1/2}$ (that we shall take as of order 1 too) the dispersion relation is of the d'Alembert type, the k^4 term in (8) may be neglected and one gets sound waves with the speed of sound $[2(1+\mu)]^{1/2}$. It is interesting to notice at this point that the usual statement that superflows are irrotational, with a harmonic velocity potential, ($\Delta\phi$=0) gets here a precise meaning. The condition $\Delta\phi$=0 is a particular form of the d'Alembert equation:

$$\frac{\partial^2\phi}{\partial t^2} - c_s^2\,\Delta\phi = 0,$$

whith c_s^2 square of speed of sound [$=2(1+\mu)$ in the present case]. This equation is the "phase equation" [20] of NLS and reduces to $\Delta\phi$=0 for a steady, or quasi steady flow; it has nontrivial solutions in a ring geometry for instance, because ϕ is a phase, so that ϕ and $\phi+2\pi$ are identical.

When k gets really big, that is much larger than $(1+\mu)^{1/2}$, then the equation for the perturbation becomes close to the free field Schrödinger equation and ω^2(**k**) close to k^4.

This splitting between two regimes for the dispersion relation will also correspond to two different regimes in the cascade of energy towards small scales. In some sense the situation might be seen as having some resemblance with the Kolmogorov picture of the turbulent cascade: the "inertial" domain would be the intermediate range of length scales between the size of the system and $(1+\mu)^{-1/2}$, and the "viscous range" the length scales much less than $(1+\mu)^{-1/2}$. In the weak turbulence theory, the nonlinear interaction between fluctuations depends crucially on their dispersion relation (that is on the frequency-wavenumber relation), whence the two different domains for the free decay of fluctuations, as a function of their wavelength.

The starting point is an expansion of a solution of (1) near the ground state $\Psi_0 e^{i\mu t}$. The perturbation is $\delta\Psi$, a function of **x** and t with a small amplitude and a small range in space. Indeed this fluctuation satisfies at the dominant order, the equation (1) linearized around $\Psi_0 e^{i\mu t}$, as given in (6).Let us define the Fourier component of $\delta\Psi$ as:

$$\delta\Psi_{\mathbf{k}} = [\delta I(\mathbf{k})]^{1/2} \exp i[\mathbf{k}.\mathbf{x}+\omega(\mathbf{k})t +\mu t],$$

Then the solution of (1) is of the form:

$$\Psi(\mathbf{x},t)= \Psi_0 e^{i\mu t}+\Omega^{1/2}\int d\mathbf{k}\ \delta\Psi_{\mathbf{k}} \quad (9),$$

where Ω is the volume of the system. Our notations are not completely rigorous, because in a finite volume **k** should take discrete values only, instead of being continuous. We have avoided however the rather cumbersome (and useless here) formalism of the large volume limit, and one can check that our final expressions are correct in this limit.

But the form of $\Psi(\mathbf{x},t)$ in (9) is not a solution of (1), because of terms in (1) wich are nonlinear in the perturbation $\delta\Psi$. Weak turbulence theory results from the expansion at the next order in the amplitude $\delta\Psi$. It is non trivial, because there is then a resonance effect allowing to create a fluctuation with wavenumber

$$\mathbf{k} =\mathbf{k}_1+ \mathbf{k}_2 \qquad (10),$$

by the nonlinear interaction between fluctuations at wavenumber $\mathbf{k}_1$ and $\mathbf{k}_2$. This is possible because, as we shall see later on, the resonance condition $\omega(\mathbf{k})= \omega(\mathbf{k}_1) +\omega(\mathbf{k}_2)$ is compatible with (10). However one must take care whether **k** is in the inertial range [k much less than $(1+\mu)^{-1/2}$] or in the "microscopic" range [k bigger or equal to $(1+\mu)^{-1/2}$]. Then the equation of weak turbulence defines the slow evolution of the intensities $\delta I(\mathbf{k})$ by this resonant interaction. It reads:

$$\frac{d\delta I_{\mathbf{k}}}{dt} = \int d\mathbf{k}_1\, d\mathbf{k}_2\, \delta(\mathbf{k}-\mathbf{k}_1-\mathbf{k}_2)\, \delta[\omega(\mathbf{k})-\omega(\mathbf{k}_1)-\omega(\mathbf{k}_2)]\ |\Psi_0|^2\, \delta I(\mathbf{k}_1)\, \delta I(\mathbf{k}_2) \quad (11),$$

In this equation the interaction coefficient $|\Psi_0|^2$ is there because one considers quadratic nonlinearities generated by the expansion of the cubic term in (1) to second order in $\delta\Psi$, so that a factor proportional to Ψ_0 remains that represents the strength of the interaction (or of the nonlinearity). Actually the correct equation is a slightly more complicated than the above form because one should add to it other combinations coming from different choices of sign for the frequency combinations. But I have chosen to write it in a form making as clear as possible its structure and scalings.

Mathematically speaking, this equation (11) is an akward object, and there is little or no hope that a sufficiently explicit solution to the general Cauchy problem can be found for it. However, one can make some progress by looking at self-similar solutions, assuming that this describes the asymptotic time behavior.

Let K be the typical wavenumber for the spectrum of fluctuations $\delta I(\mathbf{k})$, a function of time, and δI a typical value for $\delta I(\mathbf{k})$. By assumption this wavenumber is much bigger than the inverse size of the system. But it can be either within the "inertial range" that is much less than $(1+\mu)^{1/2}$ or within the microscopic range, that is of the order or bigger than $(1+\mu)^{1/2}$. This changes a lot the final result, through the estimation of $\delta[\omega(\mathbf{k})-\omega(\mathbf{k}_1)-\omega(\mathbf{k}_2)]$ in the integrand of (11). We shall consider first the most simple case, that is K within the microscopic range, where the dispersion relation is approximately the one of the ordinary free field Schrödinger equation : $\omega^2(\mathbf{k})\approx k^4$. Then the equality $\omega(\mathbf{k})=\omega(\mathbf{k}_1)+\omega(\mathbf{k}_2)$ is compatible with (10) if the scalar product $\mathbf{k}_1.\mathbf{k}_2$ is zero. Under these conditions, the right hand side of (11), as a function of K and of δI scales like $K^{D-2}\,|\Psi_0|^2\,\delta I^2$, D number of dimensions of space, although the left hand side scales as $\frac{\delta I}{t}$. Whence by combining those two results the scaling:

$$\delta I \sim \frac{K^{2-D}}{t\,|\Psi_0|^2} \quad ,$$

From the Wiener-Khintchin theorem, and with our definitions, the correlations of $\delta\Psi$ are related to $\delta I(\mathbf{k})$ as:

$$\delta I(\mathbf{k})= (2\pi)^{-D/2}\int d\mathbf{r}\; e^{i\mathbf{k}.\mathbf{r}} <\delta\Psi^*(\mathbf{r}_0)\delta\Psi(\mathbf{r}_0+\mathbf{r})>$$

From which, the amplitude of the fluctuation $\delta\Psi$ scales as $(K^D\delta I)^{1/2}$. This, together with the above estimate for δI, gives:

$$\delta\Psi \sim \frac{K}{t^{1/2}|\Psi_0|}$$

But from the energy conservation, one must have $K\,\delta\Psi \sim \varepsilon^{1/2}$, where ε is the density in space of the difference between the actual energy H and the ground state energy $H_0(N)$. This expresses that this energy difference is almost completely represented by the "kinetic energy" of small scale fluctuations $\delta\Psi$.

Combining all this, one obtains:

$$\delta\Psi \sim \frac{\varepsilon^{1/4}}{t^{1/4}|\Psi_0|^{1/2}} \qquad (12.a),$$

that is consistent with all the assumptions made to derive it and yields too:

$$K \sim (\varepsilon t)^{1/4}|\Psi_0|^{1/2} \qquad (12.b).$$

This gives the scaling laws for the late decay, since one expects that the average wavenumber reaches this microscopic domain only after it has decayed through the inertial domain. Let us study now the decay through the inertial domain. The difference with the other case is in the form of the dispersion relation to be used in evaluating $\delta[\omega(\mathbf{k})-\omega(\mathbf{k}_1)-\omega(\mathbf{k}_2)]$ in (11). Now we have to take $\omega(\mathbf{k})\approx[2(1+\mu)]^{1/2}k$. However, it will appear shortly that this dominant order is not sufficient for our purpose. We then have to take the next order term in the expansion of $\omega(\mathbf{k})$ near k=0:

$$\omega(\mathbf{k})\approx[2(1+\mu)]^{1/2}k\left[1+\frac{k^2}{4(1+\mu)}+o(k^4)\right]$$

A first integration over $\mathbf{k}_2$ in (11) yields:

$$\frac{d\delta I_{\mathbf{k}}}{dt}=\int d\mathbf{k}_1\, \delta[\omega(\mathbf{k})-\omega(\mathbf{k}_1)-\omega(\mathbf{k}-\mathbf{k}_1)]\ |\Psi_0|^2\, \delta I(\mathbf{k}_1)\, \delta I(\mathbf{k}-\mathbf{k}_1) \qquad (13).$$

In the long wavelength limit the resonance condition for the frequencies [i.e. the equality $\omega(\mathbf{k})=\omega(\mathbf{k}_1)+\omega(\mathbf{k}-\mathbf{k}_1)$] implies that the vectors $\mathbf{k}$ and $\mathbf{k}_1$ are collinear, but this does not contribute to the integral because this configuration has a negligible weight (see remarks at the end on this point). At the next order we have to consider these two vectors as almost collinear: given $\mathbf{k}$, then $\mathbf{k}_1$ is to be decomposed into its main Cartesian component, $k_{1,z}$ along $\mathbf{k}$, $0<k_{1,z}<k$, and into another small component, perpendicular to $\mathbf{k}$ and called $\Delta\mathbf{k}_1$.Then, under this approximation:

$$\delta[\omega(\mathbf{k})-\omega(\mathbf{k}_1)-\omega(\mathbf{k}-\mathbf{k}_1)]$$

$$\approx\delta\{[2(1+\mu)]^{1/2}[k-|\mathbf{k}-\mathbf{k}_1|-|\mathbf{k}_1|]+\frac{1}{2}[(1+\mu)]^{-1/2}(k^3-|\mathbf{k}-\mathbf{k}_1|^3-|\mathbf{k}_1|^3)\}$$

$$\approx\frac{1}{k\sqrt{3}}\delta\{|\Delta\mathbf{k}_1|-\left[\frac{3}{2(1+\mu)}\right]^{1/2} k_{1,z}(k-k_{1,z})\}.$$

Now one may use the same kind of estimation as before, to get the power laws for the decay. Given K, the length of $\Delta\mathbf{k}_1$ is of order $\frac{K^2}{(1+\mu)^{1/2}}$, as it results from the above estimate of $\delta[\omega(\mathbf{k})-\omega(\mathbf{k}_1)-\omega(\mathbf{k}-\mathbf{k}_1)]$. For a finite system of size L, this is consistent iff the distance between different quantized values of $\mathbf{k}$ is much less than this estimate for $\Delta\mathbf{k}$, which holds if the inequality $K^2>>(1+\mu)^{1/2}L^{-1}$ is true. This condition is to be considered together with the one defining the inertial subrange:

$$(1+\mu)^{1/2}>>K>>L^{-1}.$$

Actually this splits this range into two subdomains:

$(1+\mu)^{1/2} >> K >> (1+\mu)^{1/4}L^{-1/2}$,where the approximation developed so far applies, because there are many normal modes in the range of variation of $\Delta\mathbf{k}$, although the other subdomain is at longer wavelength defined by:

$$(1+\mu)^{1/4}L^{-1/2} >> K >> L^{-1}.$$

In this domain, one cannot any longer consider the modes as having a continuous distribution. Instead, one should treat them as a discrete set. This is a challenging problem: many modes might interact in this domain, but one cannot use the weak turbulence approximation, at least in its usual form, because the set of wavenumber is no more continuous. See remarks at the end on this point. From now on I shall deal in the subrange where the set of wavenumbers can be considered as continuous, that is whenever the first inequality $(1+\mu)^{1/2} >> K >> (1+\mu)^{1/4}L^{-1/2}$ is satisfied.

Now the integral $\int d\mathbf{k}_1\, \delta[\omega(\mathbf{k})-\omega(\mathbf{k}_1)-\omega(\mathbf{k}-\mathbf{k}_1)]\ |\Psi_0|^2\, \delta I(\mathbf{k}_1)\, \delta I(\mathbf{k}-\mathbf{k}_1)$ is , up to numerical constants,

$$\int d|\Delta\mathbf{k}_1|\ |\Delta\mathbf{k}_1|^{D-2} \int dk_{1,z}\, \delta[\omega(\mathbf{k})-\omega(\mathbf{k}_1)-\omega(\mathbf{k}-\mathbf{k}_1)]\ |\Psi_0|^2\, \delta I(\mathbf{k}_1)\, \delta I(\mathbf{k}-\mathbf{k}_1),$$

which is of order $(1+\mu)\delta I^2[\frac{K^2}{(1+\mu)^{1/2}}]^{D-2}$. Writing now that this is of the same order as $\frac{\delta I}{t}$, one obtains:

$$(1+\mu)\ \delta I\ [\frac{K^2}{(1+\mu)^{1/2}}]^{D-2} \sim \frac{1}{t}.$$

As in the previous case the conservation of energy in the cascade imposes that the product $\delta I\, K^2$ is constant and equal to ε, which yields:

$$K \sim (1+\mu)^{D''}(\varepsilon t)^{D'} \qquad (14).$$

with $D'=\frac{1}{6-2D}$ and $D''=\frac{4-D}{2(6-2D)}$, and the scaling for $\delta\Psi$ follows from this last relationship together with $\delta\Psi\, K \sim \varepsilon^{1/2}$.

Notice that the theory becomes inconsistent for D bigger than 3, since equation (14) would predict the wavenumber K to decrease in the course of time instead of increase. At D=3, the marginal case, this theory is not accurate enough, and would require probably to take into account logarithms, as usual in this sort of marginal situation. I do not believe that this signals an inverse cascade for D>3, but merely that at higher space dimensions the basic assumptions of this derivation become wrong. Most likely, the assumption of weak turbulence becomes invalid in the sense that the fluctuations in real space loose their regularity so that the Fourier components of the spectrum are no more independent Gaussian variables. These fluctuations should get a fractal support in real space, and the Wiener-Khintchin theorem should not apply anymore, because of the lack of homogeneity of the fluctuations on average.

Assuming that all the assumptions made are true (and practically thus D=2), this shows that, for this system, the formal reversibility of the dynamics is not inconsistent with the tendency to reach a sort of ground state. Indeed not all initial data tend to this ground state. The equation (4) may be considered too as a nonlinear eigenvalue equation in μ, that is the frequency of a purely oscillating solution of (1). In 1D, this equation has explicit and real solutions, given by elliptic functions. One can check that they are actually many discrete branches of those solutions, appearing as μ increases under fixed geometric conditions (the Wigner"no level crossing rule" does not seem to have been investigated for this nonlinear case, although it constitutes an interesting question). This condition of a fixed N restricts a whole continuous branch to a single solution, but more than one solution may remain at fixed N and geometry, even though they are countably many of them. However these solutions are quite special because they require well defined values of the initial energy, given N. It is possible too that they are linearly unstable, but for the ground state, as are unstable the same kind of "excited" solutions of the Newell-Whitehead-Segel amplitude equations[11]. I hope to explore this point in the future.

Let us now discuss some point raised by the previous developments.

1) The schema we propose for the long time evolution of solutions of NLS is inspired by Kolmogorov idea [12] on dissipation in turbulent flows: a cascade takes place from the large scales (L in our notations) down to the Kolmogorov scale where the molecular viscosity takes over. It is well known too that in real turbulence, this is complicated by intermittency. As explained by Kolmogorov, this is because the transfer of energy from a scale to a lower scale is a nonlinear random process, and thus the whole cascade itself is like a large number of iterations of random multiplicative process (although the extension of these random multiplicative process to more general nonlinear process by Derrida[13] should be relevant therein). This leads to log-normal distributions[14] for the intermittent fluctuations, and perhaps to a change in the scaling laws of the first Kolmogorov theory. We suggest that a similar kind of intermittency is present in this model for space dimensions large enough.

2) It is tempting to extrapolate from this a sort of nonlinear damping relevant for superflows, as one might say that, in physical terms, the cascade to the infinitesimally small scales should stop at a molecular length scale, and then become a sort of dissipation. This schema is perhaps valid in a qualitative sense, but cannot be so in a quantitative sense. This is because the dissipation mechanism we have put into evidence depends in a highly nonlinear fashion on the fluctuations near the ground state, and this does not seem to be compatible with any standard form of molecular damping terms, depending linearly on the macroscopic fields. Moreover it seems difficult to add such a damping term without dropping the property of mass conservation, as well as the possibility of permanent superflows (see below). We face in some sense the same problem as when trying to define a turbulent viscosity [15]. A more subtle question is equally relevant to this point. We alluded in the beginning to the point that one can see a general connection between the present class of problem and the one of the blackbody radiation. It is well known that the Jeans phenomenon (that is the divergence of the total energy at fixed temperature for a classical field) there is stopped at small scales by the quantum character of the fluctuations. Thus one might wonder how the cascade of energy toward the small scales is stopped by quantum phenomena. Indeed those quantum phenomena are not represented in our approach, even though the NLS equation looks a bit like the Schrödinger equation: from the point of view of field theory this is an equation for a <u>classical</u> field. The quantum version would replace the c-number field Ψ by an operator algebra defined [18] by equal time commutation relations of the form:

$[\Psi(\mathbf{x}, t), \Psi^*(\mathbf{x}',t)] = C\ \delta(\mathbf{x}-\mathbf{x}')$,

where C is a constant introduced in order to keep the dimensionless form of NLS, and δ (.) is the Dirac distribution. Within this picture, the cascade toward infinitely small scales is stopped by quantum phenomena, as described by a quantum version of the weak turbulence theory.

3) We left a bit undefined till now what we had in mind when considering a "nonequilibrium" initial situation, but for the fact that, given N, the energy is larger than in the ground state. This may be realized in many different ways. In general this initial condition should not correspond to a nonlinear eigenstate, as these eigenstates are discrete in a finite system, and for a given N. One expects that an excited state will represent something like a large scale flow in superfluid Helium, without normal component. Then, from our considerations, this flow can decay, if it is not an exact eigenstate, by the nonlinear interactions of short wavelength fluctuations. This picture is different from the familiar one, due to Landau, according to whom the flow speed should exceed a critical value to generate elementary excitations, and then decay by quantum transitions [16], as in an ordinary quantum system described by a linear Schrödinger equation. Our picture would take more into account the fundamentally nonlinear character of the basic equation. A way of maintaining the system out of equilibrium is to add a time dependent external field, that is to change (1) into:

$$i\frac{\partial \Psi}{\partial t} = \Delta\Psi + \Psi - |\Psi|^2\Psi + v(\mathbf{x},t)\Psi \quad (15),$$

where v(x,t) is a real external potential. If this potential truly depends on space (variable **x**) and time (variable t), then the energy $H\{\Psi\}$ is no more a constant of the motion, although $N\{\Psi\}$ still is. As suggested in [19] this time dependent v(.,.) could be an imposed in-plane oscillation of a superfluid ring. If v(x,t) depends on **x** on large scales only, then an equilibrium spectrum should exist allowing to transfer the injected energy down to the very small scales. This equilibrium spectrum should be a stationnary solution of the kinetic equation (11), supporting a constant flux of energy from the large to the small scales. Let us try to analyse this by following a method very similar to the Kolmogorov approach to scaling in fully developed turbulence. First we assume that the right hand side of the kinetic equation (11) describes a transfer from a range of wavenumber to another one. This amounts to replace it by a term of the form $\frac{\partial J(k)}{\partial k}$, where J(k) is precisely this flux. This means that J(k) scales as k times the right hand side of (11). In the inertial subrange, one has:

$J(k) \approx k\ (1+\mu)\ [\delta I(k)]^2 \left(\frac{k^2}{(1+\mu)^{1/2}}\right)^{D-2}$, although in the microscopic subrange, one has:

$J(k) \approx k\ (1+\mu)\ [\delta I(k)]^2\ k^{D-2}$. Notice that, as expected, the two expressions for J(k) become the same in the range $k \approx \frac{k^2}{(1+\mu)^{1/2}}$. In the present case, one does not expect the relevant parameter for the Kolmogorov scaling to be J(k), but instead the flux of energy, that is $J^\varepsilon = k^2 J(k)$, that should be a constant, independent of k. In physical terms, this flux J^ε would be proportional to an apparent dissipation per unit volume. From the above expression for J(k), one gets the following power spectra for the fluctuations $\delta I(k)$:

In the inertial subrange, $\delta I(k) \approx [J\varepsilon]^{1/2}\, k^{1/2-D}\, (1+\mu)^{D/4-1}$, and in the microscopic range $\delta I(k) \approx [J\varepsilon]^{1/2}\, k^{-(1+D)/2}\, (1+\mu)^{-1/2}$.

As expected, the total energy stored in this spectrum is infinite, since this is proportional to $\int dk\, k^{D+1}\, \delta I(k)$, that diverges at k large. But the total mass stored in this spectrum is infinite too, because the integral for N, $\int dk\, k^{D-1}\, \delta I(k)$ also diverges at large k. This means that no time independent spectrum can transport the energy to the small scales at constant N. Most likely, this means that for long times, the solution of (15) will be more and more turbulent at any scale, so that, a sizeable part-if not most of-the mass will be contained in those small scales, as one would expect in some sense from a system continuously heated.

4) A major difference between our approach of the decay and the usual one in the superfluid litterature is that our dispersion relation is of the decay type in the long wavelength limit and allows three modes interaction. According to Landau [16], the dispersion relation for elementary excitations (≈the linearized perturbations around the ground state) in superfluid Helium 4 is such that $\omega(\mathbf{k}) \approx c_s k(1-\gamma k^2)$, where γ is positive, although we have the opposite sign here [it is worth pointing out that dissenting views have been presented for what concerns real superfluid Helium 4, as in reference 17]. To model such a relation of the non-decay type (γ positive)within the NLS framework, one might add to NLS higher order space derivatives, to obtain:

$$i\frac{\partial\Psi}{\partial t} = \Delta\Psi + \xi\, \Delta^2\Psi + \Psi - |\Psi|^2\Psi \quad (16),$$

where ξ is a coefficient with the dimension of an inverse square length, of order 1 in the dimensionless unit we use. The dispersion relation (8) is changed into:

$$\omega^2(\mathbf{k}) = 2(1+\mu)(k^2-\xi k^4) + (k^2-\xi k^4)^2$$

and now the long wavelength expansion of $\omega(\mathbf{k})$ reads:

$$\omega(\mathbf{k}) \approx [2(1+\mu)]^{1/2} k\, [1-\gamma k^2 + o(k^4)]$$

where $\gamma = \frac{1}{2}[\xi - \frac{1}{2(1+\mu)}]$. From this, by taking ξ large enough, it is possible to have a positive γ, and thus a nondecay spectrum. With such a spectrum, the relations $\omega(\mathbf{k}) = \omega(\mathbf{k}_1) + \omega(\mathbf{k}_2)$ and $\mathbf{k} = \mathbf{k}_1 + \mathbf{k}_2$ cannot be satisfied simultaneously in the long wavelength limit. Now four modes interaction becomes the dominant nonlinear process of decay in the inertial subrange. This kind of interaction takes place by resonance between fluctuations with wavenumber $\mathbf{k}$, $\mathbf{k}_1$, $\mathbf{k}_2$, $\mathbf{k}_3$, such that $\mathbf{k}+\mathbf{k}_1 = \mathbf{k}_2 + \mathbf{k}_3$ and $\omega(\mathbf{k}) + \omega(\mathbf{k}_1) = \omega(\mathbf{k}_2) + \omega(\mathbf{k}_3)$. This is possible: consider the vectors $\mathbf{k}$ and $\mathbf{k}_1$ as given, then the condition on frequencies implies that, in the longwavelength limit, the sum $k + |\mathbf{k}_1|$ is equal to the sum $|\mathbf{k}_2| + |\mathbf{k}_3|$, although the vectors $\mathbf{k}+\mathbf{k}_1$ and $\mathbf{k}_2 + \mathbf{k}_3$ are equal. The locus of the end of the vector $\mathbf{k}_2$ is an ellipsoid with foci at 0 and $\mathbf{k} + \mathbf{k}_1$. Now one can use the same type of estimate as before to get the self-similar form for the decay of the spectrum in the inertial subrange. The weak turbulence equation reads(again in an oversimplified form, but correct as far as the general structure is concerned as well as the scaling properties):

$$\frac{d\delta I_k}{dt} = \int dk_1 \int dk_2 \int dk_3 \; \delta(k+k_1-k_2-k_3) \; \delta[\omega(k)+\omega(k_1)-\omega(k_2)-\omega(k_3)] \; \delta I(k_1) \; \delta I(k_2) \; \delta I(k_3) \quad (17).$$

Notice that this equation is formally independent on the "ground state" wavefunction Ψ_0. This is because the nonlinear term in (1) is cubic. A more general form of this interaction term is possible however. It would be the second derivative with respect to Ψ of the nonlinear coefficient of Ψ in an extended form of NLS (one where $\Psi|\Psi|^2$would be replaced by $\Psi V(|\Psi|^2)$, V(.) even function with real values). Let c_s be the sound speed. Thus the right hand side of (17) scales like $K^{2D-1} \frac{\delta I^3}{c_s}$ and by comparing this with the left hand side one gets a first relationship:

$\delta I \sim [\frac{c_s K^{2D-1}}{t}]^{1/2}$. The second relationship between the scaling quantities (K, t, and δI) comes again from the energy conservation:

$K \, \delta\Psi \sim \varepsilon^{1/2}$ and $\delta\Psi \sim (K^D \, \delta I)^{1/2}$. Whence the final result: $K \sim (\frac{\varepsilon^2 t}{c_s})^{1/5}$. Notice that this is independent on the dimension D. However it does not hold for the one dimensional (and integrable)case, because other invariants than the Noether invariant with different scaling laws forbid any self similar decay.

5) Till now we claimed that the generic solution of NLS at D>1 decays to the ground state by throwing the excess of energy to small scales fluctuations, this picture being inspired by what one knows of the 2D Euler dynamics [4]. Another more general, and less restrictive scenario is possible however. Starting from an initial excited state, with large scale structures only, it seems possible that for long times the solution will split into a part with wavenumber in the nondecaying subrange $[L^{-1}, L^{-1/2}(1+\mu)^{1/2}]$ plus noise at shorter wavelength and thus in a decaying subrange. Once this splitting would have occured, there would not be any way to decay for the large scales in the non decay subrange. If this is true, the ultimate state could well be an excited state, in the nondecaying subrange, plus noise. The excited state could be then viewed as a kind of permanent superflow. It is a difficult question whether this "superflow" corresponds to a single frequency solution or to a solution with a more complex time dependence. I incline to believe that it cannot have a too complex time dependence, because this would lead for long times through Arnol'd diffusion to a decay to small scales. In particular, one should take care that it is empirically known that the range of application of the KAM theory becomes narrower and narrower as the number of degrees of freedom increases.This kind of question is certainly worth exploring by direct numerical simulations.

6) We neglected in the equation of weak turbulence the contribution of exactly parallel wavenumbers. This kind of contribution cannot be treated in the framework of weak turbulence, because, for weakly dispersive waves, the interaction becomes strong and cannot be treated perturbatively. The random phase approximation, that is needed to derive the weak turbulence

equations, loses its validity because of the formation of coherent structures in real space. In the present case, these coherent structures are shock waves in a sense to be made more precise below [ordinary schock waves are not possible in dissipationless models, as equation (1)].

To show that such shock waves exist, at least in a loose sense, it is more convenient to rewrite the equation (1) for the phase and the modulus. Let us pose $\Psi(\mathbf{x},t)=r(\mathbf{x},t)\exp i\phi(\mathbf{x},t)$, where ϕ and r are two scalar real fields. Then equation (1) is equivalent to the pair :

$$\frac{\partial r}{\partial t} = \nabla(r\nabla\phi) \quad (18.a),$$

$$-r\frac{\partial \phi}{\partial t} = \Delta r-(\nabla\phi)^2+ r - r^3 \quad (18.b).$$

Looking at the derivation of the dispersion relation $\omega(\mathbf{k})$, one can check that the term responsible of the dispersion effect [that is of the $\mathbf{k}^4$ term in (8)], originates from the term Δr on the right hand side of (18.b). This can be verified directly as follows: the intrinsic length scale can be eliminated from the equations (18), only if Δr is dropped in (18.b): without this term the equations are invariant under the transform $\phi \to \sigma\phi$, $t \to \sigma t$, $\mathbf{x} \to \sigma\mathbf{x}$, $r \to r$, σ arbitrary real non zero number. This shows also that this term is negligible in the long wavelength limit. But, as we have seen, the dispersion [that is the correction to $\omega(\mathbf{k})$ beyond the d'Alembert order] plays a crucial role in the weak turbulence theory. So we can be certain that we are investigating other phenomena by dropping this term, as we shall do now. Let us transform (18) into a pair of equations for r and the irrotational field $\mathbf{\Phi} = -\nabla\phi$:

$$\frac{\partial r}{\partial t} + \nabla(r\Phi) = 0,$$

$$\frac{\partial \Phi}{\partial t} + \nabla\left(\frac{\Phi^2}{r} + 1 - r^2\right) = 0.$$

In one dimension of space, these equations are in a standard quasilinear form, and so can generate shocks from smooth initial conditions. Those shocks are eliminated by the dispersive effects represented by the neglected Δr term in (18.b), and so would generate instead "solitons". This is a way of transferring the energy in a coherent way (as opposed to the random process described in the weak turbulence approximation), although this process is efficient in the inertial range only, because the dispersion becomes of order 1 in the microscopic subrange. In this inertial range, one should compare the time scale for this coherent decay process with the one predicted by weak turbulence. The time scale for the formation of shocks is of the order of a coherence length, say λ, divided by an estimate of the fluctuation δc_s of the speed of sound over this length scale: $t_s \sim \frac{\lambda}{\delta c_s}$. The fluctuation of the sound speed is related itself to the fluctuation of the wavefunction Ψ, through $c_s^2 = 2(1+|\Psi|^2)$, which yields $t_s \sim \frac{1}{\varepsilon\lambda}$, if one writes the kinetic energy density as $\varepsilon = \frac{|\delta\Psi|^2}{\lambda^2}$. The late decay for shock in conservative systems is by the emission of solitons, as opposed to the usual dissipation inside the shock layer for dissipative systems. The typical size of these solitons is related to the other parameters through the scalings of the Korteweg-de Vries balance between weak dispersion and weak nonlinearity. In the present case, this gives for the thickness of the solitons $\delta x_s \sim (\delta c_s)^{-1/2}$. The weak turbulence dynamics is the dominant effect in the cascade to small scales if it is more efficient, that is faster, than the shock steepening to transfer energy to the scale δx_s after the delay needed for this steepening. In two dimensions, and from (14) (and by taking μ as of order 1), the central wavenumber of the weak turbulence spectrum is at $K(t) \sim (\varepsilon^2 t)^{1/2}$, which gives $K(t_s) \sim (\frac{\varepsilon}{\lambda})^{1/2}$ at the shock steepening time. At the same time, the shock will begin to transfer energy to scales of order δx_s, that is to wavenumber $K'(t_s) \sim (\delta c_s)^{1/2} = (\varepsilon\lambda)^{1/2}$. The weak turbulence mode of decay will be more efficient in general, because λ is by assumption of the order of the initial large scale, a large number then, and that $K'(t_s)$ is much larger than $K(t_s)$, ε being finite. However at very small ε (the exact limit is for ε less or equal to $\lambda^{-1/2}$), the shock steepening is the dominant process. Indeed the situation might be more complicated than that, because the shock steepening might be operating at later times, when the typical scale λ at a given time cannot be taken anymore as large. But if this happens when the length scales are of order 1, one does not expect that the sound dispersion relation is applicable anymore. Notice however that it could be that the shock steepening plays a role in some part of space to transfer energy to small scales, although it can be neglected on average.

Aknowledgement:

It is a pleasure to thank Alan Newell, Vincent Hakim and Kjartan Emilsson for interesting conversations on the present topic. I am also indebted to Jean Ginibre for communicating relevant references.

REFERENCES:

[1] Yu.N. Grigoriev, Z. Priklad.Matem. i Tekn. Fis. **3**, 27-38 (1983); Yu.N. Grigoriev and V.B. Levinsky Z. Priklad.Matem. i Tekn. Fis. **5**, 60-68 (1986).

[2] R. Robert, C.R. Acad. Sciences, **309**, 757-760 (1989); R. Robert and J.Sommeria, "Statistical equilibrium state for two-dimensional flows" preprint, submitted to the Journal of Fluid Mechanics.

[3] J.Miller, Phys. Rev. Letters **65**, 2137 (1990).

[4] G.F.Carnevale, J. McWilliams, Y. Pomeau, J.B. Weiss, W.R. Young, "Evolution of vortex statistics in two dimensional turbulence" preprint (January 1991).

[5] L.C.Young, Annals of Math. **43**, 84-103 (1942)

[6] W.F. Vinen in "Superconductivity", vol. 2, 1167-1232, edited by R.D. Parks, M. Dekker New-York (1969).

[7] J.Ginibre and A.Velo, J. of functional Analysis **32**, 33-71 (1979); H. Brezis and T. Gallouet, Nonlinear Analysis TMA, **4**, 677-681 (1980); "Introduction aux problèmes d'évolution semilinéaires", Ed. T. Cazenave and A. Horaux, Mathématiques et Applications, SMAI (1990).

[8] B.J. Lemesurier, G.C. Papanicolaou, C.Sulem and P.L. Sulem; Physica D**32**, 210 (1988).

[9] V.E.Zakharov and A.B. Shabat, Sov Phys. JETP **34**, 62-69 (1972).

[10] A.A. Galeev and R.Z.Sagdeev, part 4 of "Basic Plasma Physics", Vol. 1, A.A. Galeev and R.N. Sudan ed.,North Holland Pub.(1984), V.E. Zakharov, ibid. Vol.2.

[11] A.C.Newell and J.A.Whithehead, J. of Fluid Mech. **38**, 279 (1969); L.A. Segel, J. of Fluid Mech. **38**, 203 (1969).

[12] A.N. Kolmogorov, Doklad. Akad. Nauk, **26**, 115-118 (1941).

[13] B. Derrida and R.B. Griffiths, Europhys. Letters **8**, 11(1989); J. Cook and B. Derrida, J. of Statistical Phys. **57**, 89 (1989).

[14] A.N. Kolmogorov, J. of Fluid Mech. **13**, 82-85 (1962).

[15] A.S. Monin and A.M.Yaglom "Statistical fluid mechanics", 2 volumes, MIT Press, Cambridge, Mass. (1975)

[16] L.D. Landau, J. Phys. USSR **5**, 71-82 (1941).

[17] H.J. Maris and W.E. Massey, Phys. Rev. Letters **25**, 270 (1970).

[18]See for instance, chapter 14 in L.I. Schiff "Quantum mechanics", McGraw-Hill Kogakusha, Tokyo (1968).

[19] B. Dorizzi, B. Grammaticos and Y. Pomeau, J. of Stat. Phys. **37**, 93-108 (1984).

[20] P. Manneville and Y. Pomeau, J. de Phys. Lettres **40**, 609 (1980).

Nonlinear Waves and the KAM Theorem: Nonlinear Degeneracies

Walter Craig
Department of Mathematics
Brown University
Providence, RI 02912

C. E. Wayne
Department of Mathematics
Pennsylvania State University
University Park, PA 16802

1. Introduction

This paper is concerned with solutions of nonlinear wave equations, and other partial differential equations that model conservative phenomena in physics and applied mathematics. As the initial value problem is increasingly well understood, the focus of our attention is on the more detailed structure of the phase space in which the evolution equations are posed. The nonlinear wave equation can be viewed as an infinite dimensional Hamiltonian system, thus it is natural to study important classes of periodic and quasiperiodic solutions in the neighborhood of equilibrium. The paper (Craig & Wayne [CW]) constructs periodic solutions for nonlinear wave equations, using a version of the Nash-Moser technique to overcome the inherent small divisor problem. In that reference, certain generic requirements of nonresonance and genuine nonlinearity are needed in the existence proof. This present paper addresses problems in which the hypotheses of genuine nonlinearity are not satisfied, where nonetheless the existence of families of periodic solutions near equilibrium are obtained. Other recent work on the subject of perturbation theory for Hamiltonian systems with infinitely many degrees of freedom include Kuksin [K], Wayne [W], Pöschel [P] and Albanese, Fröhlich and Spencer [AFS].

Some of the more interesting aspects of our approach to these problems are the ties between partial differential equations, Hamiltonian mechanics, and localization theory of mathematical physics. Indeed, the central estimates in this work were pioneered by Fröhlich and Spencer [FS] in the study of the Green's function for random Schrödinger operators. The nonlinear wave equation is not the only equation of interest which has Hamiltonian structure, for which results on periodic and quasiperiodic solutions are of interest. We expect the techniques of [CW] and of this paper to extend to the nonlinear Schrödinger equation, versions of the KdV equation and other problems with infinitely many degrees of freedom, for which the equilibrium solution is an elliptic stationary point. Moreover we expect the analysis of quasiperiodic solutions to be similar to the analysis of periodic solutions in these resonant cases, and plan a further publication on this subject.

This paper describes the construction of periodic solutions of the nonlinear wave equation

$$\partial_t^2 u = \partial_x^2 u - g(x,u) \ , \qquad 0 \le x \le \pi \ , \tag{1.1}$$

where the solution $u(x,t)$ satisfies either periodic or Dirichlet boundary conditions at $x = 0, \pi$. The nonlinear term $g(x,u)$ is taken analytic, with the Taylor expansion in the variable u given by

$$g(x,u) = g_1(x)u + g_2(x)u^2 + g_3(x)u^3 + \cdots . \tag{1.2}$$

Well known examples are the sine-Gordon equation

$$\partial_t^2 u = \partial_x^2 u - b^2 \sin(u), \tag{1.3}$$

and the φ^d-nonlinear Klein-Gordon equation,

$$\partial_t^2 u = \partial_x^2 u - b^2 u + u^{d-1}. \tag{1.4}$$

All of the above partial differential equations can be considered as Hamiltonian systems with infinitely many degrees of freedom. Indeed, we may define the Hamiltonian

$$H(p,u) = \int \frac{1}{2}p^2 + \frac{1}{2}(\partial_x u)^2 + R(x,u)\, dx \ , \tag{1.5}$$

with $\partial_u R(x,u) = g(x,u)$. Denoting $z = (u,p)^T$, Hamilton's canonical equations read

$$\dot{z} = J\nabla H(z), \tag{1.6}$$

where J denotes the standard symplectic matrix. The methods of this paper are perturbative – we construct solutions near the equilibrium point $u = 0$. For the wave equation (1.1) $z = 0$ is elliptic, thus by analogy with finite dimensional problems one expects that the construction of quasiperiodic solutions encounters small divisor problems, and a form of the KAM theorem would be used. In fact the small divisor problem arises even in the construction of periodic solutions, as the presence of infinitely many degrees of freedom introduces a dense set of resonances.

In the reference [CW] the existence theory for periodic solutions is discussed, under hypotheses of nonresonance and genuine nonlinearity. The results are essentially that there is an open dense set $\mathcal{G}$ of nonlinearities such that for $g(x,\cdot) \in \mathcal{G}$, there exist families of periodic solutions of (1.1). The character of these families is typically that of a Cantor set foliated by invariant circles — a situation reminiscent of the conclusion of the KAM theorem for quasiperiodic solutions in finite dimensional Hamiltonian perturbation theory. In the results of [CW], the good set $\mathcal{G}$ depends only upon $g_1(x), g_2(x)$, and $g_3(x)$, the *3-jet* of the nonlinearity $g(x,\cdot)$.

In the present paper we extend the results of [CW] to cases which are equally nonresonant, but which are nonlinearly degenerate. These problems fail to satisfy the 'twist condition' of the previous results, thus the present work enlarges the class $\mathcal{G}$ of nonlinearities for which an existence theorem holds. For example, consider the nonlinear term

$$g(x,u) = g_1(x)u + g_M(x)u^M + \cdots \ . \tag{1.7}$$

For $M > 3$, g is not in the set $\mathcal{G}$, for the curvature of any approximate solution branch will vanish. Among other situations this appears for the nonlinear φ^d Klein-Gordon equation with $d > 4$. We show in this paper that under more subtle conditions of nondegeneracy, again families of periodic solutions of the wave equation (1.1) can be constructed. These conditions depend upon the coefficient $g_1(x)$ of course, and if M is odd, upon the first nonzero coefficient $g_M(x)$ of the nonlinear term. If the first nonzero coefficient $g_M(x)$ has even order, then the existence criterion depends upon a certain subset of the $(2M-1)$-*jet* of $g(x,u)$, that is, upon certain of the coefficients $\{g_1(x),\cdots,g_{2M-1}(x)\}$.

We feel that all these results are quite general, and will extend to other equations and to the construction of quasiperiodic solutions as well. We point out that in the study of quasiperiodic solutions, the analysis of higher order nonlinear degeneracies has not been carried out, even in the case of finite dimensional Hamiltonian systems in the neighborhood of an elliptic stationary point.

Acknowledgements: The authors would like to thank the Université de Genève, the Université de Paris 6, MSRI–Berkeley and Oxford University for their hospitality, and the National Science Foundation and the Alfred P. Sloan Foundation for their support of our research.

2. Results

It is instructive to solve the equation linearized about $u = 0$,

$$\partial_t^2 v = \partial_x^2 v - g_1(x)v \quad . \tag{2.1}$$

This is done by the elementary method of separation of variables. Let $\{(\psi_j(x),\omega_j^2)\}_{j=1}^{\infty}$ be the complete set of eigenfunction—eigenvalue pairs for the linear Sturm-Liouville operator

$$L(g_1)\psi = (-\frac{d^2}{dx^2} + g_1(x))\psi = \omega^2\psi,$$

imposing the proper boundary conditions, ($\psi(0) = \psi(\pi) = 0$ in the Dirichlet case, and $\psi(x) = \psi(x+\pi)$ in the periodic case.) We will assume that all ω^2 are positive with little loss of generality. Then a periodic solution to (2.1) is given by

$$\begin{aligned} v(x,t) &= r\cos(\Omega t + \xi)\psi_j(x) \\ \Omega &= \omega_j \quad . \end{aligned}$$

The general solution of (2.1) is given by sums of these solutions

$$v(x,t) = \sum_{j=1}^{\infty} r_j \cos(\omega_j t + \xi_j)\psi_j(x),$$

parametrized by angles $\{\xi_j\}_{j=1}^{\infty}$, and amplitudes $\{r_j\}_{j=1}^{\infty}$, (action variables $\{r_j^2\}_{j=1}^{\infty}$.) These are not usually periodic, but typically quasiperiodic if at most finitely many amplitudes r_j are nonzero, and in general they are almost periodic, unless a full set of rational relations (infinitely many) exist among the frequencies $\{\omega_j\}$. Thus it is a natural question to pose whether some of these periodic (or quasiperiodic, or almost periodic) solutions persist for the nonlinear problem.

Hypothesis: (i) Let $g(x,u)$ be π periodic in x, and analytic in the strip $\{|\mathrm{Im}\ x| < \overline{\sigma}\}$ and in u in some neighborhood of the origin. In the case of Dirichlet boundary conditions we also ask that g be odd in the (x,u)–plane.

(ii) We assume that in (1.7), $M > 3$.

The cases $M = 2, 3$ were discussed in reference [1]. Define $m = M - 1$ if M is odd, and $m = (\min\{R; M < R \leq 2M-1,\ R \text{ odd, and } g_R(x) \neq 0\} - 1)$ if M is even. If no such R exists, set $m = 2M - 2$.

Theorem 2.1. *There exists a generic set $\mathcal{G}_M$ such that if $g \in \mathcal{G}_M$ then there are uncountably many small periodic solutions to the nonlinear equation (1.1). Furthermore*

(i) The solutions are analytic in a smaller strip $\{|\mathrm{Im}\ x| < \overline{\sigma}/2\}$.

(ii) The solutions are close to the linear periodic solutions, and form a Cantor set foliated by circles. More precisely, there is a small r_0 and a Cantor set $C \in (-r_0, r_0)$ such that if $r \in C$ then there is an angle ξ such that

$$\begin{aligned} |u(x,t;r) - r\cos(\Omega(r)t+\xi)\psi_j(x)| &\leq Cr^M, \\ |\Omega(r) - \omega_j| &\leq Cr^m. \end{aligned} \tag{2.2}$$

(iii) The good set $\mathcal{G}_M$ is open and dense. If M is odd, $\mathcal{G}_M$ depends only upon the coefficients $g_1(x)$ and $g_M(x)$. If M is even, it depends upon $g_1(x)$ and $g_R(x)$, for the minimum R odd, $M < R < 2M-1$, $g_R(x) \neq 0$. If there is no such R, then $\mathcal{G}_M$ depends upon $g_1(x), g_M(x)$, and $g_{2M-1}(x)$.

For an exact description of the topology in which $\mathcal{G}_M$ is dense, see [CW] section 6.

An immediate corollary applies to a specific choice of nonlinearity. Consider the φ^d Klein-Gordon equation (1.4) on the interval $[0,\pi]$. For $d = 4$ this is addressed in [CW], however for $d > 5$ it fails to satisfy the hypothesis of genuine nonlinearity of that paper. When periodic boundary conditions are imposed, the problem can be reduced to an analysis of the phase plane for a solution $u(x - ct)$. When Dirichlet conditions are imposed this is not the case. For d even, Theorem 2.1 applies, giving the following result.

Corollary 2.2. *For an open set of parameters b^2 of full measure, (1.4) has nonlinearity within the good set $\mathcal{G}_M$, and therefore there exist families of periodic solutions, as described in Theorem 2.1.*

This particular dependence of the condition of genuine nonlinearity, and the power m on the coefficients, is natural in terms of the Birkhoff normal form for a dynamical system in the neighborhood of an elliptic stationary point. That is, odd terms in the Hamiltonian (even terms of the nonlinearity) are generically nonresonant, and do not enter the normal form at highest order. Even terms of the Hamiltonian (odd terms of the nonlinearity) are generically resonant, affecting the normal form and the frequency of the solution at highest order. Furthermore, the next to highest order corrections appear at order $2M - 1$.

We remark here that for any $g_2(x), g_3(x), \ldots$, if $g_1(x) = 0$ then the conditions of nonresonance of [CW] are violated. Indeed both the Dirichlet problem and the periodic problem are infinitely resonant, as the equation linearized about $u = 0$ is

$$\partial_t^2 v = \partial_x^2 v,$$

which has an infinite dimensional null space, spanned respectively by the functions $\{\sin(\ell x)e^{\pm i\ell t}\}$, $\{\cos(2\ell x)e^{\pm 2i\ell t}, \sin(2\ell x)e^{\pm 2i\ell t}\}$. Problems which violate the nonresonance condition, with a finite but possibly large null space will be addressed in a subsequent paper. Other than solutions with rational period that are obtained by global variational methods [B,R], the infinitely resonant case has not been addressed, so far as we know.

3. A nonlinear lattice system

We will take the point of view of embedding a circle into phase space, in such a manner that it is invariant with respect to the flow determined by the wave equation (1.1). Denoting an embedded circle by

$$\begin{aligned} S(x,\xi) &= \sum_{j=1}^{\infty} s_j(\xi)\psi_j(x) \\ s_j(\xi) &= s_j(\xi + 2\pi), \end{aligned} \tag{3.1}$$

it will be invariant under flow by the wave equation, and traversed with frequency Ω, if $S(x,\xi)$ satisfies

$$\Omega^2 \partial_\xi^2 S - \partial_x^2 S + g(x,S) = 0. \tag{3.2}$$

To treat the spatial and temporal variables on an equal footing, one expands s_j in Fourier series

$$S(x,\xi) = \sum_{\substack{j=1 \\ k=-\infty}}^{\infty} \tilde{s}(j,k)\, e^{ik\xi}\psi_j(x).$$

If $S(x,\xi)$ satisfies (3.2), the coefficients of this eigenfunction expansion of S satisfy an equation over the lattice, $(j,k) \in \mathbf{Z}^+ \times \mathbf{Z}$,

$$\begin{aligned} 0 &= (\omega_j^2 - \Omega^2 k^2)\tilde{s}(j,k) + W(\tilde{s})(j,k) \\ &= V(\Omega)\tilde{s}(j,k) + W(\tilde{s})(j,k) \quad . \end{aligned} \tag{3.3}$$

We call this the 'mode interaction equation' of the nonlinear problem (3.2). The term $V(\Omega)$ is diagonal in the given basis, while $W(\tilde{s})$ is nonlinear, and at least of order M for small $\tilde{s}$. Linearizing about $\tilde{s} = 0$, we have

$$V(\Omega)\varphi = 0,$$

with solutions $(\varphi, \Omega) = (\delta(j_0, \pm k_0), \omega_{j_0}/k_0)$ corresponding to a periodic solution of (2.1). The point spectrum of $V(\Omega)$ is typically dense in the real line, in particular it accumulates at zero; this is often called the phenomenon of small divisors. The fact that point spectra of the linearized problem approach zero is the fundamental difficulty of the problem. The technique that is presented in [CW] and this paper shows that the geometry of the lattice sites associated with the small divisors also plays an important role in the existence theory.

This lattice equation has certain symmetries which are relevant to the problem. Let $x = (j,k) \in \mathbf{Z}^+ \times \mathbf{Z}$, and write $\overline{x} = (j,-k)$. Then S is real if and only if $\widetilde{s}(\overline{x}) = \overline{\widetilde{s}(x)}$. The equation respects this condition, for $\overline{V(\Omega)(x)} = V(\Omega)(\overline{x})$, and $\overline{W(\widetilde{s}(x))} = W(\overline{\widetilde{s}(x)})$. Additionally there is the symmetry of an autonomous system; for $x = (j,k)$, define $T_\xi \widetilde{s}(x) = e^{ik\xi}\widetilde{s}(x)$. This is a unitary operator on $\ell^2(\mathbf{Z}^+ \times \mathbf{Z})$. The lattice equation is covariant with respect to T_ξ, indeed T_ξ commutes with $V(\Omega)$, and

$$T_\xi W(\widetilde{s})(x) = W(T_\xi \widetilde{s})(x).$$

Other group actions may also respect the equation (1.1), however these will not be addressed in this paper.

The existence theory is started by solving an approximate problem, given by projection of (3.3) onto a finite subregion of the lattice; $B_0 = \{x \in \mathbf{Z}^+ \times \mathbf{Z}; |x| \leq L_0\}$. The approximate problem is solved under conditions of linear nonresonance. Fix a constant $\tau > m + 3$.

Definition 3.1. *Define $\omega \equiv \omega_{j_0}/k_0$. The frequency sequence $\{\omega_j\}_{j=1}^{\infty}$ is (d_0, L_0)–nonresonant with ω if $L_0 >> |j_0| + |k_0|$, and*

(i) for all $0 < |j| + |k| \leq L_0$,

$$|k - \omega j| \geq \frac{d_0}{(|j| + |k|)^\tau}.$$

(ii) For all $(j,k) \neq (j_0, \pm k_0)$, with $|j| + |k| \leq L_0$,

$$|\omega_j^2 - \omega^2 k^2| \geq d_0.$$

Note: If a sequence of $L_0 \to \infty$, with $d_0 = o(L_0^{-1/2})$, then an open dense set of coefficients $g_1(x)$ are (d_0, L_0)–nonresonant with ω for some L_0. This is a result from [CW].

Writing $\Pi_0 V(\Omega) = V_0(\Omega)$, and $\Pi_0 W(\Pi_0 \widetilde{s}) = W_0(\widetilde{s})$, the approximate equations on $\ell^2(B_0)$ are written

$$V_0(\Omega)\widetilde{s} + W_0(\widetilde{s}) = 0. \tag{3.4}$$

Then the linearized equation about $\widetilde{s} = 0$ is simply

$$V_0(\Omega)(\delta\widetilde{s}) = 0. \tag{3.5}$$

This linear operator has a nontrivial null space for $\Omega = \omega = \omega_{j_0}/k_0$. Since the problem is (d_0, L_0)–nonresonant, the null space is two dimensional, spanned by $\varphi(p) = p\delta(j_0, k_0) + \overline{p}\delta(j_0, -k_0)$, with $p \in \mathbf{C}$. Let $N = \{(j_0, k_0), (j_0, -k_0)\}$, the support of the null vectors, and define orthogonal projections Q onto $\ell^2(N)$, and $P = (\mathbf{1} - Q)$. Equation (3.4) is solved via a Lyapounov-Schmidt decomposition.

$$P\big(V_0(\Omega)u_0 + W_0(\varphi(p) + u_0)\big) = 0 \tag{3.6}$$

$$Q(V_0(\Omega)\varphi(p) + W_0(\varphi(p) + u_0)) = 0$$
$$u_0 = Pu_0 \tag{3.7}$$

Define spaces that account for exponential decay of sequences; $\mathcal{H}_\sigma = \{u \in \ell^2(\mathbf{Z}^+ \times \mathbf{Z}); \|u\|_\sigma^2 \equiv \sum_{x \in \mathbf{Z}^+ \times \mathbf{Z}} e^{2\sigma|x|}|u(x)|^2 < \infty\}$. These form a scale of Hilbert spaces, $\mathcal{H}_\sigma \subseteq \mathcal{H}_{\sigma-\gamma}$ for all $0 \le \gamma \le \sigma$. We ask of the nonlinear term that $W \in C^\omega(\mathcal{H}_\sigma : \mathcal{H}_{\sigma-\gamma})$ for all $0 < \gamma \le \sigma < \overline{\sigma}$, with norms

$$\|W(u)\|_{\sigma-\gamma} \le \frac{C_W}{\gamma^{M+1}}\|u\|_\sigma^M$$
$$\|D_u W(u)v\|_{\sigma-\gamma} \le \frac{C_W}{\gamma^{M+1}}\|u\|_\sigma^{M-1}\|v\|_{\sigma-\gamma}$$
$$\|D_u^2 W(u)[w,v]\|_{\sigma-\gamma} \le \frac{C_W}{\gamma^{M+1}}\|u\|_\sigma^{M-2}\|w\|_\sigma\|v\|_{\sigma-\gamma} \quad .$$

The Taylor expansion of W takes the form $W(u) = W^{(M)}(u) + W^{(M+1)}(u) + \cdots$, where the term $W^{(J)}$ is J–multilinear in u. We will assume that $W^{(J)}$ is J–multilinear and symmetric in u, although the symmetry is not essential for the existence theorem. The lattice nonlinearity that comes from the nonlinear wave equation satisfies the above conditions.

Lemma 3.2. *For $r_0^m < (d_0/3L_0^2), \rho_0 = r_0$, the equation (3.6) has a solution $u_0(x;p,\Omega)$ which is analytic in a complex ρ_0–neighborhood of the set $\mathcal{N}_0 \equiv \{(p,\Omega); \|p\| < r_0,\ |\Omega - \omega| < r_0^m\}$. Furthermore, for $\overline{\sigma}/2 < \sigma_0 < \overline{\sigma} - 1/L_0$ there is an estimate*

$$\|u_0(x;p,\Omega)\|_{\sigma_0} \le \|p\|^M \frac{3C_W L_0}{d_0}.$$

This solution is covariant with respect to the translations T_ξ,

$$T_\xi u_0(x;p,\Omega) = u_0(x;T_\xi p,\Omega)$$

(where by notational abuse we denote rotations in the p–plane also by T_ξ.)

These sequences form a family of embedded circles, parametrized by $(\|p\|,\xi,\Omega)$, which are solutions of the approximate problem (3.6).

To finish the approximate bifurcation problem, equation (3.7) is also solved. This is in the form of a mapping, taking $(p,\Omega) \in \mathcal{N}_0 \to \mathbf{R}^2$. The zero set of the mapping consists locally of the Ω axis $\{p = 0\}$, and a surface $(p, \Omega_0(p))$ given as a graph over a neighborhood of zero in $\ell^2(N)$. A simple analysis of the Taylor expansion of this mapping determines that

$$\Omega_0(p) = \omega + \lambda_0^{(m)}\|p\|^m(1 + o(\|p\|)). \tag{3.8}$$

A straightforward perturbation expansion, which is left to the reader, will determine the constant $\lambda_0^{(m)}$. If M is odd, then $m = M - 1$ and

$$\lambda_0^{(m)} = \frac{1}{2k_0^2\omega}\frac{\langle\varphi(p)|W_0^{(M)}[(\varphi(p))^M]\rangle}{\|\varphi(p)\|^{M+1}} \quad . \tag{3.9}$$

When M is even, take R to be the least odd index, $M < R \leq 2M-1$, such that $W_0^{(R)} \not\equiv 0$. If $R < 2M-1$, then $m = R-1$, and

$$\lambda_0^{(m)} = \frac{1}{2k_0^2\omega} \frac{\langle \varphi(p) | W_0^{(R)}[(\varphi(p))^R] \rangle}{\|\varphi(p)\|^{R+1}} \quad . \tag{3.10}$$

If $R = 2M-1$, or there is no such R, then $m = 2M-2$, and the perturbation theory determines first that

$$u_0^{(M)}(x; p, \Omega) = -(PV_0(\Omega)P)^{-1} P(W_0^{(M)}[(\varphi(p))^M]) \quad ,$$

and then

$$\begin{aligned} \Omega_0(p) =& \omega + \frac{1}{2k_0^2\omega} \frac{\langle \varphi(p) | W_0^{(2M-1)}[(\varphi(p))^{2M-1}] \rangle}{\|\varphi(p)\|^2} \\ &+ \frac{M}{2k_0^2\omega} \frac{\langle \varphi(p) | W_0^{(M)}[(\varphi(p))^{M-1}, u_0^{(M)}] \rangle}{\|\varphi(p)\|^2} + o(\|p\|^{2M-2}) \quad . \end{aligned} \tag{3.11}$$

This perturbation analysis generalizes the formal results of [KT], regarding solutions of the nonlinear Klein-Gordon equation.

The analog of the 'twist condition' of [CW] is a condition on the nonvanishing of the coefficients $\lambda_0^{(m)}$. This will ensure that the dependence of the frequency of the solution upon the amplitude is sufficiently nondegenerate.

The full nonlinear equations (3.3) are also considered in a Lyapounov-Schmidt decomposition

$$P\big(V(\Omega)u + W(\varphi(p) + u)\big) = 0, \tag{3.12}$$

$$Q\big(V(\Omega)\varphi(p) + W(\varphi(p) + u)\big) = 0. \tag{3.13}$$

The approximate solution u_0 of (3.6) is a close approximation to the full equation (3.12), for it satisfies the estimate

$$\|P\big(V(\Omega)u_0 + W(\varphi(p) + u_0)\big)\|_{\sigma_0 - \gamma_0} \leq \frac{C_W \|p\|^M}{\gamma_0^{M+1}} e^{-\gamma_0 L_0}. \tag{3.14}$$

However, to adjust this approximate solution to a full solution involves the small divisor problem. The exact solution is obtained not over all of the parameter region $\mathcal{N}_0$, but on a closed Cantor subset $\mathcal{N} \subseteq \mathcal{N}_0$, on which the resonances of the problem are under better control. The solutions are obtained using Newton iteration steps in conjunction with approximations of the lattice $\mathbf{Z}^+ \times \mathbf{Z}$ by an increasing family of finite subdomains $B_n = \{x \in \mathbf{Z}^+ \times \mathbf{Z}; |x| \leq L_0 2^n\}$. To state the existence result, fix $1/2 < \eta < 1$.

Theorem 3.3. *Assume that the sequence $\{\omega_j\}_{j=1}^{\infty}$ is (d_0, L_0)–nonresonant with ω for $d_0 \geq L_0^{-\eta}$, for L_0 sufficiently large. Then there is a constant r_0, a sequence $u(x; p, \Omega) \in \mathcal{H}_{\overline{\sigma}/2}$ which is C^∞ on $\mathcal{N}_0 = \mathcal{N}_0(r_0)$, and a Cantor subset $\mathcal{N} \subseteq \mathcal{N}_0$ such that for $(p, \Omega) \in \mathcal{N}$, u is a solution of (3.12). Furthermore*

$$\|u - u_0\|_{\overline{\sigma}/2} \leq C\|p\|^M e^{-\gamma_0 L_0/2}. \tag{3.15}$$

The second bifurcation equation (3.13) can also be solved, giving a C^∞, T_ξ-invariant solution surface $(p, \Omega(p))$ in addition to the trivial branch of solutions $p = 0$. This solution surface is close to the approximate surface $(p, \Omega_0(p))$, however unless we specify further conditions it will not necessarily intersect the remaining set $\mathcal{N}$, and (3.12) and (3.13) will not be simultaneously satisfied. We ask that in addition to being (d_0, L_0)- nonresonant, the approximate nonlinear problem satisfies a **twist condition.** Then the surface $(p, \Omega(p))$ will intersect $\mathcal{N}$, giving rise to solutions of the full problem (3.3). For the following we fix $0 < \nu < (1 - \eta)$.

Theorem 3.4. *If the approximate problem (3.4) also satisfies the quantitative twist condition* $|\lambda_0^{(m)}| \geq L_0^{-\nu}$, *then the solution surface* $(p, \Omega(p))$ *of (3.13) intersects* $\mathcal{N}$. *Define* $\mathcal{C} = \{0 < r < r_0; \|p\| = r,\ (p, \Omega(p)) \in \mathcal{N}\}$, *the set for which a solution of the full problem is obtained. Then* meas $(\mathcal{C}) > 0$, *and is in fact of order* r_0.

The intersection points correspond to analytic solutions of the nonlinear wave equation (1.1), through their eigenfunction expansion. This proves Theorem 2.1 of the previous section.

Through exact or near resonance, the Cantor set $\mathcal{C}$ may not have $r = 0$ as an accumulation point, for $(p, \Omega) = (0, \omega)$ may be too resonant, and not in $\mathcal{N}$. However if the frequency sequence $\{\omega_j\}_{j=1}^\infty$ is fully nonresonant with ω, then $\mathcal{C}$ does accumulate at zero, and in addition there is an estimate of its density nearby. Let $\overline{\tau} > m + 3$ and $\overline{\alpha} > \overline{\tau} + 1$ be fixed.

Theorem 3.5. *Suppose that a* (d_0, L_0)*-nonresonant sequence* $\{\omega_j\}_{j=1}^\infty$ *satisfies the conditions of full nonresonance.*

(i) For all $0 < |(j, k)| < \infty$,

$$|k - \omega j| \geq \frac{d_0}{(|j| + |k|)^{\overline{\tau}}}.$$

(ii) For all $(j, k) \neq (j_0, \pm k_0)$,

$$|\omega^2 k^2 - \omega_j^2| \geq \frac{d_0}{(|j| + |k|)^{\overline{\alpha}}}.$$

Define $\mathcal{C}(r_1) = \mathcal{C} \cap [0, r_1]$. *Then there is an exponent* $\overline{\mu}$ *such that*

$$\text{meas}\,(\mathcal{C}(r_1)) \geq r_1(1 - C r_1^{\overline{\mu}})$$

for all $0 < r_1 < r_0$.

One can additionally make an estimate of the size of $\overline{\mu}$, there are similar estimates in [CW].

4. Proof of Theorem 3.3

The proof is via a modified Newton iteration scheme, similar to the Nash-Moser method. The major difference is the presence of a null space, and the sensitive parametric dependence of the approximate solutions and the linearized operator. Thus during the iteration, an acceptable set of parameters must be chosen as well, resulting ultimately in the Cantor set $\mathcal{N}$ on which the first bifurcation equation (3.12) is solved. The second bifurcation equation is a finite dimensional mapping. The zero set corresponding to a nontrivial solution is given by a graph $(p, \Omega(p))$, which gives a relationship between the action and the frequency of a solution, called the **frequency map**. This exhibits one of the differences of the present technique from the more classical versions of the KAM theorem, in which the problem is assumed nonlinearly nondegenerate, the frequency map is performed first, and only then does the analysis of the invariant sets take place.

An outline of the iteration is as follows. We choose:

(1) A sequence of length scales $L_n = L_0 2^n$ which define the approximating domains $B_n = \{|x| \leq L_n\}$ which exhaust $\mathbf{Z}^+ \times \mathbf{Z}$.

(2) A sequence of tolerances for small divisors (small eigenvalues) $\delta_n = L_n^{-\alpha}$, for a suitable $\alpha > 0$.

(3) A sequence of lengths $\ell_n = L_n^\beta$ over which linear resonances are decoupled.

(4) A sequence $\gamma_n = c_0/(n+1)^2$ which governs loss of exponential decay of the approximate solutions throughout the the iteration.

(5) And a rapidly convergent sequence $\epsilon_n = \epsilon_0^{\kappa^n}$, for $1 < \kappa < 2$, which will bound the error terms during the iteration.

The size of the error is dominated by a rapidly convergent sequence as the iteration scheme has quadratic errors; this is the usual phenomenon with the Nash-Moser technique.

The major issue to contend with is the invertibility of relevant linearized operators. Let $B \subseteq \mathbf{Z}^+ \times \mathbf{Z}$ be a subdomain of the lattice. We define the **Hamiltonian operator** on $\ell^2(B)$ by

$$H_B(p, \Omega; u) = \big(V(\Omega) + D_u W(\varphi(p) + u)\big)_B \ .$$

The subscript B denotes the restriction of the operators to $\ell^2(B)$. Invertibility depends crucially upon the small spectra of the operator $V_B(\Omega)$, as the following result demonstrates.

Lemma 4.1. *Let $A \subseteq \mathbf{Z}^+ \times \mathbf{Z}$ be a domain such that $|V(\Omega)(x,x)| > d_0$ for all $x \in A$. Then for $r_0^{m-1}/d_0 << 1$ the Green's function*

$$G_A(x,y) = \big(V(\Omega) + D_u W(\varphi(p) + u)\big)_A^{-1}(x,y)$$

satisfies the estimate

$$\|G_A\|_{\sigma_0} \leq \frac{C}{d_0} \ .$$

We call a lattice site $s \in \mathbf{Z}^+ \times \mathbf{Z}$ **singular** if $|V(\Omega)(s,s)| < d_0$, and regular otherwise. Connected regions of singular sites are called singular regions. The wave equation has singular regions consisting of either isolated sites, or else two adjacent sites, $S = \{s_1, s_2\}$, with $s_1 = (j,k)$, $s_2 = (j+1,k)$. We will consider local Hamiltonians defined on neighborhoods of singular regions. Let $S \subseteq B_{n+1} \backslash B_n$ and $C_{\ell_{n+1}}(S) = \{x; \mathrm{dist}(x, S) \leq \ell_{n+1}\}$. We will be concerned with the operators $H_S(p, \Omega; u_n)$ and $H_{C_{\ell_{n+1}}(S)}(p, \Omega; u_n)$.

The proof of Theorem 3.3 is by induction on the following statements.
$(n.1)$ There is a sequence $u_n(x; p, \Omega) = u_0(x; p, \Omega) + \sum_{j=0}^{n} v_j(x; p, \Omega)$ in $\ell^2(B_{n+1})$, which is C^∞ on $\mathcal{N}_0$, analytic in a $\delta_{n+1} r_0 / L_n^2$ complex neighborhood of $\mathcal{N}_{n+1}$ such that

$$\|P(V(\Omega)u_n + W(\varphi(p) + u_n))\|_{\sigma_n} \leq \|p\|^M \epsilon_n$$
$$\|v_n\|_{\sigma_n - \gamma_n} \leq \frac{C_G^n \epsilon_n}{\delta_{n+1} \gamma_n^s} \|p\|^M$$

for some fixed constant s.
$(n.2)$ There exists a closed domain $\mathcal{N}_{n+1} \subseteq \mathcal{N}_n \subseteq \cdots \mathcal{N}_0$ with the following properties.
(i) If $(p, \Omega) \in \mathcal{N}_{n+1}$, and $S_1, S_2 \subseteq B_n^c$ are any two singular regions, then

$$\mathrm{dist}(S_1, S_2) > 2\ell_{n+1} \quad .$$

(ii) If $(p, \Omega) \in \mathcal{N}_{n+1}$, and S is a singular region in $B_{n+1} \backslash B_n$, then

$$\mathrm{dist}(\mathrm{spec}(H_S(p, \Omega; u_n)), 0) > \delta_{n+1}$$
$$\mathrm{dist}(\mathrm{spec}(H_{C_{\ell_{n+1}}(S)}(p, \Omega; u_n)), 0) > \delta_{n+1} \quad .$$

(iii) Any C^∞, T_ξ–invariant surface $\Omega(p) = \omega + \lambda \|p\|^m (1 + o(\|p\|))$, with $|\lambda| > L_0^{-\nu}$ intersects $\mathcal{N}_{n+1}$ with nonzero measure;

$$\mathrm{meas}(\{r \in [0, r_0); \|p\| = r, (p, \Omega(p)) \in \mathcal{N}_{n+1}\}) \geq r_0(1 - C r_0^\mu) \quad .$$

A consequence of $(n.2)(i)(ii)$ is that the Green's function for any $E \subseteq B_n \backslash N$ is controlled on the parameter region $\mathcal{N}_{n+1}$.

Lemma 4.2. *Let A be a nonsingular region, and $E \subseteq (B_{n+1} \backslash N) \cup A$. The Green's function satisfies*

$$\|G_E(p, \Omega; u_n)\|_{\sigma_n} \leq \frac{C_G^n}{\delta_{n+1} \gamma_n^s} \quad ,$$

and under perturbations of u_n of size $\|u - u_n\|_{\sigma_n - \gamma_n} \leq \|p\|^M \epsilon_n / \delta_{n+1} \gamma_n^s$,

$$\|G_E(p, \Omega; u)\|_{\sigma_n - 2\gamma_n} \leq \frac{2C_G^n}{\delta_{n+1} \gamma_n^s} \quad .$$

Proof. The proof is the same as in [CW], Section 5. The arguments involve the decoupling of the local Hamiltonians at singular regions of $B_{n+1} \backslash N$. As long as the spectra of the local Hamiltonians are controlled, and the singular regions are sufficiently separated, resolvant expansions can be employed to recover the full Green's function. □

Induction step $(n.1)$ will follow from $((n-1).2)(i)(ii)$ and Lemma 4.2. Indeed the Newton iteration step is

$$v_{n-1} = -G_{B_n \backslash N}(p, \Omega; u_{n-1})\big(V(\Omega)u_{n-1} + W(\varphi(p) + u_{n-1})\big)_{B_n \backslash N} \quad ,$$
$$u_n = u_{n-1} + v_{n-1} \quad .$$

With this definition of u_n the Taylor remainder theorem will exhibit a quadratic error, and the error due to domain truncation will be exponentially small if some decay is sacrificed. We again refer to [CW] for details of the convergence proof.

The remaining task is to realize a large set of parameters $\mathcal{N}_{n+1} \subseteq \mathcal{N}_n$ such that $(n.2)$ is satisfied. Conditions (i) and (ii) decrease the size of $\mathcal{N}_n$, while condition (iii) requires that it be sufficiently large, and further satisfy certain geometrical properties related to the order of contact of the nonlinear degeneracy. Central to the verification of this induction step is a lemma on eigenvalue perturbation theory for the local Hamiltonians. For the case M even we introduce an additional hypothesis on the lattice nonlinearity W, a restriction on the self-interaction of the system within a singular region. It will always be satisfied for the nonlinear wave equation. The case M odd has no such requirement.

Hypothesis: If $z, w \in S$ a singular region, then for $M \leq J < R$,

$$\langle \delta(z) | D_u^J W(0)[(\partial_p \varphi(p))^{J-1}, \delta(w)] \rangle = 0 \ .$$

Consider a self adjoint operator $H(a)$ depending upon a parameter a, and suppose an eigenvector–eigenvalue pair $(\psi(a), e(a))$ of $H(a)$ is smooth. Then

$$\partial_a e(a) = \langle \psi(a) | \partial_a H(a) | \psi(a) \rangle \quad , \tag{4.1}$$

which is known as the Feynman–Hellman formula.

Lemma 4.3. *Let $(\psi(p,\Omega), e(p,\Omega))$ be an eigenvector–eigenvalue pair for $H_{C(S)}(p,\Omega)$. Then*

$$|\langle \psi(p,\Omega) | \partial_\Omega H_{C(S)}(p,\Omega) | \psi(p,\Omega) \rangle| \geq C_1 L_n^2 \quad . \tag{4.2}$$

For (p,Ω) satisfying $(n+1.2)(i)$,

$$|\langle \psi(p,\Omega) | \partial_p H_{C(S)}(p,\Omega) | \psi(p,\Omega) \rangle| \leq C_2 \|p\|^{m-1}. \tag{4.3}$$

Let $e(p,\Omega)$ be an eigenvalue of a local Hamiltonian (labeled by ordering), and Z be the set in $\mathcal{N}_0$ on which $e(p,\Omega)$ vanishes. Z is given by a graph $(p, \Omega_Z(p))$, and if $(p_1, \Omega_Z(p_1)), (p_2, \Omega_Z(p_2))$ are nearby points satisfying $(n.2)(i)$, then

$$|\Omega_Z(p_2) - \Omega_Z(p_1)| \leq \frac{C_3}{L_n^2} \big| \|p_2\|^m - \|p_1\|^m \big| \tag{4.4}$$

The proof of this is similar to Lemma 4.14 of [CW].

This result allows us to control the excisions of parameters in order to satisfy $(n.2)(iii)$. Consider a T_ξ invariant surface $(p, \Omega(p))$, with $\Omega(p) = \omega + \lambda \|p\|^m (1 + o(\|p\|))$, and $|\lambda| > L_0^{-\nu}$. Let $S \subseteq B_{n+1} \backslash B_n$ be a singular region, and $e(p,\Omega)$ an eigenvalue of a local Hamiltonian $H_{C(S)}$. Suppose that for some p_1, $e(p_1, \Omega(p_1)) = 0$, and that $(p_1, \Omega(p_1))$ satisfies $(n.2)(i)$. We are concerned with nearby points on the surface $(p, \Omega(p))$. In order to inductively construct the next set $\mathcal{N}_{n+1}$ a δ_{n+1}/L_n^2–neighborhood of Z is excised from $\mathcal{N}_n$. If the point $(p, \Omega(p))$ is excised in this process, then

$$\begin{aligned}\frac{\delta_{n+1}}{L_n^2} &\geq |\Omega(p) - \Omega_Z(p)| \\ &\geq |\Omega(p) - \Omega(p_1)| - |\Omega_Z(p_1) - \Omega_Z(p)| \\ &\geq (|\lambda/2| - C_3/L_n^2)\big|\|p_2\|^m - \|p_1\|^m\big| \quad .\end{aligned}$$

Hence any p such that $\|p - p_1\|^m > (4/|\lambda|)(\delta_{n+1}/L_n^2)$ is not excised, and $|e(p, \Omega(p))| > \delta_{n+1}$. Lemma 4.3 provides the main result needed to verify the induction statement $(n.2)(iii)$, and with some patience the convergence proof will follow.

References

[AFS] Albanese, C. and Fröhlich, J. and Albanese, C. Fröhlich, J. and Spencer, T.: Periodic solutions of some infinite-dimensional hamiltonian systems associated with non-linear partial difference equations: Parts I and II. Commun. Math. Phys. **116**, 475-502 **119**, 677-699 (1988).

[B] Brezis, H.: Periodic solutions of nonlinear vibrating strings and duality principles. Bull. AMS **8**, 409-426 (1983).

[CW] Craig, W. and Wayne, C.E.: Preprint. (June 1991).

[E] Eliasson, H.: Perturbations of stable invariant tori. Ann. Sc. Super. Pisa, Cl. Sci. **IV Ser. 15**, 115-147 (1988).

[FS] Fröhlich, J. and Spencer, T.: Absence of diffusion in the Anderson tight binding model for large disorder or low energy. Commun. Math. Phys. **88**, 151-184 (1983).

[KT] Keller, J. and Ting, L.: Periodic vibrations of systems governed by non-linear partial differential equations. Commun. Pure Appl. Math. **19**, 371-420 (1966).

[K] Kuksin, S.: Perturbation of quasiperiodic solutions of infinite-dimensional linear systems with an imaginary spectrum. Funct. Anal. Appl. **21**, 192-205 (1987). Conservative perturbations of infinite - dimensional linear systems with a vector parameter. *ibid.* **23**, 62-63 (1989).

[P] Pöschel, J.: Small divisors with spatial structure in infinite dimensional hamiltonian systems. Commun. Math. Physics **127**, 351-393 (1990).

[R] Rabinowitz, P.: Free vibrations for a semilinear wave equation. Commun. Pure Appl. Math. **30**, 31-68 (1977).

[W] Wayne, C. E.: Periodic and quasi-periodic solutions of nonlinear wave equations via KAM theory. Commun. Math. Phys. **127**, 479-528 (1990).

Homoclinic Structures in Open Flows

Vered Rom-Kedar
The James Franck Institute and the Department of Mathematics
The University of Chicago, Chicago, Illinois 60637

Abstract: We study the homoclinic tangle associated with the phase space flow of a particle in a cubic potential, subject to small and temporally periodic forcing. We construct a bifurcation diagram describing the changes in the Birkhoff signature of the tangle as the strength and frequency of the forcing are varied. From this diagram we find special regions in the parameter space for which we can approximate some of the properties of the flow. For example, we approximate the escape rates from the vicinity of the homoclinic tangle and the elongation rate of segments of the unstable manifold. The methods we use can be easily applied to many time periodic, near integrable *open flows*. These are flows which contain a single loop of broken separatrices, so that once a particle escapes the vicinity of the homoclinic tangle it never comes back. This study is essentially analytical; we use solutions of the autonomous system and do not compute the trajectories of the chaotic, time dependent flow.

Section 1. Introduction

Flows are often visualized by passive traces in the fluid. The large scale structures one then observes are the signature of the unstable manifolds of hyperbolic invariant sets. In fact, the stable and unstable manifolds are the appropriate extensions of streamlines to unsteady flows [1]. This observation arises by considering the dynamical system which governs the motion of fluid particles, $\dot{x} = u(x,t)$. Further, it is established in [2], [3], and [4] that the properties of the stable and unstable manifolds supply both qualitative and quantitative information regarding the transport properties of a large class of dynamical systems. Hence, the ideas and methods developed in these papers are applicable in many fields of physics.

In [3], we developed analytical methods for estimating some of the properties of chaotic maps and flows using the geometrical structure of the homoclinic tangle. Here, we report some preliminary results regarding the application of these methods to a specific physical example. We view this work as a first step in an examination of their usefulness and accuracy. We believe this study demonstrates their usefulness; we obtain approximate results using seconds of a workstation time, whereas other methods require typically hours of supercomputer time. A more comprehensive study, which includes numerical simulations and error analysis, aimed at examining the accuracy of these methods, is underway.

The methods presented here can be easily applied to any near integrable, two dimensional, area preserving, time periodic *open flow*. These are flows of the form:

$$\begin{aligned} \frac{dx}{dt} &= \frac{\partial H_\varepsilon(x,y,t)}{\partial y} \\ \frac{dy}{dt} &= -\frac{\partial H_\varepsilon(x,y,t)}{\partial x} \end{aligned} \tag{1.1}$$

where $H_\varepsilon(x, y, t) = H_\varepsilon(x, y, t+T)$ is analytic in ε near $\varepsilon = 0$ and $H_0(x, y, t) \equiv H_0(x, y)$. In addition, the flow (1.1) satisfies the following three assumptions:

(1.A) *The system (1.1) possesses an hyperbolic periodic orbit* $p(t) = (x(t), y(t), t)$ *of period* T.

(1.B) *For* $\varepsilon \neq 0$ *the stable (resp. unstable) manifold of* p *have one part which intersects the unstable (resp. stable) manifold transversly. The orbits along which the stable and unstable manifolds intersect are called* homoclinic orbits.

(1.C) *The other part of the stable (resp. unstable) manifold of* p *extends to infinity, possessing no homoclinic orbits.*

The simplest interesting example of a system of the form (1.1), satisfying (1.A-1.C), is the phase space flow of a particle in a forced cubic potential, with the Hamiltonian:

$$H_\varepsilon(x, y, t) = \frac{1}{2}y^2 + (\frac{1}{2}x^2 - \frac{1}{3}x^3)(1 + \varepsilon\cos(\omega t)) \tag{1.2}$$

where ε and ω are the two non-dimensional parameters, measuring, respectively, the strength and frequency of the forcing. Here we study this example in detail. This problem has direct applications in mechanics and chemistry and may also be considered as the normal form of a more complicated Hamiltonian system. Another example of a flow satisfying these assumptions is that of inviscid fluid particles in the vicinity of an axisymmetric vortex ring, where the core of the ring is small and its shape is a nearly circular ellipse.

The most significant assumption we make on (1.1) is (1.C). It implies that there is only one tangle of the stable and unstable manifolds, and hence there is no mechanism for re-entrainment. This is the crucial assumption made in [3]. This constraint excludes interesting dynamical systems such as the forced Duffing Eq., the forced pendulum, etc. For simplicity of notation we also assume in (1.A - 1.C) that the homoclinic tangle is associated with the intersection of the stable an unstable manifolds of a single hyperbolic periodic orbit $p(t)$. The results can be easily extended to different cases such as the the motion of inviscid fluid particles in the vicinity of an oscillating vortex pair (OVP), studied in [1].

As the theories for periodically forced autonomous systems suggest, one should start by reducing the continuous time system (1.1) to a two dimensional area preserving map. Formally, it is a well known procedure: one introduces a Poincaré section in time to obtain the required map [5],[6], [1]. Indeed, the transport theory is based upon the geometry of the manifolds in such a Poincaré section. However, in most cases, for $\varepsilon > 0$, one can only compute this Poincaré map numerically. Moreover, even if the map is found explicitly, it is unclear how to extract the structure of the manifolds analytically. Previous studies of the transport properties of time periodic flows and two dimensional maps consist of extensive numerical simulations in which exact or approximate trajectories of the chaotic system are calculated (e.g. [1], [2] and [7-9]). Here, instead of finding the Poincaré map explicitly, we use the Whisker map ([10], [11], [3]) to approximate the geometrical properties of the structure of its manifolds. In this way, we isolate regions in the parameter space (ε, ω) for which we can approximate the topology of the manifolds with simpler structures. For these simpler structures we use the methods developed in [3] to estimate properties of the flow such as the transport rates and topological entropy.

We stress that our approach is to approximate the topology of the homoclinic tangle of the Poincaré map by a similar topological structure. The Whisker map serves only as a tool to find the properties of that structure. This is to be contrasted with previous studies (e.g. [7], [10] and [11]) in which the Whisker map itself is used to approximate the dynamics of the perturbed continuous system.

This paper is organized as follows; in section 2 we argue that (1.2) gives rise to a system satisfying (1.A-1.C). Then, we set up the notation for classifying the homoclinic tangles and define precisely the quantities we estimate in this paper. We end this section with a definition of some of the geometrical parameters of the homoclinic tangle. In section 3 we compute the Whisker map for (1.1) with the Hamiltonian (1.2) and use it to estimate the geometrical parameters which were defined in section 2. The results are summarized in a bifurcation diagram, which describes the dependence of the geometrical parameters on (ε, ω). In section 4, we summarize some of the methods developed in [3] for estimating the development of tangles in specific region of phase space. Then, we use this construction to estimate the exponential growth rate of line elements for our example. Finally, we estimate the escape rates for these tangles using the Whisker map and the methods developed in [3].

Section 2. The Geometrical Properties of the Homoclinic Tangle.

a. The Open Flow Property. First, we verify that for $\varepsilon \neq 0$ the Hamiltonian (1.2) give rise to a system which satisfies the assumption (1.A-1.C). Writing the system (1.1) explicitly for our example, we obtain:

$$\begin{aligned} \frac{dx}{dt} &= y \\ \frac{dy}{dt} &= x(x-1)(1+\varepsilon\cos(\omega t)) \end{aligned} \tag{2.1}$$

For all ε, the points $(0,0)$ and $(1,0)$ are fixed points of (2.1). When $\varepsilon = 0$, the fixed point $(1,0)$ is hyperbolic, hence for ε sufficiently small $(1,0)$ remains hyperbolic and assumption (1.A) is satisfied. For the unperturbed system assumption (1.C) clearly holds. It is not hard to prove that for ε sufficiently small the perturbed system must satisfy (1.C) as well. Finally, for $\varepsilon = 0$, the parts of the stable and unstable manifolds which do not extend to infinity coincide. To prove that (1.B) holds for $\varepsilon \neq 0$ we calculate the Melnikov function and verify that it has simple zeroes [5],[6]. First, we solve for the unperturbed homoclinic orbit, $q_0(t)$,

$$q_0(t) = \Big(1 - \frac{3}{2}\operatorname{sech}^2(t/2), -\frac{3}{2}\operatorname{sech}^2(t/2)\tanh(t/2)\Big) \tag{2.2}$$

where, with no loss of generality, we choose $q_0(0) = (-\frac{1}{2}, 0)$. Using (2.2) we compute the Melnikov function $M(t_0)$:

$$M(t_0) = \int_{-\infty}^{\infty} y\,(-x+x^2)\cos(\omega(t+t_0))\Big|_{(x,y)=q_0(t)} dt = C(\omega)\sin(\omega t_0), \tag{2.3}$$

where

$$C(\omega) = \frac{9}{2}\int_0^{\infty} \frac{\sinh(t/2)}{\cosh^5(t/2)}(1 - \frac{3}{2}\operatorname{sech}^2(t/2))\sin(\omega t)dt. \tag{2.4}$$

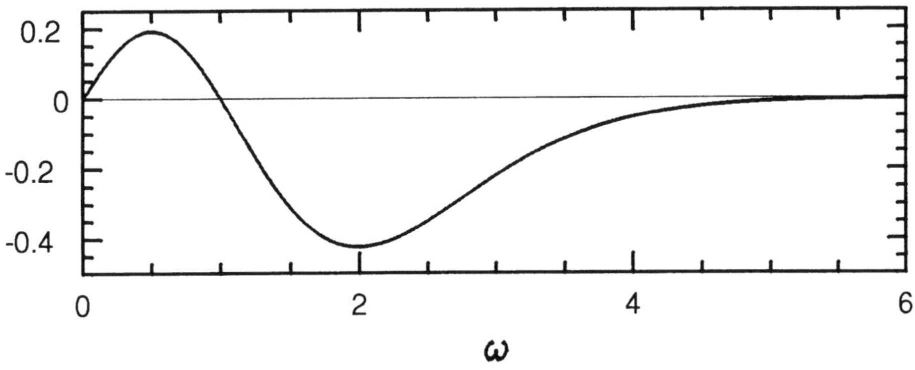

Figure 1. C(ω), The Maximal Magnitude of the Melnikov Function.

The function $C(\omega)$ is plotted in figure 1. Notice that $C(\omega) = 0$ when $\omega = 1$ and when $\omega \to 0, \infty$. Therefore, the analysis in this paper applies to small ε values and all finite values of ω, excluding neighborhoods of 0 and 1. We discuss the possible behavior near these special values of ω in the conclusion section.

b. Qualitative description of the tangle and the transport problem. So far we have verified that for ε sufficiently small the system (2.1) satisfies the required assumptions. We introduce a Poincaré crossection in time and define the Poincaré map, F, as the return map to this crossection. For ε sufficiently small F has a hyperbolic fixed point $p = (1, 0)$, and its stable and unstable manifolds intersect transversly, constituting a homoclinic tangle such as the one drawn in figure 2. In this paper we study the properties of this tangle.

We call a homoclinic point q_0 a *primary intersection point (pip)* if the segments of the stable and unstable manifolds connecting the fixed point p to q_0, denoted by $S[p, q_0], U[p, q_0]$ respectively, intersect only in p and q_0 (see figure 2). The pip orbit of q_0 is the set $\{q_i\}$, $i \in Z$, where q_i is the ith image of q_0 under the map.

Since the Melnikov function has two simple zeroes every period of the perturbation (Eq. (2.3)), the Poincaré map F has exactly two pip orbits, denoted by q_i and p_i in figure 2. We denote the region bounded by $S[p, p_0]$ and $U[p, p_0]$ by $\hat{S}$. We define the segments of W_+^u and W_+^s with end points p_i, q_i as follows:

$$J_i = S[q_i, p_i] \qquad J_i' = S(p_{i+1}, q_i) \qquad K_i = U[q_i, p_{i+1}] \qquad K_i' = U(p_i, q_i) \qquad (2.5)$$

Clearly, $M_n = F^n(M_0)$ where M stands for any of the segments J, K, J', K'. Let D_r denote the region bounded by J_r and K_r' and E_r the region bounded by K_r and J_r'. We will call these regions "lobes".

Eqs. (2.1) and (2.3) imply that for $\omega > 1$, at the Poincaré section $t = 0$, p_0 is located on the x-axis and the orientation of the manifolds at p_0 is as depicted in figure 2. However, when $\omega < 1$ this Poincaré map is symmetric about q_0; the same structure appears if one considers the Poincaré section at $t = \pi/\omega$. For simplicity of presentation, we will limit our discussions to the case $\omega > 1$, and quote the results for $\omega < 1$.

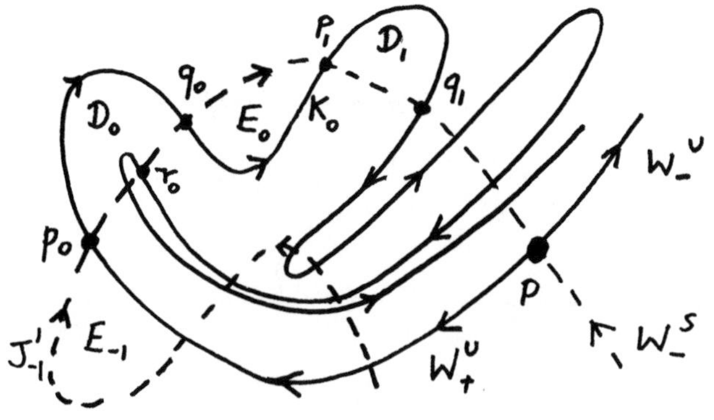

Figure 2. The Homoclinic Tangle
q_i and p_i are pip orbits. r_i is not.

Denoting the parts of the stable and unstable manifolds of F which satisfy (1.B) by W_+^s and W_+^u respectively, it follows that $W_+^u = \bigcup_{n=-\infty}^{\infty} K_n \cup K'_n$, and $W_+^s = \bigcup_{n=-\infty}^{\infty} J_n \cup J'_n$. Denoting the other parts of the manifolds by W_-^s and W_-^u respectively, the assumption (1.C) implies that W_-^s and W_-^u do not contain any homoclinic points. It is not hard to prove that for ε sufficiently small and $n > 0$ the sequences $\{K'_n\}$ and $\{J'_{-n}\}$ accumulate along W_-^u and W_-^s respectively and contain no homoclinic points as well. This is the main assumption made in [3] regarding the structure of the manifolds. We call a flow (a map) which satisfy this assumption an *open flow (map)*.

For open flows, an initial distribution of points in the vicinity of the homoclinic loop will eventually be carried to infinity. In may applications it is important to obtain the rate at which the initial distribution disperses. For example, in Chemistry this may correspond to the dissociation rates [12] and in the fluid mechanics context it corresponds to the effective diffusion rate in that region. Another quantity of interest is the boundary length; if one considers reacting fluids, the amount of product is proportional to the length of the boundary. Finally, when the particles respond to the stretching exerted by the flow (e.g. polymers suspended in a fluid), the distribution of stretching rates experienced by line elements may be important. We define the following measures of these properties for an initial uniform distribution of initial conditions in the region $\hat{S}$:

1. The amount of phase space area originating in $\hat{S}$ which escapes at the nth iteration:

$$c_n = \mu(F^{n-1}(\hat{S}) \cap \hat{S}) - \mu(F^n(\hat{S}) \cap \hat{S}),$$

   where $\mu(A)$ denotes the area of the set A.

2. The amount of phase space area originating in $\hat{S}$ which stays in $\hat{S}$ after the nth iteration:

$$R_n = \mu(F^n(\hat{S}) \cap \hat{S}).$$

3. The length of the boundary of $F^n(\hat{S})$,

$$L_n = \sum_{-\infty}^{n-1} [L(K_j) + L(K'_j)] + \sum_{n}^{\infty} [L(J_j) + L(J'_j)].$$

   where $L(K)$ denotes the arc length of the segment K.

4. The distribution of the stretching rates along the boundary of $F^n(\hat{S})$,

$$\beta(n) = \frac{L(K_n \cap D_0)}{L(K_0 \cap D_{-n})}.$$

It is easy to show that the above quantities are independent, up to a shift in n, of the definition of the "origin" of the orbit p_i and of the particular Poincaré section one chooses. It follows from the form of (2.1) that one can choose a Poincaré section for which $p_0 = (x_0, 0)$ and the map is symmetric with respect to a reflection about the x-axis with a time reversal (see [2] for a general discussion). Such a choice is both elegant and efficient computationally.

From the above quantities, we may extract some interesting asymptotic information, for example:

1. The area of the invariant set in $\hat{S}$ is given by R_∞.
2. The asymptotic behavior of c_n for large n; in particular, it is of interest to find whether c_n decay exponentially or as a power law in n (see [8] for a discussion).
3. The topological entropy, which may be estimated by the asymptotic exponential growth rate of $L(K_n)$,

$$\lambda = \lim_{n \to \infty} \frac{1}{n} \ln L(K_n).$$

4. The growth rate of the averaged stretching rate

$$\beta = \lim_{n \to \infty} \frac{1}{n} \ln \beta(n).$$

If n is not too big, one can compute the above quantities numerically. In [1] we have shown that the computation of c_n and R_n can be reduced to the computation of the escape rates, e_n, defined by

$$e_n = \mu(F^n E_0 \cap D_0).$$

In fact, it is easy to show that for open flows

$$c_n = \mu(D_0) - \sum_{j=1}^{n-1} e_j \,, \qquad R_n = \mu(\hat{S}) - \sum_{j=1}^{n} c_j. \tag{2.6}$$

The above results are exact and supply a major reduction in computation efforts of the transport rates. In [2] it has been proven that similar formulae hold for two dimensional maps which are not open maps, and for non area preserving maps. However, the dependence on numerical computation is a major obstacle for computing the escape rates in applications; for the two-dimensional case, the computational approach is feasible, but too expensive for completing, say, a bifurcation diagram describing the change in the behavior of the transport rates. Moreover, it is inappropriate for estimating the asymptotic behavior. In higher dimensions, the computational approach is bound to fail. The methods presented here carry over to higher dimensions in some special cases, in the same fashion as presented in [4].

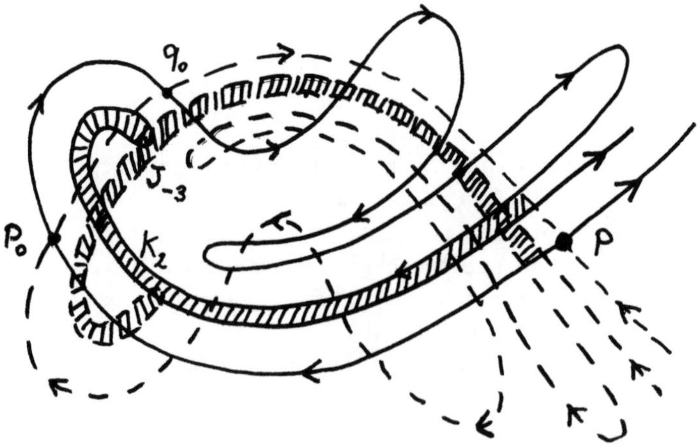

Figure 3. A type (l, m, k, 0) trellis.
$l = 2$, $m = k = 3$.

In this paper we estimate the escape rates, their asymptotic behavior and the topological entropy for special regions in the parameter space. In these regions, similar analysis supplies estimates for all the other quantities mentioned above.

c. The Geometrical Parameters, Tangles and Trellises. The quantities e_n, L_n and $\beta(n)$ depend sensitively on the manner in which the arcs J_n intersect the arcs K_m for $n, m \geq 0$. In general, given the structure of $J_0 \cap K_j$, $j = 1, \ldots, n$, one can calculate the *minimal* number of homoclinic points in $J_0 \cap K_{n+1}$ and their ordering along the stable and unstable manifolds. However, $J_0 \cap K_{n+1}$ may intersect in other points, which we call spontaneous intersection points. One can imagine that there exist some parameter values for which the arcs do not develop any spontaneous intersection points for all $j > n$. For these parameter values we have some hope for estimating e_n, L_n and $\beta(n)$ accurately using information regarding the initial development of the manifolds. Then, we would like to argue that in a neighborhood of these parameter values our estimates are still reasonable. This is the basic idea behind the methods developed in [3]. There, we use two sets of families of initial developments, namely two sets of geometrical specification for the structure of $J_0 \cap K_j$, $j = 1, \ldots, n$, for some finite integer n. The first one, due to Easton [13], has the initial structure plotted in figure 2; the arcs K_j intersect J_0 for the first time when $j = l$, and $J_0 \cap K_l$ contains exactly two homoclinic points ($l = 2$ in figure 2). For $j > l$, $J_0 \cap K_j$ is determined by the rule that no spontaneous intersection points are allowed. Easton called the tangles which obey these minimal rules type l trellises. In [3], we developed methods for estimating e_n, L_n and $\beta(n)$ for the type l trellises. In the next section we determine the regions in phase space for which (2.1) has the same initial development as a type l trellis, and in sections 4 we estimate the escape rates and the topological entropy in these regions. A tangle belonging to the second family of initial configurations, called the type $(l, m, k, 0)$ trellises in [3], is shown in figure 3. In the figure, $l = 2$, $m = 3$ and $k = 3$. For this family, the intersection of K_l with J_0 produces four homoclinic points, and the tip of K_l is contained in D_{-m}. Similarly, the tip of J_{-l} is contained in E_k. In

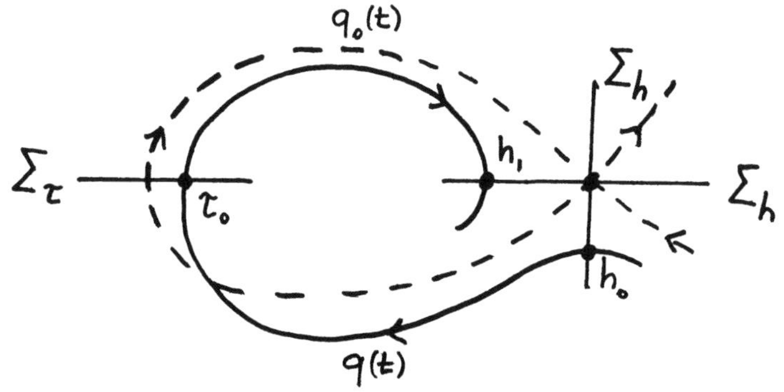

Figure 4. The Geometry of the separatrix map.
– – – – – Unperturbed separatrices, ———— An orbit.

figure 3 we hatched the lobes E_2 and D_{-3}. Each of these lobes contains the tip of the other lobe. As $m, k \to \infty$ the type $(l, m, k, 0)$ trellises approach the type l trellis. This feature allows one to examine how the topological and quantitative properties change as the parameters of the problem vary. This investigation suggests that near the type l trellises the quantities e_n, L_n and $\beta(n)$ vary smoothly. Hence, our claim that the estimates should hold in a region of non vanishing area in parameter space has some theoretical support.

Our first objective is to classify a given tangle and decide whether it is close to one of the "minimal" tangles, the type l trellises or the type $(l, m, k, 0)$ trellises. First, we define the *structure indices* l, m, k as follows:

1. The structure index l is given by the minimal value of j for which $K_j \cap J_0 \neq \emptyset$.
2. The structure index m is given by the minimal value of j for which $C_l \cap J_{-j} \neq \emptyset$, where C_l denotes the tip of the arc K_l, emanating from J_0.
3. The structure index k is given by the minimal value of j for which $K_j \cap B_{-l} \neq \emptyset$, where B_{-l} denotes the tip of the arc J_{-l}, emanating from K_{-1}.

By definition, the type l trellises have a structure index l and the type $(l, m, k, 0)$ trellises have structure indices l, m, k respectively. In addition, the critical intersection set of the type l trellis, $J_0 \cap K_l$, contains exactly two homoclinic points. Similarly, the critical intersection sets $J_0 \cap K_l$, $C_l \cap J_{-m}$ and $K_k \cap B_{-l}$ of the type $(l, m, k, 0)$ trellis contains exactly four, two and two homoclinic points respectively. We use the properties of the trellises to approximate the properties of tangles which have the same structure indices and the same number of homoclinic points in the critical intersection sets. In the next section we use the Whisker map to estimate the boundaries of the regions in parameter space in which the tangles satisfy these conditions.

Section 3. The Structure Indices.

In this section we define and compute the Whisker map for (2.1). Then we use it to construct a bifurcation diagram describing the dependence of the structure indices on ε and ω.

a. The Whisker Map. We define the separatrix map, W, as the return map of the energy and time variables (h_n, τ_n) to the crossections Σ_h and Σ_τ respectively (see figure 4):

$$
\begin{aligned}
W :&(h_n, \tau_n) \to (h_{n+1}, \tau_{n+1}) \\
&q(\tau_n) \in \Sigma_\tau \\
&h_n = H_\varepsilon(q(t^*), t^*), \quad q(t^*) \in \Sigma_h, \quad \tau_n < t^* < \tau_{n+1},
\end{aligned}
\tag{3.1}
$$

where $q(t)$ is a solution to Eq. (1.1). In the neighborhood of the separatrix the crossections Σ_h and Σ_τ are transverse to the unperturbed trajectories. Therefore, for ε sufficiently small, the separatrix map is well defined there. With no loss of generality, we assume that $h \ll 1$ near the separatrix. The Whisker map is defined to be the leading order approximation in ε and h to the separatrix map. Chirikov [10] realized that the Whisker map can be calculated explicitly as

$$
\begin{aligned}
h_{n+1} &= h_n + \epsilon M(\tau_n) \\
\tau_{n+1} &= \tau_n + T(h_{n+1})
\end{aligned}
\tag{3.2}
$$

where $M(t)$ is the Melnikov function and $T(h)$ is the period of the unperturbed orbit with energy h. In section 9 of [3], we derive the Whisker map from the separatrix map and estimate the magnitude of the next order terms, namely the error committed by replacing (3.1) by (3.2).

For our example, the Melnikov function is given by Eq. (2.3). The period of the unperturbed periodic orbits of (2.1), $T(h_0)$, is given by:

$$
T(h_0) = \sqrt{6}\int_b^a \frac{dq}{\sqrt{(q-1)^2(q+\frac{1}{2}) + 3h_0}} = \frac{2\sqrt{6}}{\sqrt{a-c}} K\left(\sqrt{\frac{b-c}{a-c}}\right) \tag{3.3a}
$$

where $K(k)$ is the complete elliptic integral of the first kind,

$$
(a-q)(b-q)(q-c) = (q-1)^2(q+\frac{1}{2}) + 3h_0, \qquad a > b > c, \tag{3.3b}
$$

and

$$
H_\varepsilon(x, y, t) = \frac{1}{6} + h_\varepsilon(x, y, t). \tag{3.3c}
$$

It follows from Eq. (2.1) that for the unperturbed problem $h_0(q_0(t)) = 0$, hence near the separatrix $|h_0| \ll 1$. Expanding (3.3) in the small parameter h_0, one obtains (with some assistance from Mathematica):

$$
T(h_0) = \ln(\frac{72}{-h_0})(1 + O(\sqrt{|h_0|})). \tag{3.4}
$$

Therefore, the Whisker map for (2.1) is given by:

$$
\begin{aligned}
h_{n+1} &= h_n + \varepsilon C(\omega)\sin(\omega\tau_n) \\
\tau_{n+1} &= \tau_n + \ln(\frac{72}{-h_{n+1}}),
\end{aligned}
\tag{3.5}
$$

where $C(\omega)$ is defined by Eq. (2.4) and is plotted in figure 1.

b. Estimating the Structure Indices. There is a simple and elegant correspondence between the value of the variables (h, τ) of the separatrix map and the geometry of the manifolds; since orbits on (or above) $S[p_0, p]$ approach p (resp. escape to infinity) their return time is not defined. Hence, if the initial energy and phase, (h_0, τ_0), of an orbit are such that $h_1 = 0$ (resp. $h_1 > 0$) then for $t > \tau_0$ this orbit belongs to $S[p_0, p]$ (resp. escapes). More generally, orbits inside $\hat{S}$ have $h < 0$ and orbits outside $\hat{S}$ have $h > 0$. Escande [11] noticed this property first, and suggested to use the Whisker map for deriving a Markov model for estimating the transport rates. Knobloch and Weiss [7] replaced the dynamics of the continuous system (1.1) by the Whisker map and computed numerically the transport rates due to modulated traveling waves. In [3] we argued that by replacing the exact separatrix map by the Whisker map one introduces error terms which grow exponentially in time. Moreover, this strategy leads to yet another numerical study, involving the Whisker map instead of the flow. Therefore, we determine the geometrical parameters of the Poincaré map using the Whisker map for a finite number of iterations, on known orbits for which the error term may be estimated. Then, we use the approximate geometrical structure to approximate the escape rates, manifold length, etc. These estimates are essentially analytical, as opposed to previous studies of transport rates. On the other hand, it is unclear how to estimate the error involved in this topological approximation. For now, we compute the geometrical parameters using the Whisker map (3.5), leaving the calculation of the error terms to future work.

The structure index l: l is defined to be the minimal integer j for which there exists an initial condition $(h_0, \tau_0) \in K_j \cap J_0$. We find the algebraic equations (h_0, τ_0) satisfy, and determine l as the minimal value for which these equations have solutions.

J_0 is characterized by initial conditions which have just past the Σ_τ crossection in the last period, therefore $0 \le \tau_0 \le T$ (where $T \equiv 2\pi/\omega$). The previous position, $(t < \tau_0)$ of such initial conditions is outside $\hat{S}$, so $h_0 > 0$, see figure 2. Finally, their future is to asymptote p, so $h_1 = 0$. We summarize these observation as:

$$J_0 = \{(h_0, \tau_0) \mid h_0 < 0,\ h_1 = 0,\ 0 \le \tau_0 \le T\} \tag{3.6}$$

and similarly we find

$$K_0 = \{(h_0, \tau_0) \mid h_0 = 0,\ h_1 < 0,\ 0 \le \tau_0 \le T\}. \tag{3.7}$$

Let $(h_0, \tau_0) \in K_0$, namely the trajectory $r(t)$, with initial energy $h_0 = 0$, passed the crossection Σ_τ at τ_0, and is in K_0 at some time t, $0 \le \tau_0 \le t \le T$, so $h_1 < 0$. We ask whether there exists (h_0, τ_0) for which $F^j r(t) \in J_0$ (then $F^j r(t) \in J_0 \cap K_j$ and the latter set is nonempty). If $F^j r(t) \in J_0$, then there exists an integer $i \ge 1$ such that $\tau_0 + (j-1)T < \tau_i \le \tau_0 + jT$ and $h_{i+1} = 0$, namely $r(t)$ is a homoclinic orbit which encircles the origin i times. We conjecture that for ε sufficiently small and ω bounded away from zero the minimal j is determined by orbits with $i = 1$. Summarizing the above arguments, we assert that l is given by the minimal integer j for which the equations

$$h_0 = 0,\quad h_1 < 0,\quad h_2 = 0,\quad 0 \le \tau_0 \le T,\quad (j + \delta(\omega) - 1)T \le \tau_1 - \tau_0 \le (j + \delta(\omega))T \tag{3.8}$$

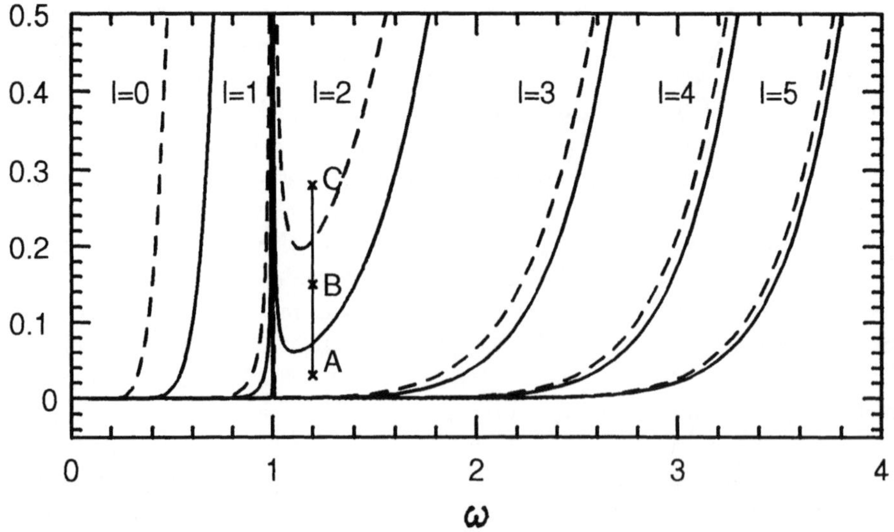

Figure 5. The Bifurcation Curves of Eq. (3.10)
——— $\varepsilon_l^b(\omega)$,– – –, $\varepsilon_l^a(\omega)$. l values are indicated to the left of the curves.

have a solution (h_0, τ_0), where

$$\delta(\omega) = \begin{cases} 1 & \text{if } \omega < 1; \\ 0 & \text{if } \omega > 1. \end{cases}$$

Using (3.2), we find that (3.8) amounts to finding a solution $0 \le \tau_0 \le T$ to:

$$M(\tau_0) = -M(\tau_0 + T(\varepsilon M(\tau_0))) \ , \quad (j + \delta(\omega) - 1)T < T(\varepsilon M(\tau_0)) \le (j + \delta(\omega))T. \quad (3.9)$$

Since $M(t)$ is a multivalued function of t, we obtain two branches of solutions, which depend on the sign of $C(w)$:

$$\sin(\omega\tau_0) = -\frac{72}{\varepsilon C(\omega)} \exp(-(j + \delta(\omega) - \frac{1}{2})T), \qquad \tau_0 \in I_\omega \qquad (3.10a)$$

or:

$$\sin(\omega\tau_0) = -\frac{72}{\varepsilon C(\omega)} \exp(2\tau_0 - (j + 2\delta(\omega))T), \qquad \tau_0 \in I_\omega, \qquad (3.10b)$$

where

$$I_\omega = \begin{cases} [\pi/\omega, 2\pi/\omega] & \text{for } \omega < 1; \\ [0, \pi/\omega] & \text{for } \omega > 1. \end{cases}$$

We equate the r.h.s. of (3.10a) to one, and find the bifurcation curves for (3.10a):

$$\varepsilon_l^a = \frac{72}{|C(\omega)|} \exp(\frac{-(2l + 2\delta(\omega) - 1)\pi}{\omega}). \qquad (3.11)$$

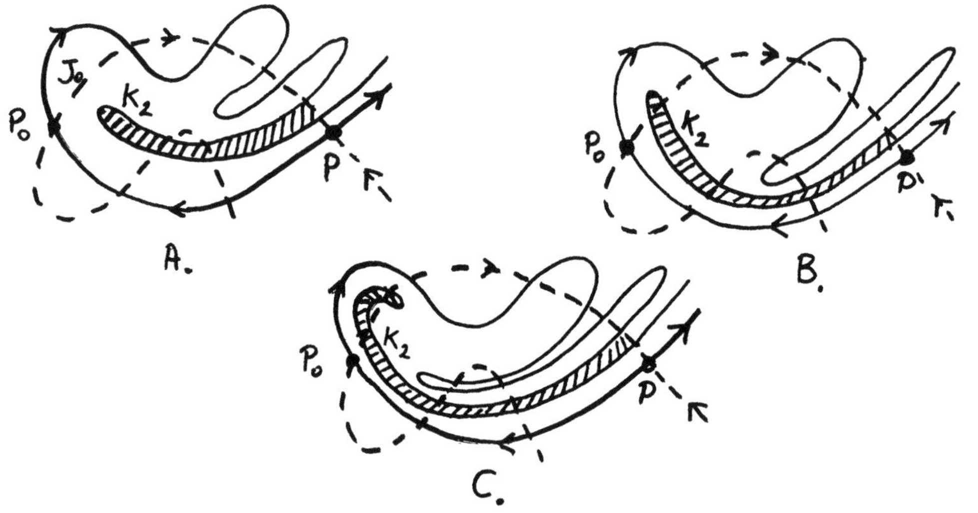

Figure 6. The geometrical interpretation of the bifurcation curves $\varepsilon_l^a, \varepsilon_l^b$.
A. $\varepsilon_3^a < \varepsilon < \varepsilon_2^b$ B. $\varepsilon_2^b < \varepsilon < \varepsilon_2^a$ C. $\varepsilon_2^a < \varepsilon < \varepsilon_1^b$.

For $\varepsilon < \varepsilon_l^a$, (3.10a) has no solutions, and for $\varepsilon \geq \varepsilon_l^a$ (3.10a) has two solutions, $\tau_0^2(l) \leq \tau_0^4(l)$. Some of the bifurcation curves $\varepsilon_l^a(\omega)$ are plotted in figure 5 (the dashed lines). We obtain the bifurcation curves of (3.10b) by requiring that the function

$$f(\tau_0) \equiv \sin(\omega\tau_0) + \frac{72}{\varepsilon C(\omega)} \exp(2\tau_0 - 2(l + 2\delta(\omega))\pi/\omega)$$

will have a quadratic zero in the appropriate interval of τ_0:

$$f(t) = 0, \qquad \frac{df(t)}{dt} = 0, \qquad t \in I_\omega,$$

and obtain

$$\varepsilon_l^b = \frac{72}{|C(\omega)|} \frac{\sqrt{1 + \frac{\omega^2}{4}}}{\frac{\omega}{2}} \exp\left(\frac{-2(l + 2\delta(\omega))\pi}{\omega} + \frac{2}{\omega} \arctan\left(\frac{\omega}{2}\right)\right), \tag{3.12}$$

where $\arctan(\frac{\omega}{2})$ is chosen to belong to I_ω. We plot the bifurcation curves $\varepsilon_l^b(\omega)$ in figure 5 (the solid lines). Clearly, $\varepsilon_l^b(\omega) < \varepsilon_l^a(\omega)$ for all ω. For $\varepsilon < \varepsilon_l^b$ (3.10b) have no solutions, whereas for $\varepsilon > \varepsilon_l^b$ it has two solutions, denoted by $\tau_0^1(l) \leq \tau_0^3(l)$. At the bifurcation point $\varepsilon = \varepsilon_l^b$, $\tau_0^1(l) = \tau_0^3(l) = \frac{1}{\omega}\arctan(\frac{\omega}{2})$. At the bifurcation point $\varepsilon = \varepsilon_l^a$, Eq. (3.10) has a cubic zero (located at $3\pi/2\omega$ for $\omega < 1$ and at π/ω when $\omega > 1$). For $\varepsilon > \varepsilon_l^a$ the four solutions to (3.10) are ordered by their initial crossing times $\tau_0^i(l) \leq \tau_0^{i+1}(l)$, $i = 1, 2, 3$. We note the peculiar property, that to leading order in ε, (3.9) has either one, two or four distinct solutions.

As one increases ε along the line ABC in figure 5, the number of solutions to (3.10) changes, and therefore the structure of the manifolds of (2.1) changes as shown in figure 6; For $\varepsilon_3^a < \varepsilon < \varepsilon_2^b$ (point A in figure 5), (3.10) has no solutions with $j = 2$, and four solutions with $j = 3$. Therefore, we estimate the structure index l of (2.1) by 3. For this value of ε, the tip of the lobe E_2 (the hatched lobe in figure 6) does not reach the segment J_0 and $K_2 \cap J_0 = \emptyset$ (figure 6.A). Increasing ε so that $\varepsilon_2^b < \varepsilon < \varepsilon_2^a$ (point B in figure 5), (3.10) has exactly two solutions with $j = 2$, and we estimate the structure index l of (2.1) to be 2. Here, the tip of E_2 crosses J_0, two homoclinic points in $K_2 \cap J_0$ are created (figure 6.B), and the tangle has the same initial development as a type 2 trellis. A further increase in ε (point C in figure 5) results in four solutions to (3.10) with $j = 2$. Then, the tip of E_2 stretches and exits D_0, giving rise to four homoclinic orbits in $K_2 \cap J_0$ (figure 6.C). In this region we expect to find subregions in which the tangles have the same initial developments as that of the type $(2, m, k, 0)$ trellises.

To summarize, in the regions of figure 5 which are bounded by a dashed line from the left and a solid line from the right, the lobe E_l intersects the lobe D_0 at exactly two homoclinic points. In these regions we approximate the topological structure of the manifolds by the simplest construction, the type l trellises. When $l = 1$ we have some theoretical justification to this procedure; the horseshoe map, which is structurally stable, defines a type 1 trellis. Hence, we expect to find an open interval, contained in $[\varepsilon_1^a, \varepsilon_1^b]$, for which the map F is topologically conjugate to the horseshoe map and its manifolds form a type 1 trellis. The hope is that even if the other trellises are not structurally stable, the properties of nearby tangles are close to theirs.

In the regions where the initial development of the tangles differs from that of a type l trellis, we need information regarding the fate of the tip of E_l. This information is supplied by calculating the structure indices m, k. For (2.1), the symmetry of the Poincaré map implies that $m = k$.

The structure index m: Following the same logic that led to (3.8), we find that an orbit contained in $K_l \cap J_{-j}$ must satisfy:

$$\begin{aligned} &h_0 = 0, \quad h_1 < 0, \quad h_2 < 0, \quad h_3 = 0, \quad 0 \le \tau_0 \le T, \\ &(l + \delta(\omega) - 1)T \le \tau_1 - \tau_0 \le (l + \delta(\omega))T, \\ &(j + \delta(\omega) - 1)T \le \tau_2 - \tau_1 \le (j + \delta(\omega))T. \end{aligned} \tag{3.13a}$$

In addition, we demand the orbit to belong to the tip of K_l, therefore

$$\tau_0^2(l) \le \tau_0 \le \tau_0^3(l) \tag{3.13b}$$

where $\tau_0^2(l), \tau_0^3(l)$ are the intermediate solutions of (3.9) with $j = l$.

m is given by the minimal value of j for which there exists an initial condition (h_0, τ_0) which solves (3.13). These conditions are meaningful only in the regions where the tip is well defined namely for $\varepsilon_l^a < \varepsilon$. Using (3.2), we find that (3.13) amounts to finding a solution τ_0 to:

$$\begin{aligned} \tau_1 &= \tau_0 + T(\varepsilon M(\tau_0)) \\ \tau_2 &= \tau_1 + T(\varepsilon M(\tau_0) + \varepsilon M(\tau_1)) \\ 0 &= M(\tau_0) + M(\tau_1) + M(\tau_2) \\ (j + \delta(\omega) - 1)T &\le T(\varepsilon M(\tau_0) + \varepsilon M(\tau_1)) \le (j + \delta(\omega))T, \end{aligned} \tag{3.14a}$$

with

$$\tau_0^2(l) \le \tau_0 \le \tau_0^3(l). \tag{3.14b}$$

Eq. (3.14) constitute a nonlinear equation for τ_0 which can be solved numerically using a Newton method. Close to the bifurcation curve $\varepsilon = \varepsilon_l^a$, we can estimate the solutions to (3.14) and exert their dependence on m; near $\varepsilon = \varepsilon_l^a$ the tip of K_l is small, therefore the distance between the zeroes $\tau_0^2(l)$ and $\tau_0^3(l)$ is small. We define θ by

$$\tau_0 = \tau_0^2(l) + \theta, \tag{3.15}$$

and assume $\theta \ll 1$. Linearizing (3.14) about $\tau_0^2 \equiv \tau_0^2(l)$, and using (3.9), we find:

$$\begin{aligned}
\theta K_2(\tau_0^2) &= -M(\tau_0^2 + T(\varepsilon M(\tau_0^2)) + \theta K_1(\tau_0^2) + T(\varepsilon\theta K_2(\tau_0^2))) \\
(j + \delta(\omega) - 1)T &\le T(\varepsilon\theta K_2(\tau_0^2)) \le (j + \delta(\omega))T \\
K_1(t) &= 1 + \frac{d}{dt}T(\varepsilon M(t)) \\
K_2(t) &= \frac{dM(s)}{ds}\bigg|_t + K_1(t)\frac{dM(s)}{ds}\bigg|_{t+T(\varepsilon M(t))}.
\end{aligned} \tag{3.16}$$

Since M is multivalued, there are two branches of solutions to (3.16)

$$\begin{aligned}
&\theta = K_3 \exp(\alpha_\pm \theta - jT) \\
\text{or:}& \\
&\theta = K_3 \exp(\alpha_\mp \theta - jT \pm \frac{1}{2}T),
\end{aligned} \tag{3.17a}$$

where the upper sign corresponds to $\omega < 1$ and the lower sign to $\omega > 1$, and

$$\begin{aligned}
B_l &= -\frac{72}{C(\omega)} \exp((-l - \delta(\omega) + \frac{1}{2})T) \\
K_3 &= \left(\frac{72}{\omega C(\omega)}\right)^2 \frac{\exp(\tau_0^2 - (l + 3\delta(\omega) - 1)T)}{\varepsilon^2 - B_l^2} \\
\alpha_\pm &= 1 - \frac{\omega}{|B_l|}(\sqrt{\varepsilon^2 - B_l^2} \pm \frac{1}{\varepsilon}(\varepsilon^2 - B_l^2)).
\end{aligned} \tag{3.17b}$$

A similar calculation for $t = \tau_0^3(l) - \theta'$, results in two equations for θ' of the same form as (3.17a), with the coefficients B_l, K_3 and $\alpha_\pm$ replaced by

$$\begin{aligned}
B_l'(\tau_0^3) &= -\frac{72}{C(\omega)} \exp(-(l + 2\delta(\omega))T + 2\tau_0^3) \\
K_3' &= \frac{72}{\omega|C(\omega)|} \frac{\exp(-\tau_0^3 + (1 - \delta(\omega))T)}{2\sqrt{\varepsilon^2 - B_l'^2} + \frac{\omega}{|B_l'|}(\varepsilon^2 - B_l'^2)} \\
\alpha_\pm' &= -1 \mp \frac{2}{\varepsilon}\sqrt{\varepsilon^2 - B_l'^2} - \frac{\omega}{|B_l'|}(\sqrt{\varepsilon^2 - B_l'^2} \pm \frac{1}{\varepsilon}(\varepsilon^2 - B_l'^2))
\end{aligned} \tag{3.17c}$$

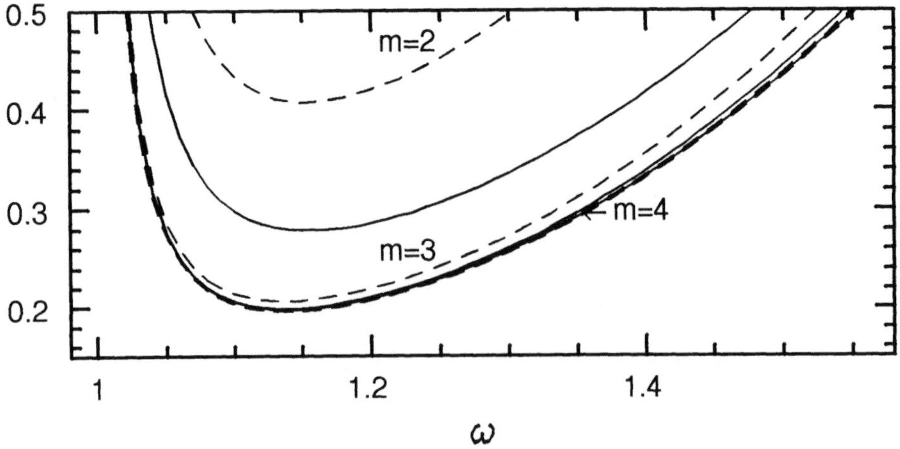

Figure 7. Approximate Bifurcation Curves for the Structure Index m. — — — $\varepsilon_2^a(\omega)$, - - - $\varepsilon_{2,m}^a(\omega)$, ——— $\varepsilon_{2,m}^b(\omega)$.

respectively. Since $\theta \ll 1$, the solutions to (3.17a) are given approximately by:

$$\begin{aligned} \theta_1 &= \frac{K_3}{\exp(jT) - \alpha_\pm K_3} \\ \theta_2 &= \frac{K_3}{\exp((j \mp \frac{1}{2})T) - \alpha_\mp K_3}, \end{aligned} \tag{3.18}$$

with similar expressions for θ'. By construction, the solutions to (3.17) are physical only when they are positive and small. Therefore, we estimate the bifurcation curves of (3.14) by counting the number of solutions which are physical and which do not overlap. For each m we obtain two approximate bifurcation curves $\varepsilon_{l,m}^b(\omega) < \varepsilon_{l,m}^a(\omega)$, where for $\varepsilon < \varepsilon_{l,m}^b(\omega)$ (3.14) has no solutions for $j = m$, for $\varepsilon_{l,m}^b(\omega) < \varepsilon < \varepsilon_{l,m}^a(\omega)$ (3.14) has exactly two solutions with $j = m$, and for $\varepsilon_{l,m}^a(\omega) < \varepsilon$ (3.14) has at least four solutions for $j = m$. In figure 7 we plot the approximate bifurcation curves $\varepsilon_{2,m}^a(\omega), \varepsilon_{2,m}^b(\omega)$ for $\omega > 1$. For $m > 4$ these curves are indistinguishable from the ε_2^a curve. In the small regions bounded between the two bifurcation curves of the structure index m, we expect to find behavior similar to the $(l, m, m, 0)$ type trellises. We note that there is a series of approximations involved in getting these curves and a more thorough analysis is needed to justify these approximations.

Section 4. The Type-l Trellises.

In section 3, we found regions in parameter space in which the tangles had the same initial development as a type-l trellis. In the first part of this section we summarize some of the methods developed in [3] for estimating the development of a type-l trellis. Then, we use this construction to estimate the exponential growth rate of the segments K_n. Finally, we estimate the escape rates for these tangles using the Whisker map and the geometrical properties of the trellises.

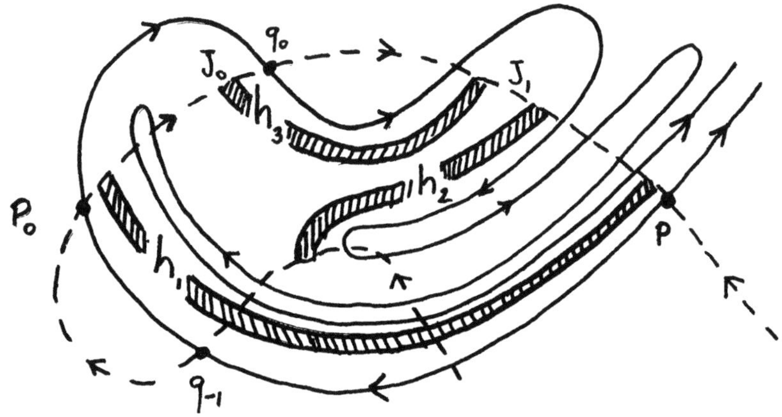

Figure 8. The states of a type 2 trellis.
The hatched strips are the members of the indicated states.

a. The Structure of a Type-l Trellis. The assumption that a type-l trellis does not develop any spontaneous homoclinic points enables one to construct a one sided symbolic dynamics which describes the dynamics of the lobes. In general, the jth image of E_0 is elongate and folded many times in $\hat{S}$. We divide this tangled lobe to several types of strips, which we call states. These states obey simple dynamical rules under the Poincaré map F. We draw typical members of the various states in figure 8 and define them as follows: Let $S = \hat{S} - E_0$. The strips of $F^n(E_0) \cap S$ which have one boundary belonging to J_0 and another belonging to J_j, $j \geq l$, belong to the state h_1. The strips of $F^n(E_0) \cap S$ with one boundary belonging to J_{-k} and the other belonging to J_{l-k} or J_{l-k+1}, where $1 \leq k \leq l-1$, belong to the state h_{k+1}. The strips of $F^n(E_0) \cap S$ which have one boundary belonging to J_0 and another belonging to J_1, belong to the state h_{l+1}. Finally, the arcs of $F^n(E_0) \cap F^{-k}(D_0)$, $k = 0, \ldots, l-1$ belong, respectively, to the state g_k, $k = 1, \ldots, l$. The arcs of $F^n(E_0) \cap F^j(D_0)$, $j \geq 1$, belong to the state g_0.

It is easy to verify that strips belonging to the above states obey the following dynamics:

$$
\begin{array}{ccccccccccc}
 & \nearrow & h_1 & \nearrow & h_l & & & & & & \\
h_1 & \longrightarrow & h_{l+1} & \longrightarrow & h_l & \longrightarrow & h_{l-1} & \longrightarrow & \cdots & \longrightarrow & h_1 \\
 & \searrow & g_1 & \searrow & g_l & \longrightarrow & g_{l-1} & \longrightarrow & \cdots & \longrightarrow & g_1 \longrightarrow g_0
\end{array}
\tag{4.1}
$$

The dynamics of the states h_i determines the folding of curves inside S. The states g_i are "passive" states, and are used for the estimates of the escape rates (see part c of this section). From (4.1) we construct the $(l+1)\times(l+1)$ transfer matrix, T_l:

$$T_l = \begin{pmatrix} 1 & 0 & 0 & \dots & 0 & 0 & 1 \\ 1 & 0 & 0 & \dots & 0 & 0 & 0 \\ 0 & 1 & 0 & \dots & 0 & 0 & 0 \\ 0 & 0 & 1 & \dots & 0 & 0 & 0 \\ \vdots & \vdots & \vdots & \ddots & \vdots & \vdots & \vdots \\ 0 & 0 & 0 & \dots & 1 & 0 & 0 \\ 0 & 0 & 0 & \dots & 0 & 2 & 0 \end{pmatrix}.$$

The number of strips of $F^n(E_0)$ $(n > 0)$ belonging to a state h_i is given by the ith component of the vector $(0,\dots,0,1,0)T_l^{n-1}$. This observation enables us to estimate $L(K_n)$, the length of the boundary of the lobe $F^n(E_0)$. The exponential growth rate of these quantities is given by $\log \lambda_{T_l}$, where λ_{T_l} denotes the modulus of the the largest root of the characteristic polynomial of T_l,

$$p_l(\lambda) = \lambda^{l+1} - \lambda^l - 2. \tag{4.2}$$

When $l = 1$, the matrix T_l is replaced by the matrix

$$T_1 = \begin{pmatrix} 1 & 1 \\ 1 & 1 \end{pmatrix}$$

and $\lambda_{T_1} = 2$. In general, λ_{T_l} is monotonically decreasing with l.

b. The Elongation Rate of The Unstable Manifold. The bifurcation diagrams of figure 5 and figure 7 may be considered as an approximate diagram of the level sets of the topological entropy; if the type-l trellises are similar to the tangles with the same initial development, then, for $\varepsilon_l^b < \varepsilon < \varepsilon_l^a$, the exponential growth rate of line elements in phase space is given approximately by $\log \lambda_{T_l}$.

In [3], we constructed the symbolic dynamics for a type $(l,m,k,0)$ trellis as well, and found that the largest eigenvalue of the matrix $T_{l,m,k,0}$ is given approximately, for $m,k > l$, by:

$$\lambda_{T_{l,m,k,0}} \approx \lambda_{T_l} + c_l \exp\{-b_l(m+k)\} \tag{4.3}$$

where c_l and b_l are computed in [3].

These findings, together with the bifurcation diagrams of figures 5 and 7, supply an estimate for the change in the topological entropy as the parameters (ε,ω) are varied. We observe that the topological entropy of the type $(l,m,k,0)$ trellises approaches that of a type l trellis exponentially in m. Using (3.18), we find that the topological entropy varies as a power law in ε near the bifurcation curve ε_a^l, with the critical index $b_l\omega/\pi$. This is a rough estimate, and a more thorough analysis is underway. We need to account for the ε dependence of the coefficients in Eq. (3.17) and we need to estimate the error terms resulting from approximating (3.14) by (3.17) and from using the Whisker map rather then the separatrix map. Furthermore, our estimates are based upon the geometry of the trellises. It is possible that the actual dependence of the topological entropy on ε is not smooth, and that the power law dependence is realized on a measure zero set of parameters. A careful numerical study is needed to resolve this question.

c. Estimate of the Escape Rates. By assigning weights to the dynamics described in (4.1), we estimate the area which escapes the region $\hat{S}$; By definition, strips belonging to the state g_1 have just escaped $\hat{S}$. In fact, e_n is equal to the area of the strips of $F^n(E_0)$ which belong to g_1. Therefore, once we determine the weights and the initial distribution of $F(E_0)$ between the states, we can construct a weighted transition matrix similar to T_l which approximates the action of the flow on the states. By construction, most of the weights in (4.1) are simply one. There are three nontrivial ($\neq 0$ or 1) weights, denoted by s_1, s_2 and s_3. s_1 (resp. s_2) measures the fraction of the area of a strip belonging to the state h_1 which maps to a strip belonging to the state h_1 (resp. h_{l+1}). Similarly, s_3 measures the fraction of a strip belonging to h_{l+1} which ends up as a strip belonging to the state h_l. The weighted transition matrix is of the form:

$$M_l = \begin{array}{c} g_0 \ldots g_l \quad h_1 \ldots h_{l+1} \\ \begin{pmatrix} L_{l+1} & 0 \\ R & W_l \end{pmatrix} \end{array}. \tag{4.4}$$

The matrix W_l realizes the dynamics on the h_i states:

$$W_l = \begin{pmatrix} s_1 & 0 & 0 & \ldots & 0 & 0 & s_2 \\ 1 & 0 & 0 & \ldots & 0 & 0 & 0 \\ 0 & 1 & 0 & \ldots & 0 & 0 & 0 \\ 0 & 0 & 1 & \ldots & 0 & 0 & 0 \\ \vdots & \vdots & \vdots & \ddots & \vdots & \vdots & \vdots \\ 0 & 0 & 0 & \ldots & 1 & 0 & 0 \\ 0 & 0 & 0 & \ldots & 0 & s_3 & 0 \end{pmatrix}$$

The matrix R administrates the transfer of areas from the h_i states to the g_i states:

$$R(1,2) = 1 - s_1 - s_2, \quad R(l+1, l+1) = 1 - s_3, \quad R(i,j) = 0 \quad \text{otherwise.}$$

Finally, L_n is an $n \times n$ transfer matrix which reflects the trivial dynamics of the g_i states:

$$L_n = \begin{pmatrix} 1 & 0 & 0 & \ldots & 0 & 0 & 0 \\ 1 & 0 & 0 & \ldots & 0 & 0 & 0 \\ 0 & 1 & 0 & \ldots & 0 & 0 & 0 \\ 0 & 0 & 1 & \ldots & 0 & 0 & 0 \\ \vdots & \vdots & \vdots & \ddots & \vdots & \vdots & \vdots \\ 0 & 0 & 0 & \ldots & 1 & 0 & 0 \\ 0 & 0 & 0 & \ldots & 0 & 1 & 0 \end{pmatrix}. \tag{4.5}$$

The lobe $F(E_0)$ has, by construction, one part which belongs to an h_l state and another small part which belongs to a g_l state. The area of this small portion is exactly e_l. Therefore, once $\mu(E_0)$, e_l and the weights s_1, s_2 and s_3 are known, we can estimate e_n as the second component of the vector v^n:

$$e_n \approx v^n(2) = v^1 M^{n-1}(2), \tag{4.6}$$

where v^1 is a vector with $(2l+2)$ components, two of which are nonvanishing:

$$v^1(l+1) = e_l, \qquad \text{and} \qquad v^1(2l+1) = \mu(E_0) - e_l \equiv w.$$

From the form of the matrix M_l, it is easy to verify that the weights s_i can be estimated by:

$$s_1 \approx \frac{e_{l+2}}{e_{l+1}}, \qquad s_2 \approx 1 - \frac{e_{l+1}}{w} - \frac{e_{l+2}}{e_{l+1}}, \qquad s_3 \approx 1 - \frac{e_{2l+1} - s_1^l e_{l+1}}{ws_2} \tag{4.7}$$

Therefore, given e_j, $j = l, l+1, l+2, 2l+1$ and $\mu(E_0)$ we can approximate the escape rates for all n. In the next section we use the Whisker map to evaluate these quantities.

d. Estimates of the Initial Escape Rates. Since the variables (h, τ) of the Whisker map are canonical variables, an area element of the phase space is given by $dhd\tau$. We estimate $\mu(E_0)$ and the e_j's by determining the values of h_n and τ_{n-1} on the boundaries of the sets they measure and integrating the area bounded by these values. Here (h_n, τ_n) is the nth image of (h_0, τ_0) under the Whisker map, (3.2). First, we evaluate $\mu(E_0)$; E_0 is bounded by the segment of the stable manifold $I_1 = \{(h_1^u(\tau_0), \tau_0) | h_1^u(\tau_0) = 0, \ \tau_0 \in I_\omega\}$ and the arc $I_2 = \{(h_1^l(\tau_0), \tau_0) | h_0^l(\tau_0) = 0, \ \tau_0 \in I_\omega\}$. Denoting the endpoints of I_ω by t_0 and t_1, we find:

$$\mu(E_0) \approx -\int_{t_0}^{t_1} h_1^l(\tau_0) d\tau_0 = -\varepsilon \int_{t_0}^{t_1} M(\tau_0) d\tau_0 = \frac{2\varepsilon |C(\omega)|}{\omega}. \tag{4.8}$$

Eq. (4.8) can be derived directly, using the geometrical interpretation of the Melnikov function [1]. It coincides with the first order approximation in ε to the difference in action of the homoclinic endpoints $(0, t_0)$ and $(0, t_1)$ [8], [14], [15].

To estimate $e_l = \mu(F^l(E_0) \cap D_0)$, we notice that the set $F^l(E_0) \cap D_0$ is enclosed by the arc $I_1 = \{(h_2^u(\tau_1), \tau_1) | h_0^u(\tau_0) = 0, \ \tau_0^1(l) \le \tau_0 \le \tau_0^3(l)\}$ and the segment $I_2 = \{(h_2^l(\tau_1), \tau_1) | h_2^l(\tau_1) = 0, \ \tau_0^1(l) \le \tau_0 \le \tau_0^3(l)\}$, where $\tau_0^1(l)$ and $\tau_0^3(l)$ are the solutions of (3.10b) with $j = l$. Since $\tau_1^3(l) < \tau_1^1(l)$, we find

$$\begin{aligned} e_l &\approx \int_{\tau_1^3}^{\tau_1^1} h_2^u(\tau_1) d\tau_1 = \varepsilon \int_{\tau_1^3}^{\tau_1^1} (M(\tau_1) + M(\tau_0)) d\tau_1 \\ &= \varepsilon \int_{\tau_1^3}^{\tau_1^1} M(\tau_1) d\tau_1 + \varepsilon \int_{\tau_0^3}^{\tau_0^1} M(\tau_0)(1 - \frac{M'(\tau_0)}{M(\tau_0)}) d\tau_0 \\ &= \varepsilon \int_{\tau_1^3}^{\tau_1^1} M(\tau_1) d\tau_1 - \varepsilon \int_{\tau_0^1}^{\tau_0^3} M(\tau_0) d\tau_0 + M(\tau_0) \Big|_{\tau_0^1}^{\tau_0^3}, \end{aligned} \tag{4.9}$$

where we used the shorthand notation $\tau_i^j \equiv \tau_i^j(l)$. Note that (4.9) contains expressions which are simple functions of the Melnikov function, evaluated at the homoclinic points. We believe that as in the case of (4.8), (4.9) supplies the leading order approximation in ε to the exact formula, given in terms of the action of the homoclinic points τ_1^1 and τ_1^3. We use Eqs. (2.3), (3.9) and (3.10) to evaluate (4.9) for our example and obtain

$$e_l \approx \varepsilon C(\omega)(\frac{1}{\omega}\cos(\omega t) + \sin(\omega t)) \Big|_{\tau_0^1}^{\tau_0^3}. \tag{4.10}$$

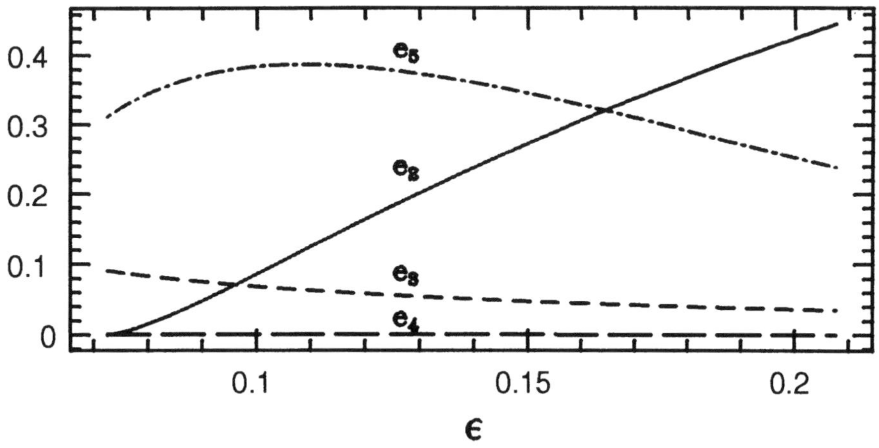

Figure 9. The Normalized Escape Rates for $\omega = 1.2$.

In a similar fashion we calculate $e_j = \mu(F^j E_0 \cap D_0)$, $l < j \le 2l$. The arches $F^j E_0 \cap D_0$ are bounded by the arcs $I_1 = \{(h_2^u(\tau_1), \tau_1)|\ h_0^u(\tau_0) = 0,\ \tau_0^1(j) \le \tau_0 \le \tau_0^2(j)\}$, $I_2 = \{(h_2^l(\tau_1), \tau_1)|\ h_0^l(\tau_0) = 0,\ \tau_0^4(j) \le \tau_0 \le \tau_0^3(j)\}$ and the segments $I_3 = \{(0, \tau_1), \tau_1^4(j) \le \tau_1 \le \tau_1^1(j)\}$, $I_4 = \{(0, \tau_1), \tau_1^2(j) \le \tau_1 \le \tau_1^3(j)\}$, where $\tau_0^1(j)$ and $\tau_0^3(j)$ are the solutions of (3.10b) and $\tau_0^2(j)$ and $\tau_0^4(j)$ are the solutions of (3.10a). Their images, τ_1^i, obey $\tau_1^2(j) < \tau_1^3(j) < \tau_1^4(j) < \tau_1^1(j)$. Therefore,

$$\begin{aligned} e_j &\approx \int_{\tau_1^2}^{\tau_1^1} h_2^u(\tau_1)d\tau_1 - \int_{\tau_1^3}^{\tau_1^4} h_2^l(\tau_1)d\tau_1 \\ &= \varepsilon \int_{\tau_1^2}^{\tau_1^1} M(\tau_1)d\tau_1 - \varepsilon \int_{\tau_1^3}^{\tau_1^4} M(\tau_1)d\tau_1 + \varepsilon \int_{\tau_0^2}^{\tau_0^1} M(\tau_0)d\tau_0 - \varepsilon \int_{\tau_0^3}^{\tau_0^4} M(\tau_0)d\tau_0 \qquad (4.11) \\ &\quad - \varepsilon M(\tau_0)\Big|_{\tau_0^2}^{\tau_0^1} + \varepsilon M(\tau_0)\Big|_{\tau_0^3}^{\tau_0^4}, \end{aligned}$$

where $\tau_i^k \equiv \tau_i^k(j)$. Using Eqs. (2.3), (3.9) and (3.10), we find that for our example

$$e_j \approx -\varepsilon C(\omega)(\frac{1}{\omega}\cos(\omega t) + \sin(\omega t))\Big|_{\tau_0^1, \tau_0^3} - 144 \exp((-j - \delta(\omega) + \frac{1}{2})T), \qquad (4.12)$$

where we use the notation $f(t)|_{a,b} = f(a) + f(b)$.

Finally, we estimate $e_{2l+1} = \mu(F^{2l+1}(E_0) \cap D_0)$. The set $F^{2l+1}(E_0) \cap D_0$ is composed of two arches, denoted by A^u and A^l. The area of the arch A^u is approximated by (4.12), with $j = 2l + 1$. Orbits belonging to the lower arch encircled the origin twice before escaping. Let t_0^i be the four solutions to (3.14a) with $j = l$ and the l in (3.14b) replaced by $2l + 1$. It follows from the geometry of the manifolds that the solutions t_0^i obey

$$\tau_0^2(l+1) < t_0^1 < t_0^2 < \tau_0^1(l), \qquad \text{and} \qquad \tau_0^3(l) < t_0^3 < t_0^4 < \tau_0^3(l+1). \qquad (4.13)$$

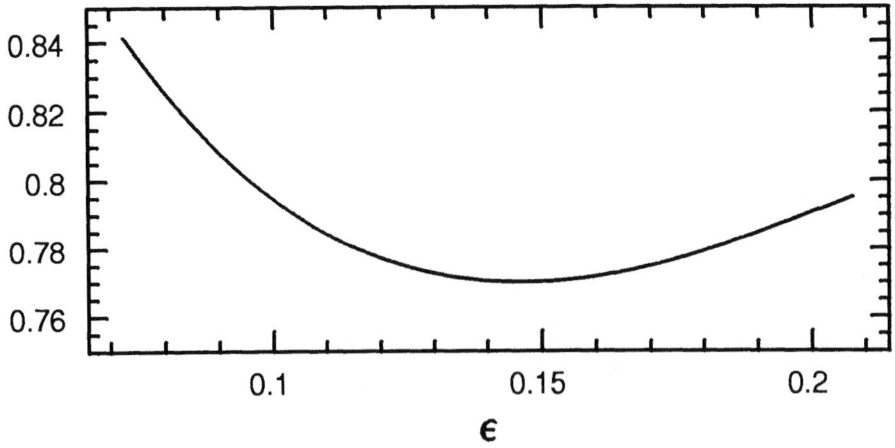

Figure 10. The Asymptotic decay Rate of The Escape Rates for $\omega = 1.2$.

and that $t_2^4 < t_2^1 < t_2^2 < t_2^3$. $h_3 \in A_l$ is bounded by the arcs $I_1 = \{(h_3^u(t_2), t_2) |\ h_0^u(t_0) = 0,\ t_0^3 \le t_0 \le t_0^4\}$ and $I_2 = \{(h_3^l(t_2), t_2) |\ h_0^l(t_0) = 0,\ t_0^1 \le t_0 \le t_0^2\}$ and the segments on which $h_3 = 0$ and t_2 varies between t_2^2 to t_2^3 and between t_2^4 to t_2^1. Therefore,

$$\mu(A^l) \approx \int_{t_2^4}^{t_2^3} h_3^u(t_2)dt_2 - \int_{t_2^1}^{t_2^2} h_3^l(t_2)dt_2 \qquad (4.14).$$

Using (3.2) we find

$$\begin{aligned}\int_{t_2^4}^{t_2^3} h_3^u(t_2)dt_2 &\approx \varepsilon \int_{t_2^4}^{t_2^3} (M(t_2) + M(t_1) + M(t_0))dt_2 \\ &= \varepsilon(2M(t_0)\Big|_{t_0^3}^{t_0^4} - M(t_1)\Big|_{t_1^4}^{t_1^3} + \sum_{i=0}^{2} \int_{t_i^4}^{t_i^3} M(t_i)dt_i).\end{aligned} \qquad (4.15)$$

The integral of h_3^l is of the same form, with the end points t_i^4, t_i^3 replaced by t_i^1, t_i^2 respectively.

To summarize, we find simple expressions for estimating $\mu(E_0)$ and the e_j's, for $j \le 2l + 1$, in terms of the solutions to Eqs. (3.9) and (3.14), which can be found numerically. In figure 9 we present the computations of these quantities, normalized by $\mu(E_0)$, for $\omega = 1.2$ and $\varepsilon_2^b < \varepsilon < \varepsilon_2^a$. This computation takes a few seconds on a Sparc workstation. The contribution of e_{2l+1} (e_5 in the figure), is quite significant, and causes a long transient behavior of the approximate e_n's, given by Eq. (4.6).

For large n, the approximate e_n's decay exponentially as $(\zeta_l)^n$, where ζ_l is the largest root of:

$$\zeta^{l+1} - s_1\zeta^l - s_2 s_3 = 0. \qquad (4.16)$$

Using (4.7-4.16), we calculate $\zeta_2(\varepsilon, \omega)$ for $\omega = 1.2$ and $\varepsilon_2^b < \varepsilon < \varepsilon_2^a$. The results are presented in figure 10.

Section 5. Summary and Conclusions.

We applied some of the methods developed in [3] to the phase space flow of a particle in a cubic potential, perturbed by a temporally periodic forcing. In section 2 we defined the structure indices, which classify the flows by the structure of their homoclinic tangles. In section 3 we used the Whisker map to compute these indices and in section 4 we estimated the escape rates using the Whisker map and the approximate structure of the manifolds.

We have demonstrated that the methods developed in [3] can be used to obtain a global description of the changes in the properties of the flow as the parameters vary. Moreover, we found that these methods are easy to apply and require negligible amount of numerical computations and programming. However, we have not addressed the question of accuracy of the methods in this paper. The use of the Whisker map introduces errors which can be estimated analytically. Their impact on our calculations must be examined carefully. The topological approximations regarding the structure of the manifolds is a more delicate issue. We do not foresee a rigorous way to justify this approximation, and hope to supply numerical evidence which will support our results. A possible theoretical resolution to this problem is to find structurally stable maps which attain a type l trellis (e.g. the horseshoe map for the $l = 1$ case). We believe that even if such maps do not exist, the results may be accurate in neighborhoods of the type l trellises, up to a measure zero set of parameter values.

The preliminary results for the specific example of the flow (2.1), are summarized by figures 5, 7 and 10; in the first two figures we find the approximate bifurcation curves for the structure indices, which supply approximate level sets for the topological entropy of the flow. In the last figure we plot the asymptotic decay rate of the escape rates as a function of ε for a fixed value of ω. The curve in not monotonic in ε, suggesting the surprising result that for some parameter values an increase in ε causes a decrease in the asymptotic escape rates.

Our results are valid for ε sufficiently small, and for finite values of ω which are bounded away from zero and one; for $\omega = 0, 1$ or infinity the Melnikov function vanishes identically, and we cannot conclude what is the structure of the homoclinic tangle using the Whisker map approximation. The limit $\omega \to 0$ is the adiabatic limit in which the homoclinic structure is different then the one considered in this paper [16]. Further analysis is required to conclude whether the ideas presented here are of any practical value in this case. In the limit $\omega \to \infty$ we expect to see exponentially small separation of the manifolds [17]. Formally, this limit corresponds to $l \to \infty$ in our analysis. However, the regular perturbation theory which we use is not valid in this limit. A generalization of the results to this case will probably require an extensive analytical effort. As for the region $\omega \approx 1$, the structure of the tangle is determined by the next nonvanishing term in the expansion series of the distance function in ε. The analysis presented here can be easily generalized to this case by replacing the Melnikov function by the appropriate higher order term. However, the calculation of the higher order terms is usually too hard and therefore impractical.

Acknowledgment: I have benefitted from discussions with Robert MacKay and Dana Hobson. This work is partially supported by NSF grant DMS8903244 and by ONR grant N00014-90J1194.

References

[1] V. Rom-Kedar, A. Leonard, and S. Wiggins [1990]. An Analytical Study of Transport, Mixing, and Chaos in an Unsteady Vortical Flow. J. Fluid Mech., **214**, 347-394.

[2] V. Rom-Kedar and S. Wiggins [1990]. Transport in Two-Dimensional Maps. Arch. Rat. Mech. Anal., **109**, 239-298.

[3] V. Rom-Kedar [1990]. Transport Rates of a Family of Two-Dimensional Maps and Flows, Physica D, **43**, 229-268.

[4] S. Wiggins [1990]. On the geometry of Transport in Phase Space, I. Transport in k-Degree-of-Freedom Hamiltonian Systems, $2 \leq k < \infty$. Physica D, **44**, 471-501.

[5] S. Wiggins [1990]. *Introduction to Applied Nonlinear Dynamical Systems and Chaos.* Springer-Verlag: New York, Heidelberg, Berlin.

[6] J. Guckenheimer and P. Holmes [1983]. *Non-Linear Oscillations, Dynamical Systems and Bifurcations of Vector Fields.* Springer-Verlag, New York.

[7] E. Knobloch and J.B. Weiss, [1987]. Mass Transport and Mixing by Modulated Traveling Waves. Phys. Rev. A, **36**, 1522.

[8] R. S. MacKay, J. D. Meiss, and I. C. Percival [1984]. Transport in Hamiltonian Systems. Physica D, **13**, 55-81.

[9] R. Camassa and S. Wiggins [1991]. Chaotic Advection in Rayleigh Bénard Flow, Phys. Rev. A. **43** (2), 774-797.

[10] B. V. Chirikov, *A Universal Instability of Many-Dimensional Oscillator Systems*, Physics Reports, **52**, No. 5, 263, 1979.

[11] D. F. Escande [1988]. Hamiltonian Chaos and Adiabaticity in *Plasma Theory and Nonlinear and Turbulent Processes in Physics* (Proc. Intl. Workshop, Kiev, 1987), edited by V. G. Bar'yakhtar, V. M. Chernousenko, N. S. Erokhin, A. G. Sitenko, and V. E. Zakharov. World Scientific: Singapore.

[12] M. J. Davis [1985]. Bottlenecks to Intramolecular Energy Transfer and the Calculation of Relaxation Rates. J. Chem. Phys., **83**, 1016-1031.

[13] R. W. Easton, *Trellises Formed by Stable and Unstable Manifolds in the Plane*, Trans. American Math. Soc., **294**, 2, 1986.

[14] D. Bensimon and L. P. Kadanoff [1984]. Extended Chaos and Disappearance of KAM Trajectories. Physica D, **13**, 82.

[15] T. J. Kaper, G. Kovacic and S. Wiggins [1990]. Melnikov Functions, Action, and Lobe Area in Hamiltonian Systems. J. Dyn. Diff. Eqns., submitted.

[16] T. J. Kaper and S. Wiggins [1990]. Lobe Area in Adiabatic Systems. Physica D, proc. for the CNLS 10th Annual Conference "Nonlinear Science: The Next Decade", submitted.

[17] P. Holmes, J. Marsden and J. Scheurle [1988]. Exponentially small Splittings of Separatices with applications to KAM theory and Degenerate Bifurcations. Cont. Math. 81, 213-244.

LARGE SCALE STRUCTURES IN KINETIC PLASMA TURBULENCE

D.F. Escande

Equipe Turbulence Plasma*, Institut Méditerranéen de Technologie,
Technopôle de Château-Gombert, 13451 MARSEILLE CEDEX 13, France

Abstract : The qualitative features of large scale structures in the simplest case of kinetic plasma turbulence, the electron beam-plasma instability, are described. The corresponding theory is still unsatisfactory, and even paradoxical. The weaknesses of the traditional approach of microscopic plasma theory are analysed. A new approach is introduced, resting on a classical mechanics technique. It starts from a N-body description of the plasma, goes through the derivation of a Hamiltonian describing the self-consistent evolution of Langmuir waves and near-resonant particles, and provides a derivation of the quasilinear equations where spontaneous emission effects are included. This derivation makes clear the physics of wave-particle interaction, and answers many question raised by the Vlasovian approach. In particular, Landau damping turns out to be a non-resonant nonlinear phenomenon of synchronization of particles with a wave. The regime where chaos is dominant is still unexplained analytically, but the results of a numerical simulation starting from the self-consistent equations agree with experiments. This new technique should find applications in other fields than plasma physics.

---

* Team of the Laboratoire de Physique des Interactions Ioniques et Moléculaires, Université de Provence, Unité de Recherche Associée au Centre National de la Recherche Scientifique n° 773.

1. INTRODUCTION

When a weak electron beam is injected into a thermal plasma, electrostatic modes at the plasma frequency (Langmuir waves) are destabilized, and grow starting from the entrance point of the beam till the instability saturates. This saturation occurs by two possible mechanisms [1] depending on the beam temperature (its velocity spread) :

- for a cold beam, beam particles are trapped in the most unstable mode, which yields a regular spatial modulation of the mode amplitude (figure 1),
- for a warm beam, beam particles are diffusively slowed down by a broad spectrum of modes, which yields a chaotic spatial modulation of the modes amplitudes (figure 2).

Both cases correspond to the formation of spatial structures in the plasma. In a true experiment a Langmuir mode is characterized by its frequency ω.

Theorists prefer the initial condition problem where a spatially uniform beam is initially present in the plasma. Then similar structures evolve in time, and a Langmuir mode is defined through its wave number k. Temporal and spatial scales are related through the group velocity of the mode of interest. The typical behaviour of the mode's field amplitude E_ω or E_k is displayed in figures 1 and 2.

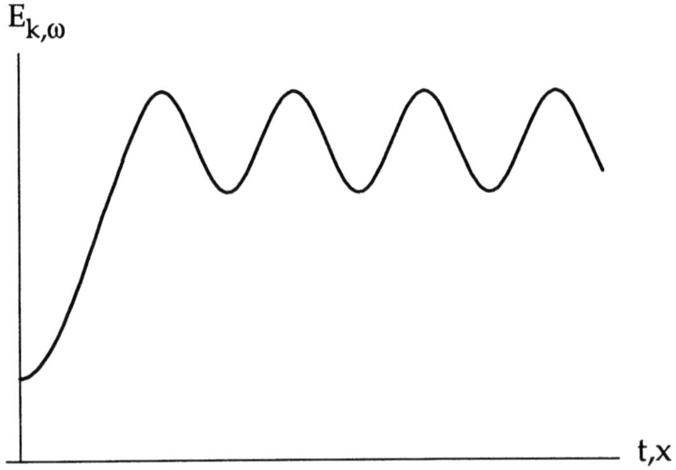

Figure 1. Spatial or temporal evolution of the electric field of one Langmuir mode in the cold beam-plasma instability.

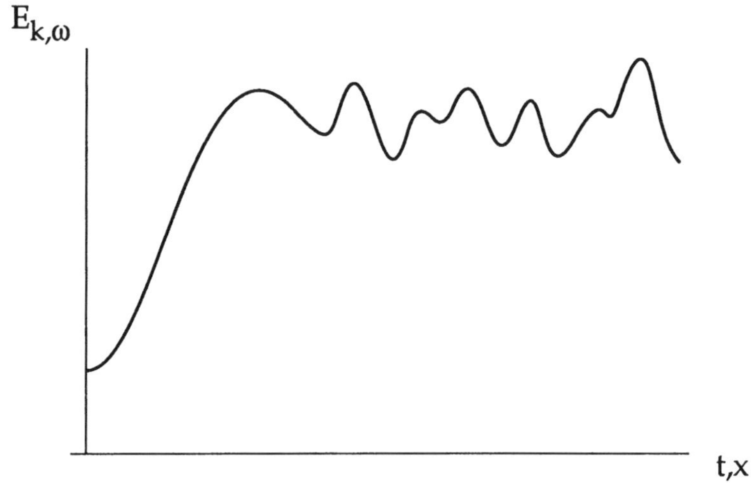

Figure 2. Spatial or temporal evolution of the electric field of one Langmuir mode in the warm beam-plasma instability.

Many questions are left unanswered in the theory of this basic phenomenon of kinetic plasma turbulence. In the beginning of the sixties the warm beam case was given a theoretical description in a so-called quasilinear theory [2] which makes correct predictions even though its assumptions are proved to be wrong both theoretically and experimentally [3]. As a more recent theory with better assumptions [4] makes predictions which are not experimentally verified [3], theorists are left with a paradox !

The saturation of the cold beam case has not yet received a fully analytical treatment [5,6]. However, an interesting set of equations (*self-consistent equations)* was derived, which describes the self-consistent coupling of Langmuir modes with resonant particles, and made possible a convincing computer simulation of the saturation [5].

The above paradox was originally tackled from a Vlasovian point of view. This motivated the Turbulence Plasma team, created in Marseille in 1988, into developing a new approach through the self-consistent equations where particles and modes played a symmetrical role, and where most of the irrelevant degrees of freedom of the full Vlasovian description were removed. This approach rapidly gave hints that it could tell more than the Vlasovian one about phenomena like Landau damping (the damping of waves by near-resonant particles). This, in turn, made desirable a more basic derivation of the self-consistent equations.

This paper reports on the present state of development of the theory, and shows several directions for future investigation it opens, in particular in other fields than plasma physics. The beam-plasma system is considered as a mechanical N-body object. Through simple physically intuitive steps, the properties of Langmuir waves for *one realization* of the plasma are derived. *Averaging over realizations* of the plasma yields the usual quasilinear equations where the effect of spontaneous emission of the particles is included. As yet, the range of validity of these equations is not improved with respect to the original derivations, but their physical content becomes transparent. In particular Landau damping turns out to be a nonlinear non-resonant effect, a fact completely obscured by the Vlasovian approach ! Furthermore, explaining the quasilinear paradox is now clearly connected with the justification of the quasilinear computation of the diffusion coefficient in the much simpler (but far from trivial !) dynamics of one electron in a prescribed set of Langmuir waves. A numerical simulation of the self-consistent equations yields results in agreement with the experiment of reference 3.

We consider here plasma turbulence from a kinetic point of view. This means we deal with particles, and we do not use a continuum mechanics approach. Collisions are considered as negligible, but this does not preclude the existence of dissipative-like effects (Landau damping). The plasma is unmagnetized, it is considered as one-dimensional, and the ions are a static neutralizing background (this last assumption corresponds to the so-called regime of weak turbulence). The large scales we consider are those larger than the Debye screening length λ_D, and are related to electrostatic electron waves (the Langmuir waves) which are coherent plasmon states.

Since this paper is intended for a broad audience, section 2 yields a simplified presentation of the features of the beam-plasma instability, and section 3 shows the outline of the theoretical results available till 1988. Section 4 presents the main stream of the new mechanical approach. For pedagogical reasons, we use Newton's law of mechanics and the Coulomb force, instead of more sophisticated (and elegant) Lagrangian or Hamiltonian formalisms. This way, the most sophisticated tool we use is the Fourier series.

2. PICTURES OF THE BEAM-PLASMA INSTABILITY

For a thermal plasma with electron thermal velocity v_T (figure 3) and plasma frequency ω_p ($\omega_p^2=ne^2/(\varepsilon_0 m)$, with n the electron density), the Langmuir waves have the Bohm-Gross dispersion relation shown in figure 4. This picture should be completed with a curve symmetrical with respect to the k axis, which corresponds to modes propagating in the negative direction. When a weak electron beam is present in the plasma, the beam's Langmuir modes couple with the plasma's ones, and the Bohm-Gross dispersion relation is modified. For a cold beam with velocity u, a narrow part of the beam modes become unstable, and simultaneously decrease their phase velocity (figure 5). For a warm beam, the plasma Langmuir modes become unstable for phase velocities in the region where the slope of the beam-plasma distribution function is positive (Landau instability, figure 6). In fact figures 5 and 6 is somewhat simplified with respect to reality. In particular, for wave numbers far from the one defined by ω=u.k, there are two beam modes corresponding to the two counterstreaming Langmuir modes of the beam. Therefore, there are always three modes with a positive phase velocity for a given wave number. In figures 5 and 6 the emphasis is on the unstable branch, and the imaginary part of the pulsation (or wave number) is missing.

f(v)

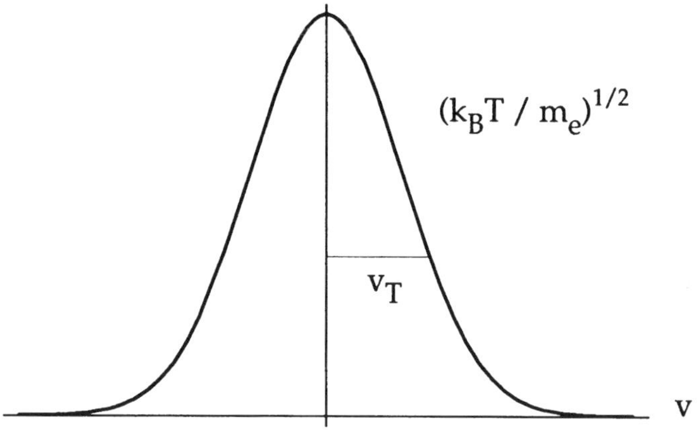

Figure 3. Maxwellian distribution function for electrons.

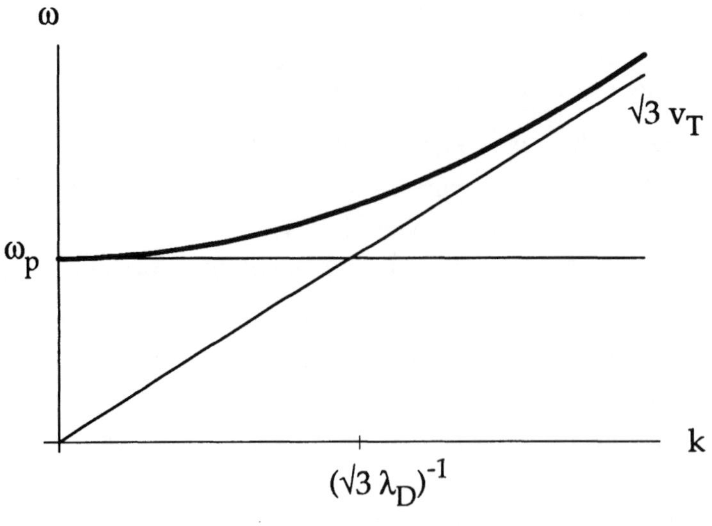

Figure 4. Bohm-Gross dispersion relation.

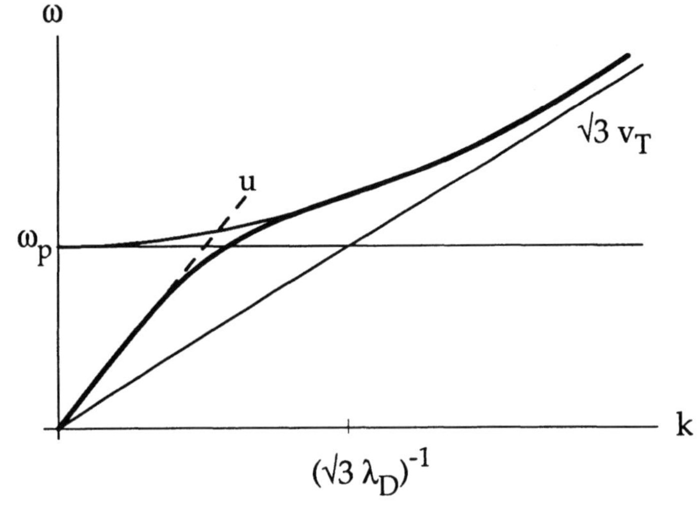

Figure 5. Dispersion relation of electrostatic waves for the cold beam-plasma instability.

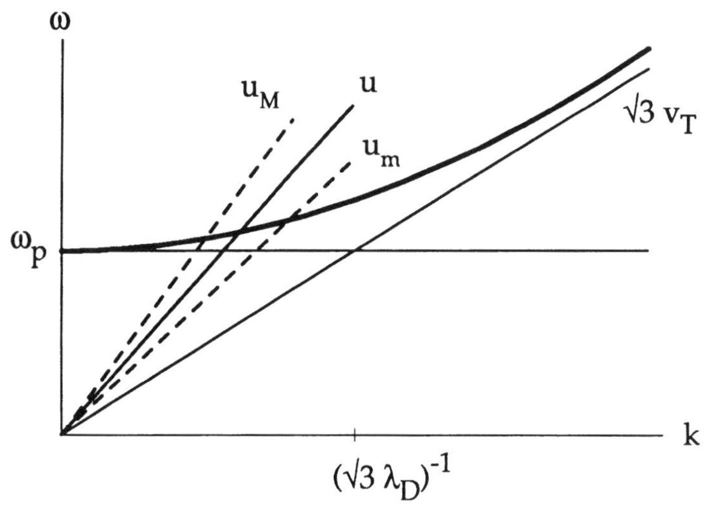

Figure 6. Dispersion relation of electrostatic waves for the warm beam-plasma instability.

For the initial value problem, it is convenient to write the mode's electric field as $E_k(t)\exp[i(kx-\omega_k t)]$, where ω_k and k are real. During the linear stage of the instability, E_k grows like $\exp(\gamma_k t)$. In the cold beam case, γ_k is very peeked as a function of k. Hence one may consider that only one mode is excited. When it saturates, the period of the oscillations of E_k in figure 1 is much larger than ω_p^{-1}. In the warm beam case, a broad spectrum of unstable modes is selected, and each mode fluctuates on scales much larger than ω_p^{-1} in figure 2. In both cases there are large-scale modulated envelopes.

Together with the saturation of the modes, there is a change in the particles' orbits in phase space. Figure 7 displays the structure of the initial phase space in the cold beam case. The initial trapping domain [7] related to the most unstable mode is clearly away from the beam particles. As shown in figure 8, when the mode amplitude grows, the beam particles are trapped in the wave troughs, and begin to rotate in phase space. For an appreciable amount of time, most particles may be considered as having the same bounce frequency, which produces a rotating bar in phase space. This bounce frequency is the one visible in figure 1.

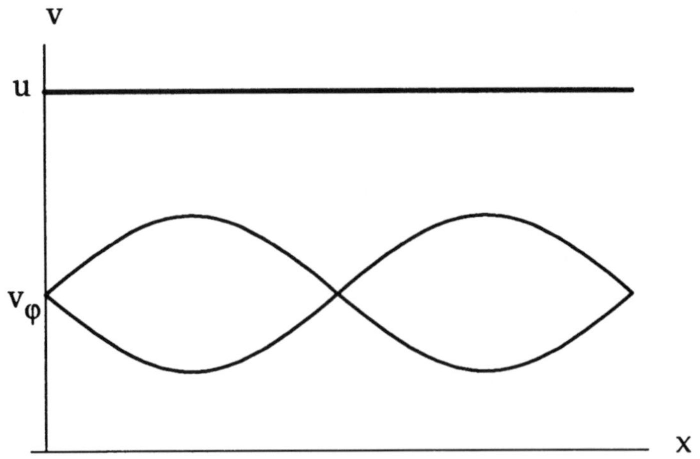

Figure 7. Initial phase space for the cold beam-plasma instability. The beam with velocity u (thick line) and the initial trapping domain of the unstable mode are displayed.

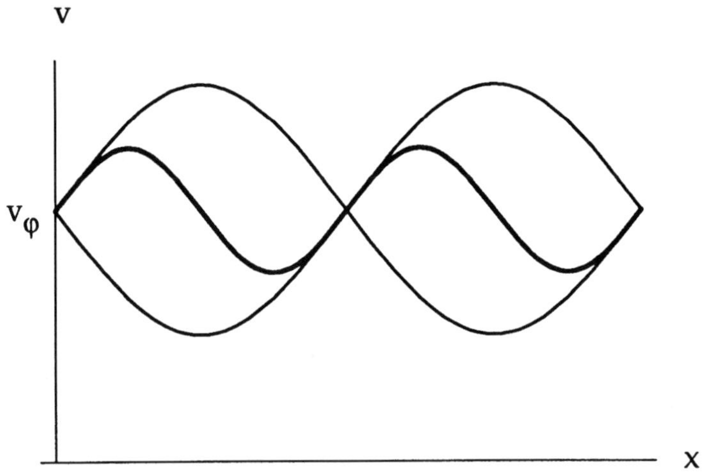

Figure 8. Final phase space for the cold beam-plasma instability. The final trapping domain of the unstable mode and trapped beam particles (thick line) are displayed.

The right part of figure 9 displays the structure of the initial phase space for the warm beam case. The trapping domains of some unstable modes are visible. Initially nearby trapping domains do not overlap, or overlap so weakly that the chaotic time scale is much larger than the linear (Landau) growth time. When the instability develops, chaos becomes dominant, and particles diffuse in phase space. This induces a slowing down of the beam particles, and the transition from a beam-like tail to a plateau, as shown in the left of figure 9.

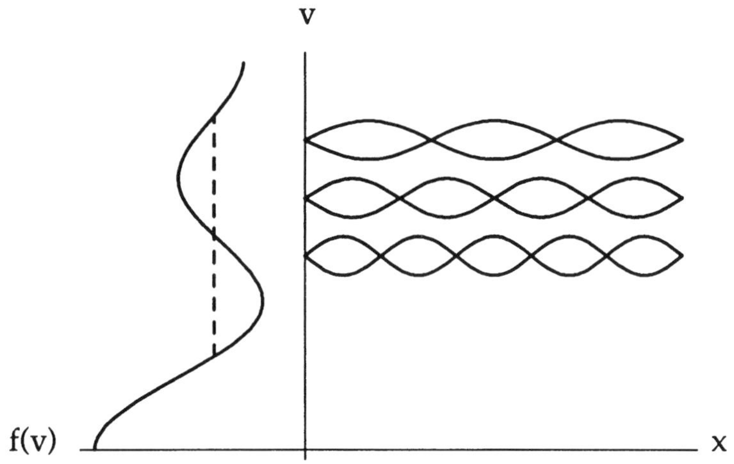

Figure 9. On the right, phase space for the warm beam-plasma instability. On the left velocity distribution function before (bump) and after (plateau) saturation.

3. THEORY TILL 1988

The above features of the beam-plasma instability have been observed in a series of experiments both in the cold [8] and warm [9] case. Whereas the qualitative understanding of both saturations is good, theory is still incomplete, and even paradoxical.

In principle, macroscopic classical physics should be descriptible as a N-body problem, i.e. as a problem of classical mechanics (this is Laplace's dream). The complexity of the problem and the chaos inherent to it forbid an explicit solution. Till now, this has excluded approaches starting from classical mechanics which is an essential root of physical intuition though (1st annoyance). It has induced the splitting of macroscopic classical physics into a series of subfields, each of it characterized by an appropriate tool accounting for the relevant dynamics of interest : Boltzmann equation, Navier-Stokes equation, etc... The traditional tool of microscopic plasma physics is the Vlasov equation

$$\frac{\partial f}{\partial t} + v\frac{\partial f}{\partial x} + \frac{qE}{m}\frac{\partial f}{\partial v} = 0 \tag{1}$$

where f is the velocity distribution function, and where the electric field is computed through the Poisson equation

$$\frac{dE}{dx} = \frac{e}{\varepsilon_0}\left(- \int f(v)\, dv + n_0 \right) \quad , \tag{2}$$

where n_0 is the ion density, and f is for electrons (one could also write a Vlasov equation for ions). The Vlasov equation is analogous to the Boltzmann equation, and is derived from the Liouville equation through the heavy (2nd annoyance) BBGKY hierarchy.

Already for linear theory, further annoyances show up. Linearizing the Vlasov-Poisson system, Fourier-Laplace transforming it, a pole hunting, and an analytic continuation yield the Landau effect but no physical interpretation (3th annoyance) of it, despite the considerable amount of work. Mysteriously (4th annoyance), there are eigenmodes for Landau instability, but none for Landau damping. When tackling the same problem with distributions, one finds a continuum of stable modes for a given wave number k : the van Kampen modes. They have not been detected experimentally, and constitute a kind of optional knowledge in plasma physics (5th annoyance). They enable the interpretation of Landau damping as a consequence of their phase mixing, but do not give its physical interpretation from the viewpoint of wave-particle interaction either (6th annoyance).

Many textbooks fill in this gap with the following calculation : the change of kinetic energy of electrons in the presence of a sinusoidal wave with constant amplitude is computed perturbatively, and is balanced with the opposite change of wave energy ; this trick yields Landau's formula for the growth/damping rate γ_L, but van Kampen's approach shows it is fundamentally wrong for the damped case (where is phase mixing ?), which is confirmed when the calculation is redone with the wave amplitude varying exponentially in time (7th annoyance). However these calculations suggest that, at least for the instability, Landau effect is due to the synchronization with the wave of particles within γ_L/k of the phase velocity [10,11]. Hence resonant particles would not participate in the effect whose resonant character is clearly suggested by the Vlasovian approach (8th annoyance) !

In a true plasma, since there is a thermal level for field modes, Landau damping cannot lead below it. Therefore, it is not a linear effect as in a Vlasov plasma (9th annoyance). Would true Landau damping be a nonlinear non-resonant effect ? This is confirmed in the next section. In order to compute the thermal level, Vlasov's equation is useless, and textbooks introduce test particles which give spontaneous emission. Putting together this emission with the Landau effect yields the thermal level. This computation is typical of microscopic descriptions in plasma physics : there is no systematic way of dealing with them, but one must require ingeniosity and experience in order to select and to combine various basic equations (10th annoyance). None of them gives a complete description of the plasma state : a continuous or a granular medium ? This state of plasma theory contributed to a fame of plasma physics which is far from excellent among theoreticians (11th annoyance, for plasma physicists only !).

Things worsen in the nonlinear regime. In the warm beam case, a theory appeared in the sixties, the quasilinear theory [2], which neglects the coupling between unstable Langmuir modes. It predicts the behaviour described in the previous section, and an experiment [9] was in good agreement with its quantitative predictions. However its assumptions were proved to be incorrect long before the saturation of the instability [12]. This motivated new efforts for a correct description of the nonlinear regime, and a realistic model predicted that the Landau growth rate, and the diffusion coefficient predicted by quasilinear theory should be renormalized by a factor of order 2 in this regime [4]. A numerical simulation indicated a smaller renormalization [13]. A quite accurate experiment was carried out in order to check these predictions. It confirmed that mode-coupling could not be neglected, but, paradoxically, it did not find any renormalization [3] (12th annoyance). The old quasilinear theory makes right predictions with wrong assumptions, when the more recent Vlasovian approach, *a priori* more correct, is not confirmed experimentally (13th annoyance) !

In the cold beam case, no fully analytic description is available. However, the analytic endeavours opened a quite interesting path. They split the beam-plasma system into a bulk (the plasma) described by the Vlasov equation, and a beam described as a set of N charged sheets per spatial period L of the beam-plasma system [5]. This gives amplitude equations for the E_k where the source is due to the N beam particles, while the particle's motion is ruled by Newton equations where the force is due to the global electric field of M active Langmuir modes. This yields the set of self-consistent equations already mentioned. Since the unstable spectrum is narrow, M=1 was chosen, and a numerical simulation gave the saturation features described in section 2. Seven years later, the Hamiltonian character of the self-consistent dynamics was made explicit by the derivation of a self-consistent Hamiltonian whose canonical equations yield the self-consistent equations [6]. The same paper introduced a rotating bar model for the description of the saturation in phase space, which accounted for the results of the previous simulation, but with 5 adjustable parameters.

4. MICROSCOPIC PLASMA PHYSICS THROUGH CLASSICAL MECHANICS

The 13 above annoyances, and especially the quasilinear paradox, directed the Turbulence Plasma team into a new approach of weak Langmuir turbulence that would be closer to mechanical intuition and could benefit from the recent progress of chaos theory. Therefore a description with a finite number of degrees of freedom was highly desirable, and the self-consistent Hamiltonian looked like a promising starting point. However, considering the flow of reasoning on Langmuir turbulence starting from basic principles, this Hamiltonian is an intermediate point in the stream.

Upstream are basic principles enabling to derive it. Its original derivation was not completely satisfactory for several reasons. When keeping in mind the mechanical character of a plasma, there should be a way avoiding to go through the infinite number of degrees of freedom of the Vlasov-Poisson system in order to get the finite-dimensional self-consistent Hamiltonian. Furthermore there should be a Lagrangian or Hamiltonian path to it, which avoids building it by induction from the equations of motion. This suggested to start with a N-body problem. This approach proves to be quite efficient, and yields a lot more information than the Hamiltonian itself. It is the topic of section 4.1.

Downstream are all the results which can be collected from it. Already the first 10 annoyances are suppressed this way, and there is a good hope to solve the quasilinear paradox from this starting point. Section 4.2 presents the results already available when chaos may be neglected. Section 4.3 gives a short account of the present results of a numerical simulation, using the self-consistent equations, of the regime corresponding to the quasilinear paradox.

A general feature of the reasoning is to avoid as much as possible statistical methods, and to consider one realization of the plasma. Upstream, the statistics is restricted to the concept of typical realization of the plasma. Downstream, the averaging over realizations of the plasma is introduced after making sure that Landau damping cannot be obtained as a Floquet exponent for one sample.

4.1. FROM N-BODY TO FIELD-PARTICLE INTERACTION [14]

As the plasma (or beam-plasma system) is considered as one-dimensional, it may be described as a set of charged sheets. Since we consider Langmuir weak turbulence, ions may be considered as a neutralizing background whose role vanishes when the plasma is considered as periodic. As a result, our starting point is a set, with spatial period L, made of N electronic sheets per period. The starting equation could quite naturally be the Lagrangian of this system, but, for pedagogical reasons, it is interesting to start with the corresponding set of N Newton equations describing the motion of charged planes under the action of the Coulomb force due to the others

$$\ddot{x}_j = -i \frac{q\sigma}{\varepsilon_0 m L} \sum_{k \in \mathbb{Z}_0} \frac{1}{k} \sum_{l \neq j}^{N} e^{ik(x_j - x_l)} \qquad j = 1,N \tag{3}$$

where σ is the charge surface density of a sheet, and other notations are standard.

We now introduce the concept of reference multibeam (section 4.1.1), collective variables (section 4.1.2), and linear eigenmodes (section 4.1.3). We then address the question of the validity of the linearization, what turns out to be an inverse problem : for a typical initial plasma, what is the best choice of reference multibeam (section 4.1.4) ? The solution to this problem motivates considering a Bohm-Gross plasma, i.e. a plasma with no particle resonant with Langmuir modes (section 4.1.5). Adding a resonant tail to the velocity distribution function leads naturally to the self-consistent Hamiltonian (section 4.1.6).

4.1.1. Reference multibeam

A reference multibeam is an ideal state of the plasma where all particles belong to a set of, say r, beams whose particles (at least two per beam) are equidistributed spatially (figure 10). This is the analogue of the unperturbed distribution function of Vlasovian approaches. The j-th particle has a position

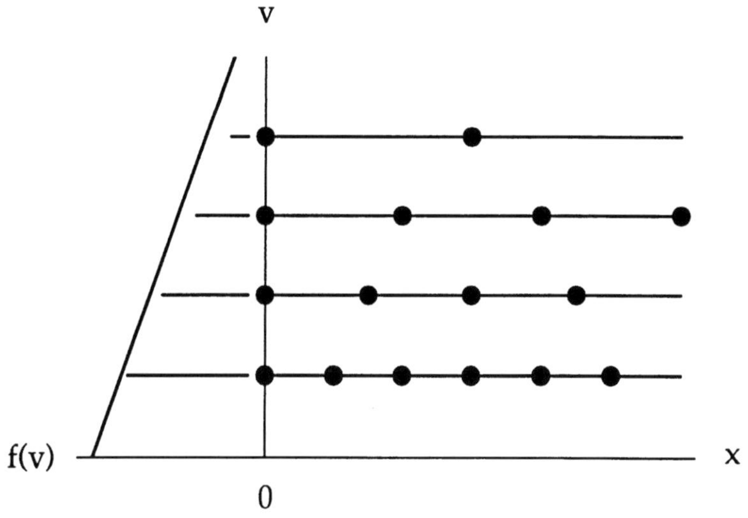

Figure 10. Reference multibeam. The particles are shown as black dots and the display is similar to that of figure 9.

$$\xi_j = \xi_{j0} + v_j t \qquad (4)$$

where v_j belongs to $\{ u_\nu \}$, $\nu = 1,r$. This definition of the reference multibeam goes one step further in the direction opened by Dawson in 1960 when he dealt with the plasma as a set of fluid beams [15].

4.1.2. Collective variables

We now consider plasmas whose particles have positions close to those of a given reference multibeam. It is natural to introduce the variables $\eta_j = x_j - \xi_j$. If particle j moves in the field of an unspecified spectrum of Langmuir modes (which may have an infinite number of components in the case of a cold plasma), its acceleration is

$$\ddot{x}_j = \sum_{k=-\infty}^{+\infty} \frac{qE_k(t)}{m} e^{i(kx_j - \omega_k t)} \quad . \qquad (5)$$

Here we perform a change of variable corresponding to the definition of N collective amplitudes a_k and their complex conjugates defined by

$$\ddot{x}_j = \sum_{|k| \leq \frac{2\pi N}{L}} a_k(t)\, e^{i(k\xi_j - \omega_k t)} \quad , \qquad (6)$$

which are reminiscent of qE_k/m, with three important differences : the number of modes is the number of particles, the position of particle j is replaced with that of its nearby reference multibeam particle, and pulsation ω_k is as yet unspecified. Equation (6) can be inverted easily to give the a_k explicitly as a function of the $\ddot{x}_j$. In the Lagrangian approach a slightly different definition of the a_k's is preferred,

$$\eta_j = \sum_{|k| \leq \frac{2\pi N}{L}} \frac{a_k(t)}{\Omega_{kj}^2} e^{i(k\xi_j - \omega_k t)} \qquad (7)$$

where

$$\Omega_{kj} = kv_j - \omega_k \, . \tag{8}$$

4.1.3. Linear eigenmodes

We now look for a_k's which are slowly varying on the time scale of the largest $(\Omega_{kj})^{-1}$. Linearizing the equations of motion in the η_j's yields

$$\frac{2i\,\omega_p^2}{N} \sum_j \frac{1}{\Omega_{kj}^3}\, \dot{a}_k + O\left(\sum_j \frac{1}{\Omega_{kj}^4}\, \ddot{a}_k\right) = a_k \left(1 - \frac{\omega_p^2}{N} \sum_j \frac{1}{\Omega_{kj}^2}\right) . \tag{9}$$

Then, a_k may be taken as constant in this equation if $D(\omega,k) = 0$, where D is the coefficient of a_k in equation (9) ; note that in the fluid limit

$$D(\omega,k) = 1 - \int_{-\infty}^{+\infty} f(v) \frac{dv}{(\omega_k - kv)^2} \, . \tag{10}$$

Therefore $D = 0$ corresponds to the famous Bohm-Gross full dispersion relation (no expansion in k has been performed to yield equation (10)) [16].

In a plasma with a thermal bulk, if there are particles with a velocity larger than $\sqrt{3}v_T$ (the minimum phase velocity in figure 4) this dispersion relation yields by continuity the beam modes related the fastest particles instead of the dispersion of figure 4. Therefore the previous definition of collective modes is not completely satisfactory. Furthermore especially without any previous knowledge in plasma physics the above linearization is, as yet, just a trick. Indeed we just considered plasmas close to a given reference multibeam. We must now address the difficult part of the problem : if we take initially at random a plasma according to a uniform distribution in positions and to a given distribution in velocities, can we typically find a nearby reference multibeam maximizing the number of wave numbers for which the above linearization is relevant ? How is the set of such wave numbers defined ?

4.1.4. Inverse problem

We take a plasma at random as just defined. In order to define a reference multibeam for this sample, there are two, *a priori* arbitrary, steps : the choice of the number r of beams and that of their velocities. For each particle there is a closest beam velocity. The particle is said to be in the velocity domain of the corresponding beam (figure 11). We then count the number of particles in each velocity domain and assign this number of particles to the corresponding beam. The last free choice is the phase corresponding to the initial position of one the beam's particle. We then number beam particles from left to right and do similarly for the particles in its velocity domain. This defines the j's and η_j's. At this point we can define collective variables as before.

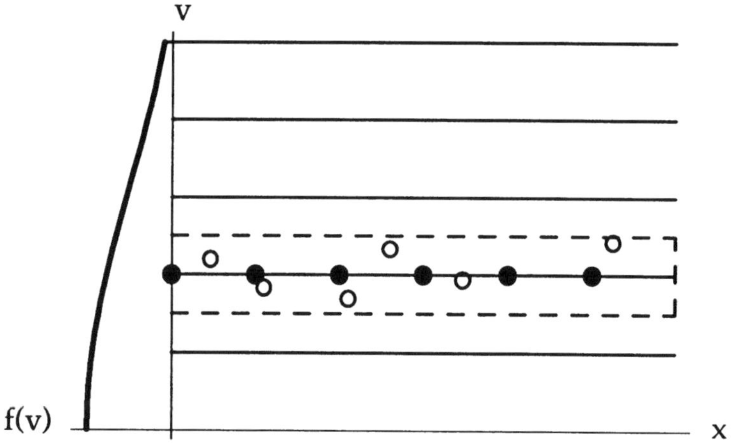

Figure 11. Definition of the number of particles in a given beam : actual particles are white dots ; reference particles are black dots.

We choose ω_k according to the full Bohm-Gross dispersion relation. If we do not linearize the equations of motion, but still make the assumption of slow variation for the a_k's, equation (9) is replaced with an amplitude equation with the same left-hand side, and a nonlinear source term in the right-hand side which incorporates all wave numbers. For the nonlinear source term to be small, k should be much smaller than the reciprocal of η_0, the typical value of η. A small value of η_0 can be obtained by putting many particles per beam. For the amplitude equation for a_k to be correct, the second derivative of a_k should be negligible initially. This can be simultaneously ensured for a number r of wave numbers by an appropriate choice of the set of phases corresponding to the beams. A large value of N enables us to have simultaneously a large number of beams and of particles per beam. Then, equation (9) turns out to be correct for k much smaller than ω_p/u_m, where u_m is the maximum velocity in the initial distribution function ($u_m > v_T$), provided the plasma density is low enough (negligible Coulomb collisions). This property enables the tail particles to play a part disproportionate with respect to their small number, what should be cured in some way.

4.1.5 Bohm-Gross plasma

As a first step, let us consider a Bohm-Gross plasma, i.e. a plasma with no tail, or, more specifically, a thermal-like plasma whose tails are cut out at $u_m < \sqrt{3} v_T$ (figure 12). Then, for a typical sample of the plasma, we recover the usual Bohm-Gross dispersion relation of figure 4 for wave numbers lower than $k_D = 2\pi/\lambda_D$, a usual condition in plasma physics. What occurs when k increases ? It is interesting to make an analogy with chains of nonlinear oscillators like Fermi-Pasta-Ulam's. In this case, mode coupling is small for large k's. When mode-coupling increases, thermalization of a mode's energy occurs on smaller and smaller time scales due to dynamical chaos [17]. One may expect something similar to occur in the plasma case. Numerical simulations of the plasma case give a clear evidence of thermalization for k above a critical wave number which is larger than k_D [18]. We expect this wave number to be typically

$$k_* = k_D \left(\frac{n k_B T}{W} \right)^{1/2} \tag{11}$$

where n is the plasma density, T its temperature, and W is the total electrostatic energy density ($W = \varepsilon_0 E^2/2$).

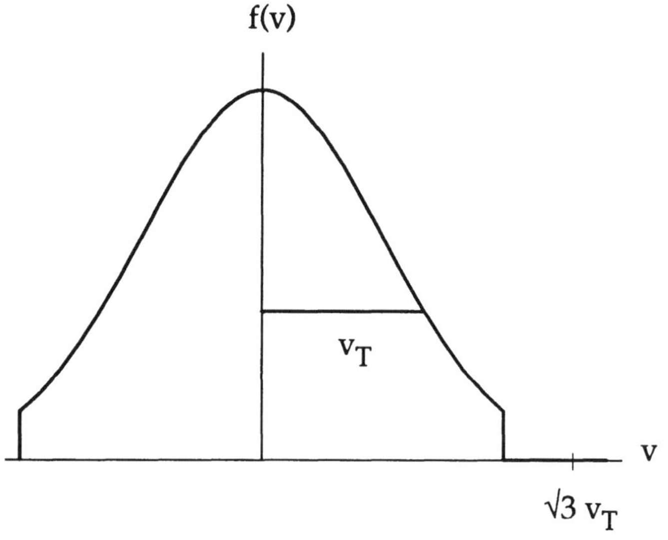

Figure 12. Velocity distribution function of a plasma with no particle resonant with Langmuir waves.

Therefore, even without Landau damping, modes with k larger than k_* cannot be observed because of thermalization. The corresponding a_k's are rapidly fluctuating quantities. If tail particles are added, this property prevents the existence of a resonant-like behaviour, and Landau damping does not play any role either.

Modes with $k>k_*$ provide a fluctuating action on plasma particles. The $k>>k_*$ part of it has been known for a long time under the name of Coulomb collisions (quite weak in one dimension). The other part has been missed by previous approaches which could not tackle the graininess of the plasma. Nevertheless it is quite living a part as well, and it could provide the missing ingredient for the explanation of Langmuir paradox : electrons would thermalize due to chaos on the scales about $(k_*)^{-1}$. At this point one could wonder whether chaos at the thermalization scale could not be an essential ingredient in the explanation of so-called anomalous transport [19] as well. After these speculations we go back to the main stream of the reasoning.

4.1.6 Plasma with a tail

The preceding step suggests that, when dealing with a plasma with N particles, some of them having velocities larger than $\sqrt{3}v_T$, one should split it into a Bohm-Gross bulk which allows the definition of collective variables, and a tail which has a number of particles n<<N. These particles have Newton equations

$$\ddot{x}_j = \sum_k a_k\, e^{i(kx_j - \omega_k t)} - \frac{iq\sigma}{\varepsilon_0 mL} \sum_k \frac{1}{k} \sum_{l\in \text{tail}} e^{ik(x_j - x_l)} \qquad (12)$$

where $|k|$ takes on N-n values in the first summation and all integer values in the second one. Similarly envelope equations for the collective variables are

$$\dot{a}_k = \frac{\omega_k^3}{2mk} \sum_{l\in \text{tail}} e^{i(kx_l - \omega_k t)} + \text{Nonlinear source } (\{a_{k'}\}) \qquad (13)$$

The ω_k in this formula corresponds to the bulk and thus differs from that in section 2. It is interesting to notice how a N-body system splits naturally into n bodies and N-n collective degrees of freedom which in turn split into Langmuir, thermalization, and Coulomb scales (large, intermediate, and small). In particular, the N-body system possesses a wave-like field, the Langmuir waves.

Equations (12) and (13) are an extended version of the self-consistent equations of reference 5. The last term of equation (12) corresponds to the beam-beam interactions, and embodies in particular effects at the beam's plasma frequency. As was done in references 5 and 6, we may neglect this term for a small tail. Moreover, tail particles are mainly sensitive to near-resonant modes, and we may restrict the summation in equation 12 to a small number of modes nearly resonant with the tail particles (this number was 1 in references 5 and 6). Self-consistency only requires the corresponding envelope equations to be considered, especially for Langmuir modes where the nonlinear coupling can be neglected in many applications. Dealing from the outset with a Lagrangian or Hamiltonian formalism makes it quite trivial that the self-consistent equations stem from the canonical equations of a self-consistent Hamiltonian. Define real variables X_j and Y_j by

$$X_j + iY_j = \frac{2im}{\varepsilon\beta_j} \quad a_k\, e^{i\omega_j t} \qquad , j = 1,M , \tag{14}$$

$$\beta_j = \sqrt{\frac{\omega_j^3}{2\,\omega_p^{\,2}}} \; , \tag{15}$$

$$\varepsilon = \sqrt{\frac{2\,e^2\,n_t}{\varepsilon_0\,N}} \; , \tag{16}$$

where n_t is the tail density. The self-consistent Hamiltonian is

$$\begin{aligned} H_{sc} &= \frac{1}{2}\sum_{l=1}^{N} p_l^{\,2} + \frac{1}{2}\sum_{j=1}^{M} \omega_j \left(X_j^{\,2} + Y_j^{\,2}\right) \\ &+ \varepsilon \sum_{l=1}^{N}\sum_{j=1}^{M} \frac{\beta_j}{k_j} \left(Y_j \sin k_j x_l - X_j \cos k_j x_l \right) , \end{aligned} \tag{17}$$

where N now stands for n, and M is the number of Langmuir modes of interest. It is useful to introduce the action-angle variables (I_j, θ_j) related to mode j

$$I_j = \frac{1}{2} (X_j^{\,2} + Y_j^{\,2}) , \tag{18}$$

$$X_j = \sqrt{2\,I_j} \;\; \cos\theta_j , \tag{19}$$

$$Y_j = \sqrt{2\,I_j} \;\; \sin\theta_j ; \tag{20}$$

with these variables the Hamiltonian becomes

$$H'_{sc} = \frac{1}{2}\sum_{l=1}^{N} p_l^{\,2} + \sum_{j=1}^{M} \omega_j I_j - \varepsilon\sqrt{2} \sum_{l=1}^{N}\sum_{j=1}^{M} \frac{\beta_j}{k_j} \sqrt{I_j}\, \cos\!\left(k_j x_l - \theta_j\right) \quad . \tag{21}$$

It is worth noticing that the energy conserved during the motion embodies the wave-particle coupling energy. It is easy to see from equation (21) that the total wave-particle momentum

$$P = \sum_{l=1}^{N} p_l + \sum_{j=1}^{M} k_j I_j \qquad (22)$$

is conserved [6].

4.2. FROM FIELD-PARTICLE INTERACTION TO QUASILINEAR EQUATIONS [20]

As we are dealing with Hamiltonian dynamics, it is natural to study one realization of the plasma (section 4.2.1) before studying ensemble averages (section 4.2.2). More preliminary results concerning this section were given in references 21 and 25.

4.2.1. Normal modes in one plasma

We may introduce reference multibeams for tail particles as well (figure 10). For such a multibeam the source term for modes at t=0 vanishes due to destructive interferences. If the modes' amplitudes are zero initially, they stay so for all times. A reference multibeam is the equivalent of a Vlasovian trivial state, spatially uniform with zero field. The dynamics of the zero-field state is here defined by one angle per reference beam. The zero-field state is therefore an invariant torus which is not a KAM torus, since its dimensionality is smaller than the number of degrees of freedom (we now assume there are at least three particles per beam).

We now restrict our attention to the cases where the beams' velocities are integer multiples of a basic one Δv. Linearizing the equations of motion about a zero field state yields a series of equations which may be cast under the form

$$\dot{U}(t) = (M_0 + M_1(t)) U(t) \tag{23}$$

where U is a N'=2(N+M) dimensional vector, M_0 is a static N'xN' matrix and M_1 is a time-periodic N'xN' matrix with a frequency $\Delta v/L$. Equation (23) defines a Floquet problem. Its general solution is

$$U(t) = V(t)\, e^{(\gamma - i\,\omega)\,t} \tag{24}$$

where γ and ω are reals defined for M=1 by

$$\gamma = (\varepsilon\,\beta)^2 \sum_{l=1}^{N} \frac{\gamma\left(p_l - \omega\right)}{\left[\gamma^2 + \left(p_l - \omega\right)^2\right]^2}\,, \tag{25}$$

$$\omega = 1 + \frac{(\varepsilon\beta)^2}{2} \sum_{l=1}^{N} \frac{\left(p_l - \omega\right)^2 - \gamma^2}{\left[\gamma^2 + \left(p_l - \omega\right)^2\right]^2}\,. \tag{26}$$

In order to simplify the formulas, they are written for $\omega_j = k_j = 1$, and index j is omitted. Most exponents are zero, but for each beam there are just four nontrivial ones of the form $\gamma_1 \pm i\omega_1$, $-\gamma_1 \pm i\omega_1$. Their symmetries are a mere consequence of the Hamiltonian (symplectic) nature of the dynamics. The four complex normal modes per beam enable altogether the construction of the classical two real modes per beam.

If $|\gamma| >> \Delta v$, one finds a solution

$$|\gamma| = \gamma_L \tag{27}$$

with the Landau growth rate

$$\gamma_L = \frac{\pi\,\varepsilon^2 \beta^2 N}{2} \frac{\partial f}{\partial p} \tag{28}$$

where f is the tail velocity distribution function (from now on $\int f(p)dp = 1$ on the domain of tail velocities) evaluated at the phase velocity of the wave. Obviously equation (27) is meaningful only if γ_L is positive. We thus find that one realization of a plasma with a positive slope of the velocity distribution function possesses unstable eigenmodes with the Landau growth rate. Their frequency turns out to be given by the usual formula with a principal part of Vlasovian theory. Particles are acted upon by a force due to the wave which produces on a particle a time-averaged acceleration

$$<\dot{p}> = \frac{(\omega - p)}{[\gamma^2 + (\omega - p)^2]^2} (\varepsilon\beta Z)^2 \tag{29}$$

where Z is the wave amplitude. This shows that the Landau instability is related to a synchronization of particles with the wave. Such a formula was alreary derived through a perturbation calculation by Kupersztych [11], and the synchronization mechanism is clearly suggested in Nicholson's book [10]. The particles whose velocity is the most affected by the wave are those with a velocity γ/k away from the wave's phase velocity (here k=1). Conversely they are the most active in the instability. Resonant particles play no role (remember that the wave amplitude is assumed to be small). This contrasts with the opposite suggestion coming from the Vlasovian approach ! The quadratic dependence of the acceleration on Z is consistent with the existence of the conserved total momentum (22). The existence of a damped mode with an opposite γ is a consequence of the symplectic character of the dynamics. With probability one the unstable modes show up for a random initial condition.

If there is no Landau instability for a given k, we just find a family of normal modes with $\gamma=0$ or γ vanishing in the limit of small Δv's. They correspond to modes already identified by Dawson [15] as corresponding to the Vlasovian van Kampen modes. For a typical initial condition and for a given k, the most strongly excited van Kampen modes are those with a velocity close to the phase velocity of the bulk mode (the one for $\varepsilon=0$). This is consistent with the Vlasovian case.

At this point we found the van Kampen modes, but no Landau damping. Is it a failure of our approach ? No, this is a mere consequence of the

Hamiltonian character of the plasma. If Landau damping was related to damped eigenmodes, symplecticity would impose the simultaneous existence of unstable modes with the opposite growth rate, which would show up with probability one ! There is no eigenmode corresponding to Landau damping, a fact already present in the Vlasovian approach which also tells us that Landau damping is the result of the phase mixing of van Kampen modes. Phase mixing is a statistical property which is missed for one realization of the plasma. We must now consider ensemble averages.

4.2.2. Ensemble averages

We would like to consider an ensemble of plasmas whose tails have a fixed distribution function, but random initial positions for the particles. Ensemble averaging on such initial conditions can be quite easily performed on the results of second order perturbation expansion in ε [21]. This yields for the wave amplitude

$$<\dot{I}_j> = 2\,\gamma_{Lj}\,<I_j> + S_j \tag{30}$$

where $\gamma_{L\,j}$ is given by equation (28), and S_j by

$$S_j = \frac{\pi\, N\, \varepsilon^2\, \beta_j^2}{k_j^3}\; f\left(\frac{\omega_j}{k_j}\right) \tag{31}$$

For the particles one finds diffusion and friction coefficients enabling one to write the Fokker-Planck equation

$$\frac{\partial f}{\partial t} = \frac{\partial}{\partial p}\left[\, D(p)\, \frac{\partial f}{\partial p}\right] + F(p)\, \frac{\partial f}{\partial p} \tag{32}$$

where

$$F(p) = -\frac{\pi \varepsilon^2 \beta^2}{\omega(p)} \tag{33}$$

and

$$D(p) = \frac{\pi \varepsilon^2 \beta^2}{p} I(p) \tag{34}$$

where the p dependence of ω and I means the choice of ω_j and I_j with j such that $\frac{\omega_j}{k_j} = p$. These equations are the usual quasilinear equations completed with the spontaneous emission of tail particles. In the case of the warm beam-plasma instability they provide the plateau formation of figure 9, before leading to a distribution function with a negative slope which anticipates the formation of a globally maxwellian distribution. Unfortunately this nice result is as yet obtained by a perturbation technique requiring the wave amplitude to stay almost constant for it to be valid. In order to describe the growth or damping of a wave, we thus need a more powerful technique we now describe.

We formally integrate the linearized particle motion of equation (23), and insert the result in the equations ruling the linearized wave motion. This provides a differential equation for the square of the wave amplitude we may average over initial realizations of the plasma. If we average over initial positions of tail particles when keeping fixed for each of them its reference particle, we violate the conditions of validity of the linearized equations. Instead, for each sample we must number the particles as we did in section 4.1.4. For N large enough and a typical sample, this warrants that linearization is justified. Since the dominant van Kampen modes are concentrated near the phase velocity of the bulk mode, the correlation time of the wave amplitude is long with respect to the inverse of the maximum Doppler frequency of the tail particles. This enables us to split averages occurring in the evolution equation of the wave, and to recover equation (30). The term corresponding to spontaneous emission contains initial conditions explicitly, and is hard to estimate this way, but it must have the same value as in equation (30), as the two derivations have a common time interval of validity. In the second derivation, as in classical derivations of Landau damping [1] or quasilinear theory [2], the time interval of validity is bounded by the time for separation of nearby orbits which is either the trapping time for regular motion, or the Lyapunov-Landau-Dupree time [22]

$$\tau_D = (k^2 D)^{-1/3} \tag{35}$$

for chaotic motion, where D is the diffusion coefficient for particles.

Equation (30) provides the Landau damping we were looking for. Because of spontaneous emission this equation also shows the well-known existence of a thermal level in the damped case. Therefore the Landau damping rate turns out to be a rate of relaxation toward the thermal level. A wave cannot be damped below this level (it would grow in the unlikely case where it would start below this level). As a result, in a true plasma, Landau damping is a nonlinear effect, which is hidden in the Vlasovian approach where the thermal level vanishes. It is also a non-resonant effect, as formula (29) can be recovered by a treatment of the particles similar to the one just described for the waves. This treatment also provides the Fokker-Planck equation (32). In the warm beam instability, many modes are simultaneously unstable. Each of them tries to synchronize particles with itself, and this results in a global beam drag. Other results can be got by the same technique, in particular the existence of a diffusion of the phase of a wave about the average value given by the Vlasovian expression with a principal part. The average behaviour may be seen on one realization of the plasma because fluctuations about the average are small when the number of particles is large [33].

4.3. TOWARD A CHAOTIC QUASILINEAR THEORY [23]

The above derivation of the quasilinear equations does not hold when chaos is dominant. This corresponds to the regime where the quasilinear paradox is present. Before tackling chaos in the self-consistent case, it is tempting to first study the much simpler (but far from trival !) case of the motion of one electron in the field of a fixed spectrum of Langmuir waves. Unexpectedly, numerical simulations revealed that for wave amplitudes about 18 times above the threshold of chaos, the diffusion coefficient corresponding to the chaotic motion is 2.4 times the quasilinear one, and it is still 10 per cent above it 100 times above threshold [24]. This regime is hereafter referred to as intermediate resonance overlap. It indicates a dependence of transport on the graininess of the turbulent spectrum, but not on the spectral density. In the limit of a continuous spectrum, transport is quasilinear in the non-self-consistent case. This enhanced value of the diffusion coefficient was missed in previous calculations because there is an initial time interval, typically given by τ_D, such that transport is quasilinear, as can be shown by perturbation theory [25].

The self-consistent equations provide a quite attractive basis for a numerical simulation of the saturation of the warm beam-plasma instability. Indeed, no computational effort is wasted in the description of the bulk plasma, and it is even possible to focus on a part of the velocity domain where the distribution function has a positive slope. In agreement with experiments, the simulation reveals a strong mode coupling, but no renormalization of the quasilinear predictions when the graininess is low [26]. For intermediate resonance overlap, an enhanced value of the growth rate with respect to γ_L is observed, in agreement with the results of a previous "Vlasovian" code [13] (the quotation marks are used because, due to discretization, the code recovers the spontaneous emission dramatically missing in a pure Vlasovian approach). So, the self-consistent case looks similar to the non-self-consistent one : except for graininess reasons, the quasilinear predictions are correct. The similarity between the low-dimensional case and the high-dimensional one, is not fortuitous. Indeed, there is good evidence that Hamiltonian dynamics with many degrees of freedom may be locally reduced to dynamics with 1.5 degree of freedom [27].

The analytical explanation of the numerical results, even in the non-self-consistent case, is not trivial. Indeed this means the understanding of the origin of decorrelations in Hamiltonian dynamics. The stretching and folding mechanism observed in phase space on the homoclinic tangle, the structure underlying chaos, is the origin of the process. This was used in [28], but the understanding of homoclinic tangles still is in infancy [29]. As yet the existence of a supraquasilinear regime is confirmed by a calculation using an approximation of the true dynamics by a mapping, and making reasonable assumptions on the chaotic dynamics [24].

5. CONCLUSION

This paper has given a simplified account of the qualitative features of large scale structures in the simplest case of kinetic plasma turbulence, the electron beam-plasma instability. The basic physics of this phenomenon is understood, but theory is still unsatisfactory, and even paradoxical. The weaknesses of the traditional approach of microscopic plasma theory were analysed. Then a new approach, relying upon a classical mechanical technique (a part of Laplace's dream), was described. It starts from a N-body description of the plasma, goes through the derivation of a Hamiltonian describing the self-consistent evolution of Langmuir waves and near-resonant particles, and provides a derivation of the quasilinear equations where spontaneous effects are included. This derivation makes clear the physics of wave-particle interaction, and answers many question raised by the Vlasovian approach. In particular, Landau damping turns out to be a non-resonant nonlinear phenomenon of synchronization of particles with a wave. The regime where chaos is dominant is still unexplained analytically, but a numerical simulation starting from the self-consistent equations yields results which agree with experiments.

Future research on the chaotic regime should benefit from the present formulation with a finite number of degrees of freedom, especially when the connection with low-dimensional dynamics becomes visible. The generalization of available results to 3 dimensions, or weak turbulence phenomena including ions, does not look difficult.

In fact the new mechanical approach should be interesting to a much broader audience than plasma physicists. First, the self-consistent equations have a typical structure, occurring in the study of electronic tubes and, in particular, free electron lasers. All reasonings held for the beam-plasma system can thus be generalized to these devices.

Second, the new approach is not restricted to Coulombian interactions and might have a more universal character. Other classical systems could thus be described by using a similar approach. How feasible is it ? It is too early to tell. However, some similarities between the plasma and fluid turbulence problems are striking. First, a new regime between the inertial Kolmogorov regime and the dissipative regime has been recently evidenced in fluid turbulence [30]. Our analysis shows the existence in the plasma of scales intermediate between the large ones (Langmuir waves) and the "dissipative" ones (Coulomb scales), where mode coupling becomes dominant and where non-dissipative thermalization processes should occur. Second, it becomes clear that a correct description of fluid turbulence should incorporate both a traditional modal description and localized objects (the vortices). In the plasma case, we saw that the degrees of freedom of the system split naturally into collective modes and particles : the N-body system is made of several objects of different natures. Indeed we even found that the collective degrees of freedom themselves split into large scales (Langmuir waves), small scales (Coulomb scales) and intermediate scales where chaos is dominant. By analogy with Landau damping one may wonder about the existence in a fluid of a damping of modes by vortices.

Similarities are also striking with other non-classical plasmas : the quark-gluon plasmas. Indeed in a high energy collision the quark-gluon plasma goes from an initial stage of weak coupling to one of strong coupling [31]. This is similar to what occurs in the warm beam-plasma instability where the initial phase may be described perturbatively, and the final one is fully chaotic. The evidence for existence of intermittency in high energy collisions [32] further supports the analogy.

Naturally, the results presented here are still preliminary. We are like archeologists digging out an ancient town. At present the main lines of the city map are visible (this was not yet the case in a previous publication [25]), but a lot of dust still needs to be removed before the main buildings be ready for a visit. Nevertheless having a tour may be interesting. Section 4.2 is the closest to the state of a traditionally publishable result.

I thank the organizers of the workshop for giving me the opportunity to present the theory as it stands without the referees' censorship (but with more risks of publishing an error !). I thank the members of the Turbulence Plasma team for the joy I have to work with them. I am grateful to M. Antoni, Y. Elskens, A. Verga, and S. Zekri who made a careful reading of the manuscript, to Y. Elskens who made numerous improvements to the style, and to him and S. Zekri for their assistance in preparing the figures.

REFERENCES

1 Chen, F.F, *Plasma physics and controlled fusion* (Plenum, New York, 1984).
2 Vedenov, A.A., Velikhov, E.P., and Sagdeev, R. Z., Nucl. Fusion **1**,82 (1961) ; Drummond, W.E., and Pines, D., Nucl. Fusion, Suppl. Pt **3**, 1049 (1962).
3 Tsunoda, S.I., Doveil, F., and Malmberg, J.H., Phys. Rev. Lett. **58**, 1112.(1987).
4 Laval, G., and Pesme, D., Phys.Fluids **26**, 52 (1983) ; Phys. Rev. Lett. **53**, 270 (1984).
5 O'Neil, T.M., Winfrey, J.H., and Malmberg, J.H., Phys. Fluids **14**, 1204 (1971).
6 Mynick, H.E., and Kaufman, A.N., Phys. Fluids **21**, 653 (1978).
7 The wave's trapping domain takes on the shape of cat's eye, the typical signature of a nonlinear resonance.
8 Gentle, K.W., and Roberson, C.W., Phys. Fluids **14**, 2780 (1971) ; Gentle, K.W., and Lohr, J., Phys. Fluids **16**, 1464 (1973).
9 Roberson, C., and Gentle, K.W., Phys. Fluids **14**, 2462 (1971).
10 Nicholson, D.R., *Introduction to basic plasma theory* (Wiley, New York, 1983) p. 87.
11 Kupersztych, J., Phys.Rev. Lett. **54**,1385 (1985).
12 O'Neil, T.M., Phys. Fluids **17**, 2249 (1974).
13 Theilhaber, K., Laval, G., and Pesme, D., Phys. Fluids **30**, 3129 (1987).
14 Part done with Mickaël Antoni.
15 Dawson, J.M., Phys. Rev. **118**, 381 (1960).
16 Bohm, D., and Gross, E.P., Phys. Rev. **75**, 338 (1949).
17 Benettin, G., Galgani, L., and Giorgilli, A., Nature **311**, 444 (1984) ; Phys. Lett. **A120**, 23 (1987).
18 Rouet, J.-L., and Feix, M.R., to be published in Phys. Fluids 1991 ; Rouet, J.-L., Thèse de l'Université d'Orléans, February 1990.
19 *Turbulence and anomalous transport in magnetized plasmas*, edited by Grésillon, D., and Dubois, M.A. (Editions de Physique, Les Ulis, France, 1987).
20 Part done with Stephane Zekri, Hanafi Derfoul, and Yves Elskens.
21 Derfoul, H., Escande, D.F., and Zekri, S., to be published in the proceedings of the "First south-north international workshop on fusion theory", Tipaza, Algeria, September 1990.
22 Dupree, T.H., Phys. Fluids **9**, 1773 (1966).
23 Part done with Isidoros Doxas, John Cary, and Alberto Verga.
24 Cary, J.R., Escande, D.F., and Verga, A.D., Phys. Rev. Lett. **65** , 3132 (1990).

25 Escande, D.F., in *Nonlinear World,* edited by Bar'yakhtar, V.G, Chernousenko, V.M., Erokhin, N.S., Sitenko, A.G., and Zakharov, V.E., (World Scientific, Singapour, 1990) pp. 817-836.
26 Doxas, I., Escande, D.F., Verga, A.D., Zekri, S., Cary, J.R., and Mantacheff, J., to be published in the proceedings of the "US-Japan Workshop on Nonlinear Dynamics and Acceleration Mechanisms ", Tsukuba, Japan, October 1990.
27 Escande, D.F., Kantz, H., Livi, R., Ruffo, S., in preparation.
28 Escande, D.F., in *Plasma Theory and Nonlinear and Turbulent Processes in Physics,* edited by Bar'yakhtar, V.G. Chernousenko, V.M. Erokhin, N.S., Sitenko, A.G., and Zakharov, V.E. , (World Scientific, Singapore, 1988) 398-430.
29 Elskens, Y., and Escande, D.F., Nonlinearity, May 1991.
30 Gagne, Y., and Castaing, B., C. R. Acad. Sci. Paris sér.2 **312**, 441-445 (1991) ; Castaing, B., Gagne, Y., and Hopfinger, E.G., Physica D46, 177-200 (1990).
31 Le Bellac, M., "Hot QCD and the quark-gluon plasma", this volume ; Kapusta, J., *Finite temperature field theory* (Cambridge University, U.K., 1989).
32 Bialas, A., and Peschanski, R., Phys. Lett. **B253**, 225 (1991) and references therein.
33 Dawson, J., Phys.Fluids, **5**, 445 (1962).

SUPERSONIC HOMOGENEOUS TURBULENCE

D. H. Porter[1], A. Pouquet[2] and P. R. Woodward[1]

1. Department of Astronomy and Minnesota Supercomputer Institute, University of Minnesota, Minneapolis, U.S.A.
2. Observatoire de la Côte d'Azur, CNRS URA 1362, 06003 Nice–cedex, France.

ABSTRACT

Supersonic turbulence is observed in the giant molecular clouds in the galactic disk and may also occur at re–entry of space shuttles. Numerical simulations in three dimensions concerning compressible homogeneous non–stationary flows using the fluid equations with the Navier–Stokes formulation, a hyperviscosity method, and new simulations of the Euler equations using the PPM code are presented. Results are for random flows at a *rms* Mach number of unity on uniform grids with periodic boundary conditions. Through visualization and analysis of the fluid simulations, we conclude that: **i)** shock waves and shock intersections play an important role in the transfer of energy from long to short wavelengths. Weak shocks survive for several τ_{ac}, where τ_{ac} is the acoustic time of the energy containing modes in the initial state. In the context of our decay problems, vorticity is concentrated in filaments, and not sheets, within a few acoustic times. Whereas in two dimensions eddies tend to slowly merge over many τ_{ac}, in 3–D vortex filaments break up into short filaments within a few τ_{ac}; **ii)** three temporal phases are identified in the evolution of the flow, onset and formation of shocks, quasi–supersonic phase with many interacting shocks, and post–supersonic phase with a velocity spectrum flatter than that of Kolmogorov.

1. INTRODUCTION

Compressible flows are observed within giant molecular clouds which are the site of star formation and supernovae explosions. Several characteristic features of these clouds remain unexplained: they resist collapse for times an order of magnitude larger than what the linear analysis for a static cloud predicts; and supersonic flows, with *rms* Mach numbers of ~ 4 persist although shock formation should lead to a rapid dissipation of kinetic energy. In the former case, internal motions are present that help support the cloud through an additional pressure term (Chandrasekhar, 1951); the formation of shocks can provide another mechanism of support (Léorat et al., 1991). In the latter case, magnetic fields are invoked to substantially reduce the rate of energy dissipation. Indeed, magnetic fields are observed both on large scales through the alignment of bipolar jets emerging from protostars, and on smaller scales; their magnitude is, in velocity units, slightly larger than the velocity of the fluid and thus dynamically important (Heiles, 1987; Lada and Shu, 1990).

Turbulence is another key element in the dynamical evolution of molecular clouds (Falgarone, 1990; Falgarone and Philipps, 1990; Scalo, 1990; Henriksen, 1990), but little is known on compressible turbulence at high Mach number. Numerical computations of homogeneous turbulent flows in three dimensions for both the decay problem (Feireisen et al., 1981; Erlebacher et al. 1987; 1990; Lee et al., 1991; Zang et al., 1991) and the stationary case (Kida and Orszag, 1991) are scanty.

Section 2 gives a review of numerical results using Navier–Stokes solvers with explicit dissipation, through a Laplacian and with an hyperviscosity method; Section 3 describes results using an Euler solver, PPM (Woodward, 1986; Woodward and Collela, 1984; Collela and Woodward, 1984) in three space dimensions, and Section 4 is the conclusion.

2. THE VISCOUS CASE

2.1 Introduction

We discuss numerical simulations using the Navier–Stokes equations, and a standard pseudo–spectral code. Let us give for reference purposes the relevant equations and the various parameters of the initial conditions. The velocity field is decomposed into its solenoidal u^s and compressional u^c components as follows:

$$\mathbf{u} = \mathbf{u}^s + \mathbf{u}^c \quad , \tag{2.1}$$

with $\nabla \cdot \mathbf{u}^s = 0$. The associated Fourier velocity spectra are denoted respectively by $E_t^v(k)$, $E_s^v(k)$ and $E_c^v(k)$ with $E_s^v + E_c^v = E_t^v = \frac{1}{2}\int \rho u^2(\mathbf{x})\, d^d\mathbf{x}$. The fluid dynamical equations for a perfect gas law are written as:

$$\begin{aligned} \partial_t \rho + \nabla \cdot (\rho \mathbf{u}) &= 0 \quad , \\ \partial_t \mathbf{u} + \mathbf{u} \cdot \nabla \mathbf{u} &= -(\gamma - 1)\frac{\nabla(\rho e)}{\rho} + \frac{1}{R\rho}\left(\nabla^2 \mathbf{u} + \frac{1}{3}\nabla(\nabla \cdot \mathbf{u})\right) \quad , \\ \partial_t e + \mathbf{u} \cdot \nabla e &= -(\gamma - 1) e \nabla \cdot \mathbf{u} + \frac{1}{\rho R}\tau : D + \frac{\gamma}{Pr R\rho}\nabla^2 e \quad , \end{aligned} \tag{2.2}$$

where ρ is the density, $e = c_v T$ the internal energy and T the temperature; $\gamma = 1.4$; the sound velocity is c (with mean value c_0):

$$c^2 = \gamma(\gamma - 1)e \quad ;$$

the Prandtl number $Pr = \mu_0 c_p / \kappa$ is equal to unity in all the viscous calculations; c_p (*resp* c_v) is the constant pressure (*resp* constant volume) specific heat; μ_0 and κ are the diffusion coefficients of the velocity and temperature. The Reynolds number is $R = \rho_0 U_0 L_0 / \mu_0$ where ρ_0 is the mean density taken equal to unity, and U_0 is the velocity at the energy–containing scale $L_0 = 2\pi / k_0$. The mean temperature is also taken equal to unity. N is the number of grid points in each direction, uniformly distributed;

$k_{min} = 1$ is the minimum wavenumber, and $k_{max} = N/2$ the maximum wavenumber. Finally, τ_{ij} is the stress tensor

$$\tau_{ij} = \mu_0 \, (-\frac{2}{3}\nabla \cdot \mathbf{u}\delta_{ij} + D_{ij}) \tag{2.3a}$$

and D_{ij} the deformation tensor

$$D_{ij} = \frac{1}{2}(\partial_j u_i + \partial_i u_j) \ . \tag{2.3b}$$

The dispersion of the density fluctuations

$$\delta\rho = (\rho - \rho_0)/\rho_0 \ ,$$

as well as the temperature fluctuations are initially between 0% and 20%, and the density contrast

$$\Delta\rho = \rho_{max}/\rho_{min} \tag{2.4}$$

is initially close to four.

The viscosity is adjusted in a dichotomic way so that no numerical noise appears. This typically requires that shocks be resolved on the order of ten grid points or more. Otherwise, oscillations will appear in their vicinity which will be particularly prominent in the velocity divergence and vorticity fields. The viscosity typically is 10^{-2} for a Mach one flow on a $N = 128^3$ grid and with initial conditions centered in the large scales ($k_0 \sim 2$). The boundary conditions are periodic in all directions. The initial conditions are that of a random flow with a spectrum of prescribed width and characteristic scale, namely:

$$E(k) \sim k^4 e^{-2(k^2/k_0^2)} \ .$$

The ratio of the compressional to total components of the kinetic energy

$$\chi = E^c/E^v \tag{2.5}$$

is yet another parameter of the initial conditions; we take $\chi \sim 0.09$.

Initially, the kinetic energy is equal to unity, and the initial *rms* Mach number $M_a = U_0/c_0$ is close to one. The helicity is defined as $H = \int \mathbf{u} \cdot \omega \, d^3\mathbf{x}$ where ω is the vorticity, and the relative helicity is:

$$\tilde{r} = H \ / < \mathbf{u}^2 >^{\frac{1}{2}} \ < \omega^2 >^{\frac{1}{2}} \ . \tag{2.6}$$

A partially implicit temporal scheme is used in conjunction with the viscous term to handle the regions of space where the minimum density ρ_{min} is particularly low, a problem that arises in three dimensions.

2.2 Navier–Stokes solver

The decay problem with adequate viscosity to resolve the shocks has been performed in three dimensions on a $N_{max} = 128^3$ grid. The main results of these Navier–Stokes simulations are now briefly summarized:

i) There are two main regimes, one subsonic in which the solenoidal modes dominate at small scales and one supersonic in which the compressional modes dominate at small scales; the transition between the two occurs for an initial *rms* Mach number of $M_a \sim 0.3$, as in two dimensions (Passot and Pouquet, 1987). This corresponds to the presence of at least one strong shock in the flow. However, when at small Mach number the density fluctuations initially are larger than $\mathcal{O}(\mathrm{M}^2)$, another regime occurs in which most of the kinetic energy is in the acoustic modes (Erlebacher et al., 1990; Bayly et al., 1991). In all cases, the equipartition between E_s and E_c in the small scales, postulated by Kraichnan (1953) on the basis of a statistical argument, does not seem to hold.

ii) Once the shocks have formed, they are the main process through which kinetic energy is dissipated into heat in the supersonic regime. As the Reynolds number increases, the turbulent transfer of energy to small scales through mode coupling will become more prominent. It should be the main dissipative process of the late acoustic phase.

iii) The temporal evolution of the Mach number displays a plateau corresponding to the replenishing of the kinetic reservoir through energy exchanges with the heat bath; this indicates that there are standing adiabatic waves present in the system. It is similar in 2-D and 3-D, at least for a few eddy turn–over times. However, for times long enough that the flow has become substantially subsonic, the Mach number may become very low in three dimensions, since no spurious conservation of vorticity will hinder the decay as is the case in 2-D. Finally, we note that acoustic oscillations are visible for long times.

One significant difference between the 2-D and 3-D Navier–Stokes results is that the density contrast is stronger in 3-D. For example, for a run with initially a Mach number $M_a = 2.2$, a Taylor wavenumber $k_T \sim 3.8$, $\chi = 0.05$ and a Reynolds number $R = 55$, the density contrast $\Delta\rho$ reaches a value close to 100, the minimum density being $\rho_{min}/\rho_0 = 0.07$, and for a similar run with $k_T \sim 1.5$ and $R = 120$, $\Delta\rho = 54$. At similar Reynolds numbers in two dimensions, $\Delta\rho$ is less than ten. It is likely that this phenomenon amplifies with the Reynolds number. Finally, inertial indices in three dimensions are not established at such low resolutions of the computation. The spectra are steeper than k^{-2}, the main cause apparently being contamination by the nearby viscous range. However, we note that at a fixed time t_0 the solenoidal spectrum is still steeper than its acoustic counterpart, indicating a slower transfer to small scales of the vortices.

To be able to reach higher effective Reynolds numbers, and thus a larger span in wavenumbers in the inertial range, adequate modeling of small–scale flows is in order.

2.3 Hyperviscosity solver

The limitation on the Reynolds number that is reasonably attainable with Navier–Stokes solvers is drastic when the three space dimensions are retained. A way around this difficulty is to resort to a model for the small scales, both in the inertial range and in the dissipation range of the flow. The one–point or two–point closures used in modeling industrial flows with complex boundaries will not be mentioned here (see Ha Minh and Vandromme, 1986).

A common simple approach to such modeling of small scales is to use a high power of the Laplacian, or a *hyperviscosity* method, corresponding to a $k^{2\eta}$ dissipation, with $\eta = 1$ for Navier–Stokes. By a "$k^{2\eta}$ dissipation" we mean that an isolated Fourier mode of wavenumber k decays at a rate proportional to $k^{2\eta}$. This decay rate has units of inverse time, which are the same for each wavenumber and not dependent on the sound or flow crossing time of a wavelength for each mode. The Green's function associated with such a diffusion operator has negative lobes, giving rise to a dissipation that is not everywhere positive in space, although the integral of the local dissipation remains positive. One may introduce on basic principles a bi–Laplacian $\sim k^4$ dissipation and impose the positivity in all space (Passot and Pouquet, 1988). The price to pay is that the dissipation operator is now nonlinear. In the simplest case, one may replace the standard dissipation function with constant viscosity μ_0 in equation (3a) by one in which the viscosity coefficient μ' depends quadratically on both the solenoidal and the compressional components of the velocity field, namely:

$$\mu' = \mu_1(I_2 - \frac{1}{3}I_1^2) \tag{2.7}$$

where $I_1 = D_{ii}$ (see equation (2.3b)), $I_2 = D_{ij}D_{ij}$, and now μ_1 is an adjustable parameter depending in particular on the grid resolution. Field–dependent viscosities as a model of small scales have already been derived, for example that based on second–order closures (Chollet and Lesieur, 1981), or that of Smagorinsky (1963) in convection in the meteorological context.

The hyperviscosity method has only been tested in two dimensions until now. It is found that the large scales are reproduced accurately (Passot and Pouquet, 1988); the small scales are noisy and an adequate filtering should be introduced, for example through a "laplacian viscous term" at a lower level, allowing it to be felt only in the close vicinity of k_{max}.

3. PPM SIMULATIONS OF 3-D COMPRESSIBLE TURBULENCE

3.1 Introduction

A very different approach consists in dealing with the Euler equations, omitting in equations (2.2) all dissipative terms, and introducing an adequate dissipative mechanism to spread shocks over about two zones. The dissipation of small scale waves is then to be regarded as caused by numerical errors in approximating the Euler equations. These errors must be very carefully controlled so that the numerical solution converges as uniformly as possible to a weak solution of the Euler equations. The Piecewise-Parabolic Method, PPM, (Woodward and Collela, 1984; Colella and Woodward, 1984; Woodward 1986) represents the present state of the art for such schemes. PPM incorporates high–order interpolation techniques augmented by monotonicity constraints in order to resolve sharp features in the flow over only about two computational zones. In addition, it tests for the presence of contact discontinuities, and if these are detected it uses specialized interpolation techniques to see that these discontinuities remain sharp as they move through the grid. Shocks are detected as well, and extra dissipation is carefully computed so that shocks will remain sharp but at the same time emit a minimum of numerically generated oscillatory waves into the flow behind them. The version of PPM used in the computations presented here is closest to that described by Woodward (1986), where these special features of the method are discussed in detail.

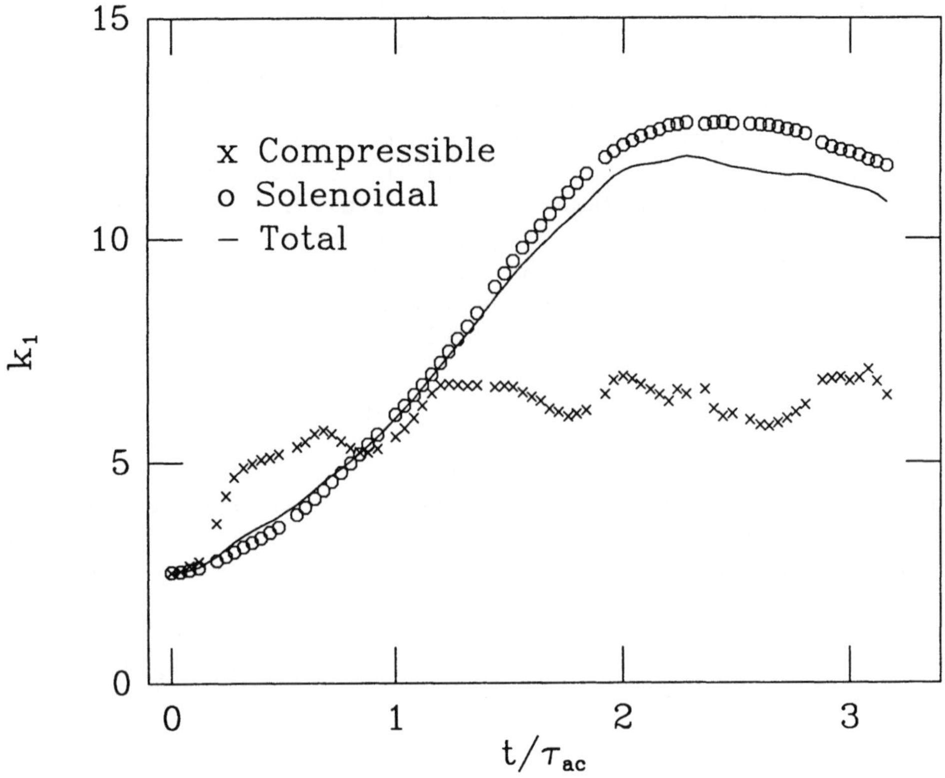

Figure 1: Temporal evolution of the characteristic wavenumber k_1 separated into its solenoidal (circles), longitudinal (crosses) and total (solid line) components. The unit time for all the figures is the acoustic time τ_{ac} of the energy–containing eddies. Note the onset of the quasi–steady supersonic phase in k_c at $t_1 \sim 0.3$ and that of the post–supersonic phase in k_s at $t_2 \sim 2.1$. All wavenumbers k_n defined in (3.1) globally evolve in a similar way, except for $n = -1$: the compressional component of the integral wavenumber k_{-1} displays clear acoustic oscillations of period τ_{ac}.

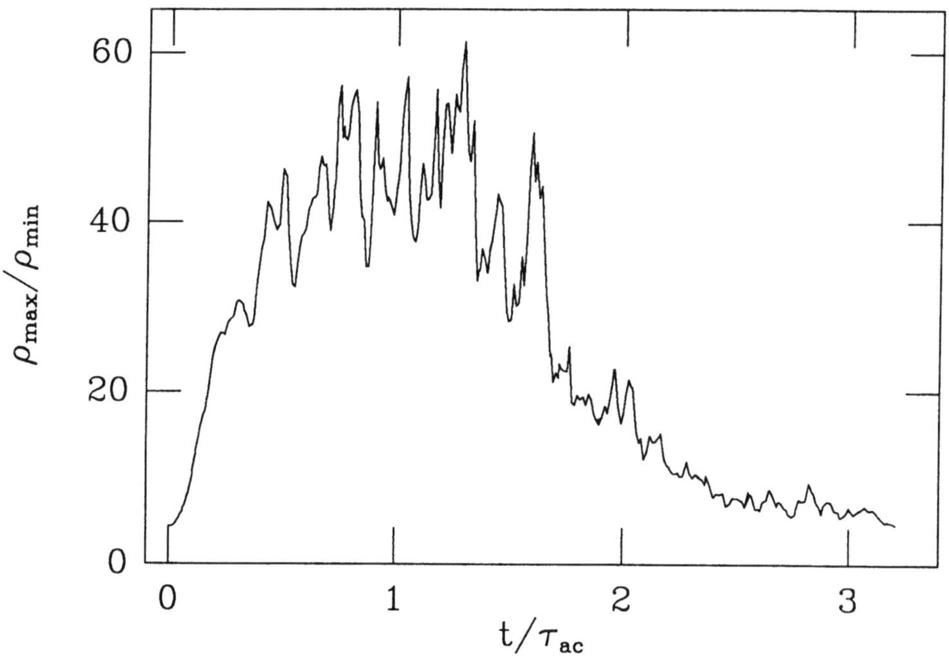

Figure 2: Density contrast as a function of time. Note the domain $t_1 < t < t_2$ during which it remains quasi-steady.

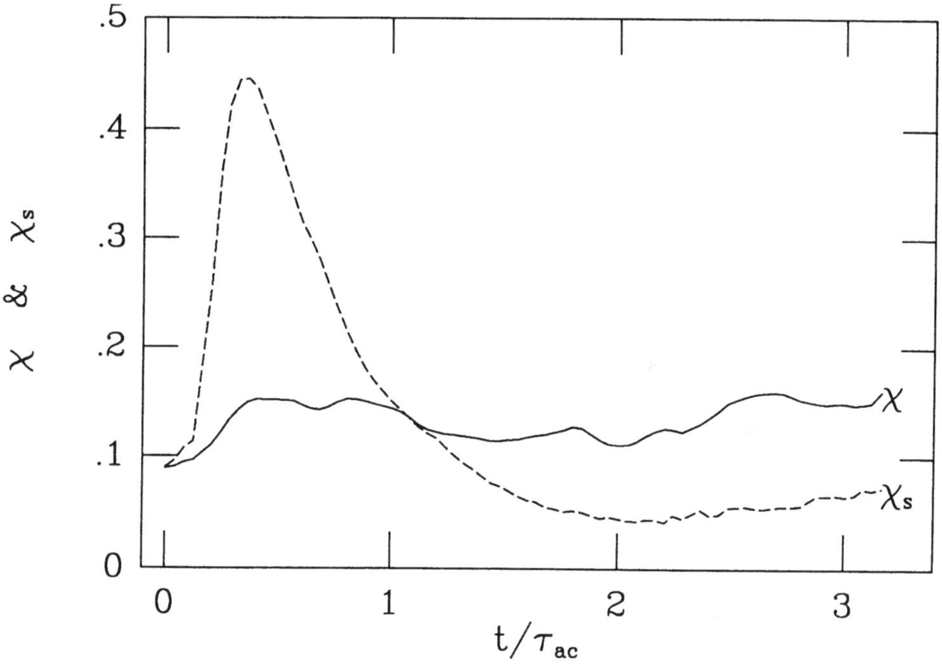

Figure 3: Large-scale (energy) and small-scale (enstrophy) ratios χ and χ_2 (see text for definitions) of the compressional part of the kinetic energy as a function of time.

We have recently undertaken a detailed comparison of PPM calculations with those of PPMNS, a Navier–Stokes solver, (Porter et al., 1991d) in the specific context of modeling homogeneous turbulent flows in the limit of high Reynolds number. The numerical viscosity of PPM has been determined for isolated sinusoidal waves (Porter and Woodward, 1991; Porter et al., 1991a); it has a wavenumber–dependency intermediate between a k^4 and k^5 law. It is important to note that the similarity in decay rates of isolated sinusoidal disturbances of long wavelength for PPM and for spectral methods with bi-Laplacian dissipation does not imply similar accuracy in modeling intermediate and small scales. Both PPM and the bi-Laplacian methods mentioned above are highly nonlinear. A simple power–law for decay rates of isolated modes is but one measure of dissipation; there are infinitely many possible dissipative forms which lead to the same decay rates of isolated modes. The treatment of very short wavelength modes in PPM involves much stronger dissipative mechanisms arising from this scheme's monotonicity constraints and its special dissipation terms which turn on only within strong shocks.

3.2 Temporal evolution of a supersonic flow

In the 3D–PPM simulation that we now present, we choose units so that the mean density ρ_0 and initial mean sound speed c_0 are both unity (see also Porter et al., 1991b & c). We set the initial *rms* fluctuations in the density, $\delta\rho = (\rho - \rho_0)/\rho_0$, pressure, $\delta p = (\rho e - \rho_0 e_0)/(\rho_0 e_0)$, and velocity, δU_0, to 20 %, 20 %, and 1, respectively. Boundary conditions are periodic and the initial conditions are identical to those given in Section 2. This numerical simulation was performed on a 256^3 mesh, and the maximum time of the computation is $t = 3.2\ \tau_{ac}$, where τ_{ac} is the acoustic time of the energy–containing scale and is the unit of time for these computations.

Let us define $k_{n,c}$ and $k_{n,s}$ as the wavenumbers associated with the compressional ($x = c$), solenoidal ($x = s$) and total ($x = t = c + s$) components of the velocity field $\mathbf{u} = \mathbf{u}^S + \mathbf{u}^C$:

$$k_{n,x} = \left\langle \int_0^\infty k^n E_x^v(k)dk \;/ \int E_x^v(k)dk \right\rangle^{\frac{1}{n}} \qquad (3.1)$$

where E_x^v are the power spectra of the velocity for the c–, s– and total components. One recovers the Taylor k_T and the integral k_I wavenumbers for $n = 2$ and $n = -1$ respectively.

The kinetic energy of a compressible flow decays inexorably as a consequence of shock formation and vortex stretching. It appears to take place in *three distinct phases* separated by short transitional periods (Porter et al., 1991b): a first period dominated by the phase–coherent mechanism of shock formation, then a compressional period of shock interactions, and thirdly, a period dominated by vortex interactions and vortical decay. This is revealed in the temporal evolution of several small–scale variables: the characteristic wavenumbers k_1, representative of all k_n except for $n = -1$ (see below), is shown in Figure 1 and the density contrast $\Delta\rho$ in Figure 2. In Figure 1, the crosses correspond to the wavenumber associated with the compressional velocity modes, the circles to the solenoidal modes, and the solid line corresponds to the wavenumber computed from the total velocity.

Indeed, we can distinguish three distinct phases of the evolution in these plots. First is a period of rapid growth, up to $t_1 \sim 0.3$. This onset phase corresponds to the formation of shocks which, as coherent nonlinear structures, feed compressional modes of all wavenumbers at the same time. At $t \sim t_1$, the density fluctuations $\delta\rho$ reach a value of 52%, the density contrast $\Delta\rho \sim 60$, and $\rho_{min}/\rho_0 = 0.14$. From t_1 onward, the compressional modes $k_{n,c}$ are roughly constant while the solenoidal components continue their longer period of initial growth. Following this phase of growth and shock formation is a supersonic phase for $t_1 < t < t_2$ with $t_2 \sim 2.1$, during which the solenoidal modes $k_{n,s}$ continue to grow, and the density contrast remains at a high value, on average 45; there are local fluctuations due to the presence of many strong shocks and to their interactions. During this second phase a roughly linear growth for $k_{n,s}$ occurs, at a rate which increases with n for $n = 1, 2, 3$, and 4. For $n > 4$, $k_{n,s}$ increases abruptly at time t_1, and then increases at a rate independent of n until time t_2. During this second phase local regions of supersonic flow can be found.

Finally we have a post–supersonic phase in which the *rms* Mach number is substantially lower than unity, and the local Mach number is smaller than one almost everywhere in space. During this third and final phase, both the density contrast $\Delta\rho$ and the density fluctuations $\delta\rho$ are much smaller ($\Delta\rho \sim 4$ and $\delta\rho \sim 0.15$ at $t_{max} = 3.2\tau_{ac}$). The characteristic wavenumbers k_n are now roughly constant in time; they scale linearly with the order n, and vary from $k_1 \sim 10$ to $k_8 \sim 50$. Furthermore, the ratio $k_{n,s}/k_{n,c}$, approximately 2 for n=2, diminishes for higher moments. The integral wavenumber k_{-1}, however, has a different evolution. Both components grow in the shock formation and supersonic phases, and start to decay again at the onset of the post–supersonic phase (Porter et al., 1991c). This can be attributed to a shift at late times to larger scales of the energy–containing range in a self–similarly decaying flow. Furthermore, large amplitude acoustic oscillations of period τ_{ac} are present in the integral scale of the longitudinal component, $k_{-1,c}$. These oscillations are also observed in $k_{n,c}$ for $n =$ 1, 2, and 4, but are much weaker in amplitude (decreasing as n increases, and in fact marginal at $n = 4$); they are in phase with the oscillations in $k_{-1,c}$.

The dominance of solenoidal modes present in the initial conditions with our choice of χ is recovered after a short period of time during which the compressional modes overpower the solenoidal ones in the small scales due to strong and rapid shock formation preceding slower vortex formation. In Figure 3 are shown both χ defined in (2.5) and χ_2 defined as

$$\chi_2 = \Omega_c/\Omega_v \tag{3.2}$$

where $\Omega_x = \int k^2 \, E^v_x(k) \, dk$ are the enstrophies of the flow; χ_2 characterizes the relative degree of compression in the small scales. The value of χ remains close to 10%, while the evolution of χ_2 (which emphasizes small scales) reflects the three characteristic temporal phases. Both χ and χ_2 tend to increase slightly during the post–supersonic phase.

After the time of shock formation the *rms* Mach number M_a, initially equal to 1.1 with excursions up to $M_{max} \sim 6.2$, decays steadily and linearly from $t = 0.4$ to $t = 2.0$. It is below unity after $t \sim 0.6$. The *rms* Mach number is equal to 0.36 at the final time of the computation. At this time the local Mach number still displays a few local excursions up to $M_{max} = 1.5$, as can be inferred from histograms of the Mach number (Porter et al., 1991c). For example, at $t = 2.96$, all of the regions of locally supersonic

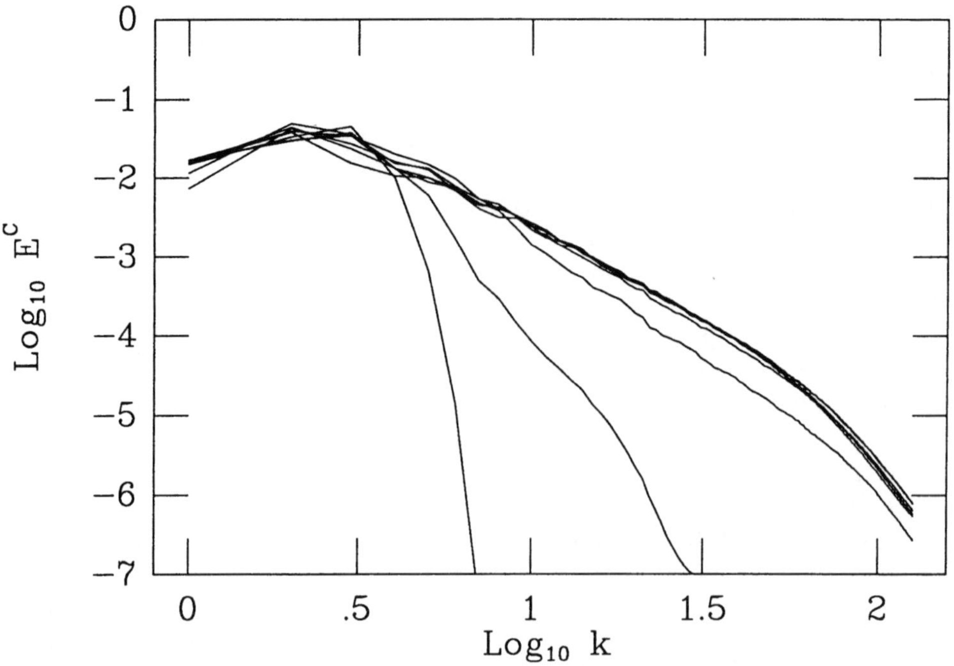

Figure 4a Growth of compressional velocity spectra in the supersonic phase for times 0, 0.12, 0.24, 0.36, 0.48, 0.64, 0.8 and 1.0 in units of τ_{ac}.

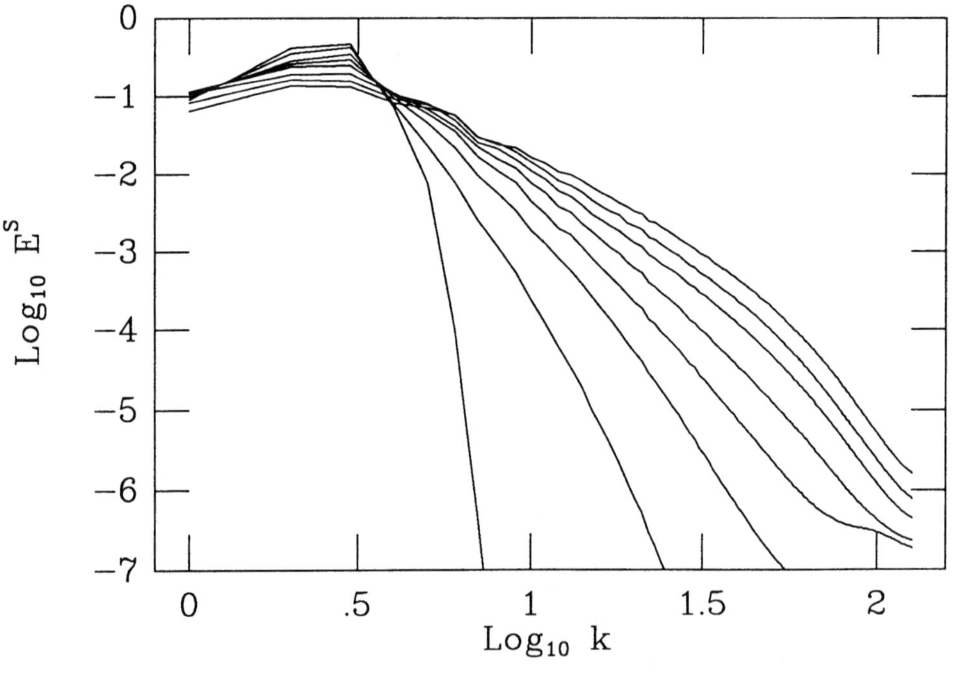

Figure 4b Growth of solenoidal velocity spectra in the supersonic phase. Note the early establishment of a self-similar spectrum for E_c while E_s continues to grow.

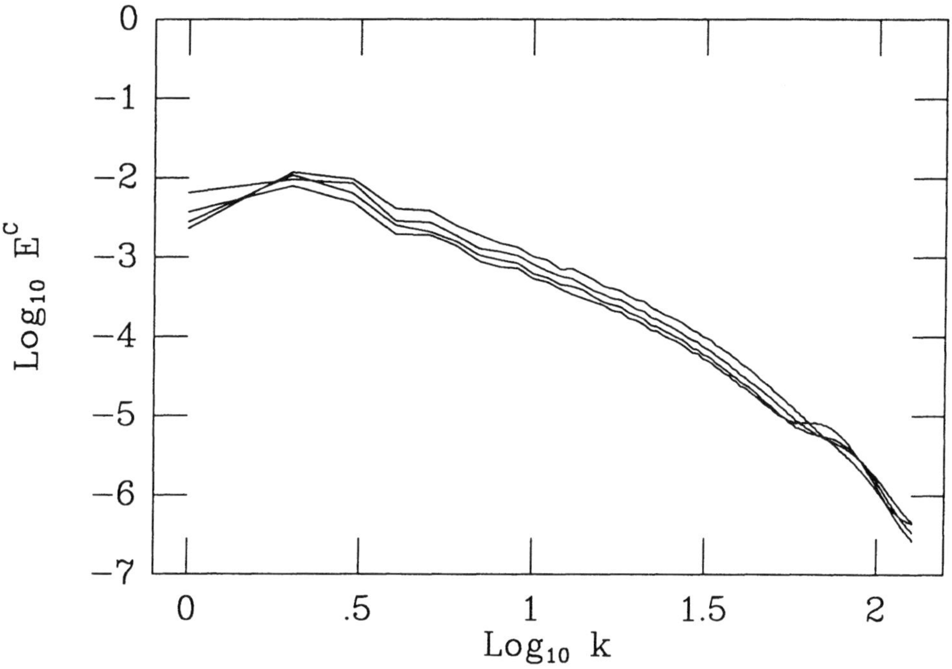

Figure 5a Compressional velocity spectra in the post–supersonic phase for times 2., 2.4, 2.8 and 3.16 in units of τ_{ac}.

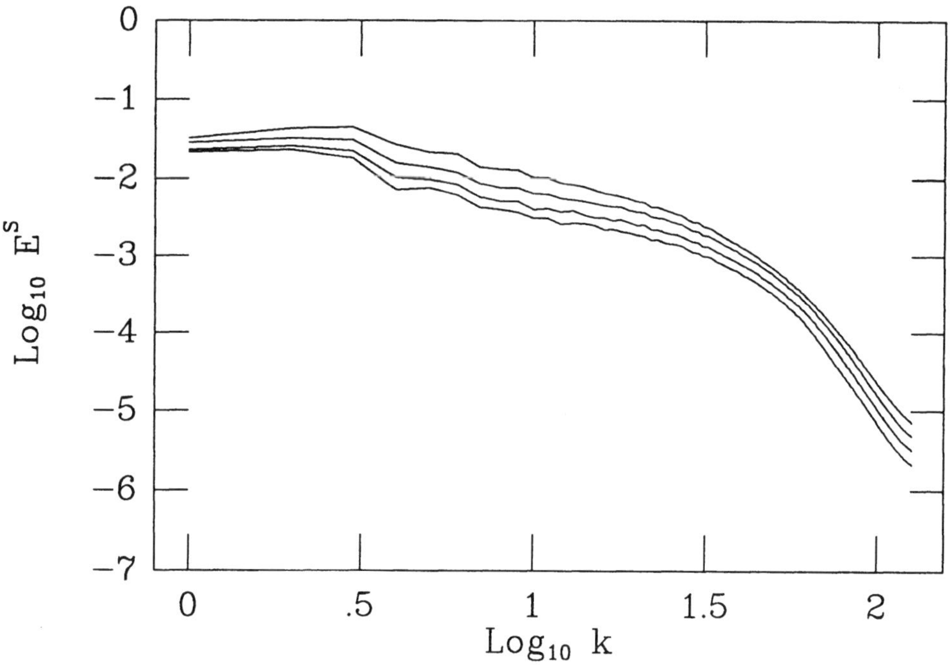

Figure 5b Solenoidal velocity spectra in the post–supersonic phase. Note the shallowness of the solenoidal spectrum.

flow occupy 0.03% of the volume. This small portion of the 16.8 million grid points is equivalent to a cubic box roughly 17 grid points on a side.

We note that the time for compressional modes to develop fully is considerably shorter than the time for the solenoidal modes to develop in the system examined here. All of the compressional modes jump simultaneously from their initial very small values (e.g., essentially zero for $k_n > 5$) to within 86% of their peak values in the time interval $[0.3t_1, t_1]$: all compressional modes are excited together when shocks form by time t_1. The solenoidal modes grow more slowly and at different times: solenoidal modes of shorter wavelengths take longer to reach their peak amplitudes than do those of longer wavelengths, indicative of a forward cascade of energy. The time for all of the solenoidal modes to develop fully seems to be linked with the eddy rotation time of the energy containing scale, which is close to t_2.

The slower growth of the solenoidal modes as compared to the compressional modes which we observe in our simulation can be explained as follows. Given our initial conditions, and for almost any supersonic and random initial conditions, the entire velocity field (not just the initially compressional component) contributes to shock formation. In general, a random pressure field is not consistent with stable eddies. Furthermore, in our 3D simulation, initial pressure fluctuations are about six times too small to support stable eddies that rotate with the initial velocities given. Therefore the initial energy in the solenoidal field is quickly converted into compressional energy, as elements of gas moving in roughly straight lines collide. In order for an eddy to equilibrate, the flow must traverse the eddy's circumference. But in order for a shock to develop, two parts of a flow moving in opposite directions at supersonic speeds need only move a fraction (less than half) of their initial separation. The distance equal to half of the energy containing scale corresponds to both a typical eddy's diameter and the typical initial separation of elements of gas that energetically collide. Both sets of motions come from the same random velocity field, so they typically have similar amplitudes. Hence, the ratio between the time of shock formation and the time of eddy formation is roughly the ratio of the distances given above, which is the ratio of the radius of a circle to its circumference; in fact, $t_2/t_1 \sim 7$.

3.3 Characteristic structures of a decaying supersonic flow

The temporal evolution of the velocity power spectra $E(k) \sim k^i$ with i the spectral index, follows that of the integrated variables we have described above. Small scales develop as the flow evolves; the compressional spectrum settles at $t \sim t_1$ and then decays in a self-similar fashion, with an index $i_c \sim -2.06 \pm 0.03$ on average. On the other hand, the evolution of the solenoidal component of the velocity is much slower. A self-similar regime occurs only after $t \sim t_2$, with an index close to $i_s \sim -0.94 \pm 0.02$. The total spectrum is dominated at all scales, except for a short time early in the quasi-supersonic phase, by its solenoidal component.

We show the temporal evolution of the spectra early in the quasi-supersonic phase ($0 < t < 1$) and in the post-supersonic phase ($2.0 < t < 3.2$) in Figures 4 and 5 respectively, separated into their compressional (a) and solenoidal (b) components. The spectral indices, once the supersonic phase is well established, are close to -2 for all three (s, c and t) spectra, as is expected for a velocity discontinuity, which is the structure that must be dominant during that time. In the last phase, the spectral

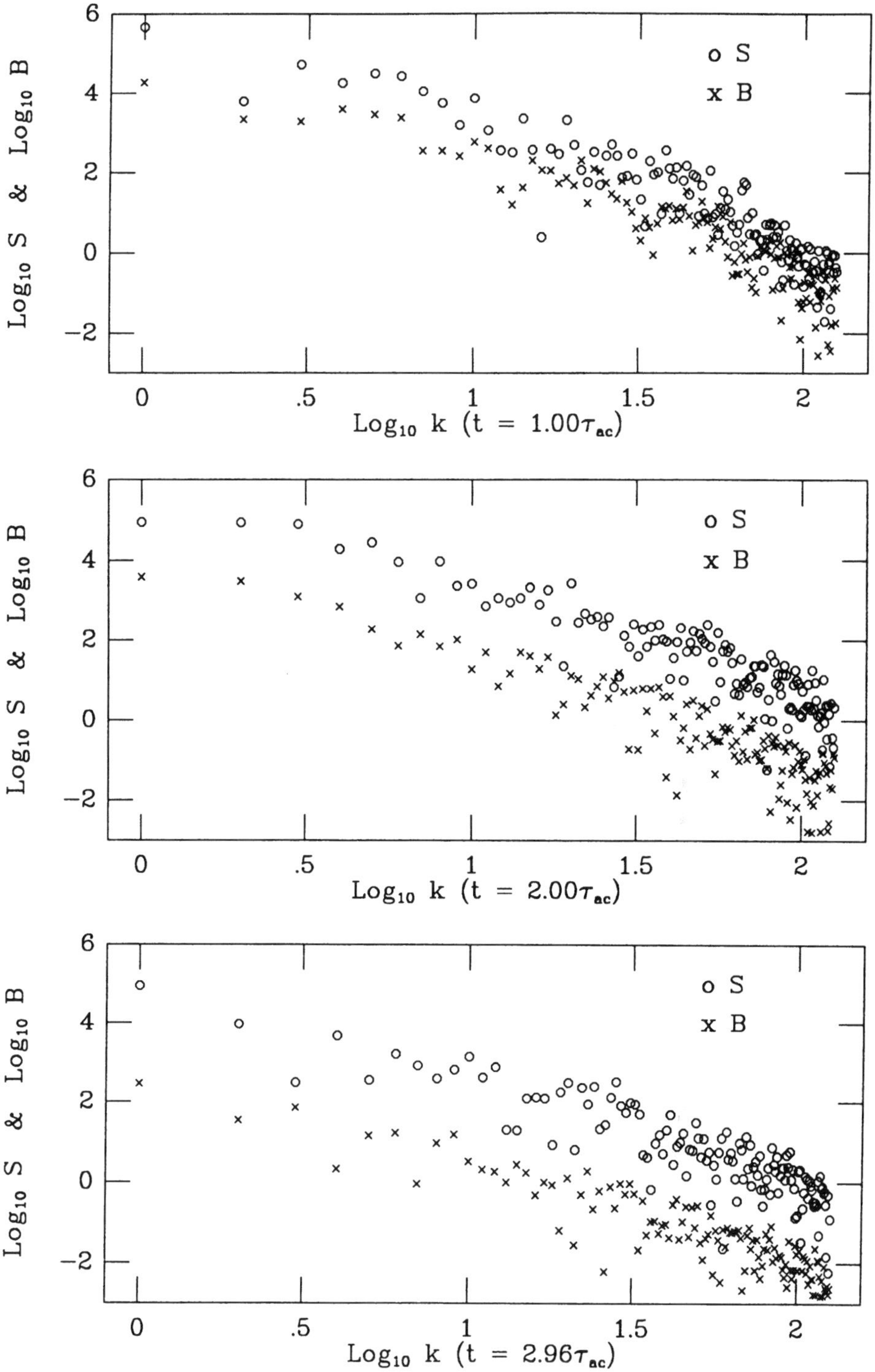

Figure 6: One-dimensional spectra of the production of vorticity through the baroclinic term **B** (crosses) and the linear term **S** (circles). Time increases from top to bottom. The baroclinic term is negligible except during the first half of the supersonic phase.

indices for the s- and c– spectra differ. Whereas the compressional component retains a rather steep spectrum, whose index is still close to but shallower than −2, the solenoidal component has an index close to −1. Since the latter dominates the total velocity, the velocity spectrum is thus quite flat during that period. This behavior may be due to hysteresis: the history of a supersonic phase in a now–subsonic flow leaves traces. This effect was already encountered in our two–dimensional computations, where filamentary structures in entropy, density and vorticity were observed at late times (Porter et al., 1991c) and there were strong local vortices superimposed onto warped vorticity sheets in the subsonic phase (Passot et al., 1988). The k^{-1} spectrum is also reminiscent of the spectrum of a passive scalar advected by an incompressible flow (Leslie, 1973).

The temporal evolution of vorticity comoving with the flow is given by

$$d\omega/dt = \mathbf{B} + \mathbf{S} \quad , \tag{3.3a}$$

with the inhomogeneous baroclinic term

$$\mathbf{B} = (\nabla\rho \times \nabla p)/\rho^2 \tag{3.3b}$$

and the terms linear in ω (stretching and compression)

$$\mathbf{S} = (\omega \cdot \nabla)\mathbf{u} - \omega\,(\nabla \cdot \mathbf{u}) \quad . \tag{3.3c}$$

From inspection of the histograms of both **B** and **S** we infer that in the early supersonic phase the baroclinic and the linear terms are roughly balanced: 90 % of the points in the flow have a value of vorticity production below 250 for the former, and 450 for the latter, values that are comparable; we obtain equivalent results at the 99 % level (1,350 *vs* 2,130). This indicates that vorticity production through shock curvature (including intersecting shocks) is as important as the linear term, at least for Mach numbers of unity and for a length of the inertial range spanning roughly one decade in wavenumbers. In order to assess the relative importance of these two terms graphically, we now examine the plots of one–dimensional (that is, wavenumbers taken in only one direction) power spectra in log–log coordinates for the norms of both the baroclinic and linear terms as they appear in the equation given above. Figure 6 shows this comparison at times 1, 2, and 2.96. We find $B < S$ most of the time and for most scales. However, the two terms are comparable during the time of initial growth of the vorticity – up to when the compression first becomes stationary ($t \sim t_1$). The linear term is mostly dominant later on due to its being proportional to the vorticity. In an initially supersonic flow, the baroclinic term acts as the principal trigger of vorticity production on small scales; intersecting shocks plant the seed of vorticity at short wavelengths which then may undergo an exponential growth for some time via the linear terms during the intermediate phase $t_1 < t < t_2$. Vortex sheets are established at shock intersections, and these sheets subsequently roll up due to the familiar shear instabilities. These vortex tubes are also unstable and are disrupted by kink instabilities, in which the phenomenon of vortex stretching plays a major role. This sequence of events is reflected in the figures 1 and 2, but it is most clearly revealed in animations of volume rendered images of the flow which we have generated in our "numerical laboratory" (Porter et al., 1991c). It would be useful as well to compare the intensity of the incompressible stretching term

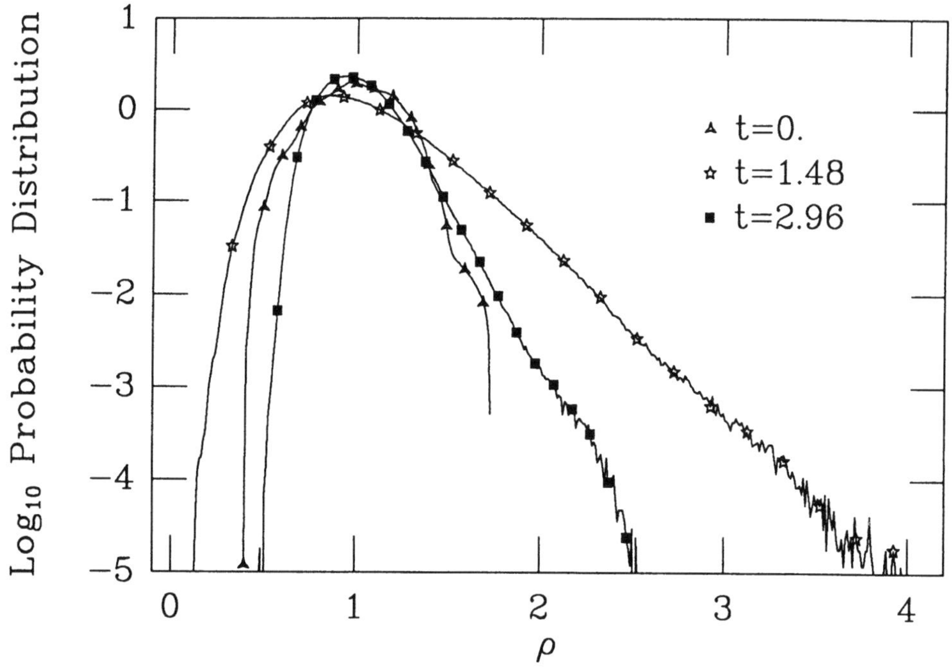

Figure 7a: Note the narrowing ot the density contrast ρ_{max}/ρ_{min}.

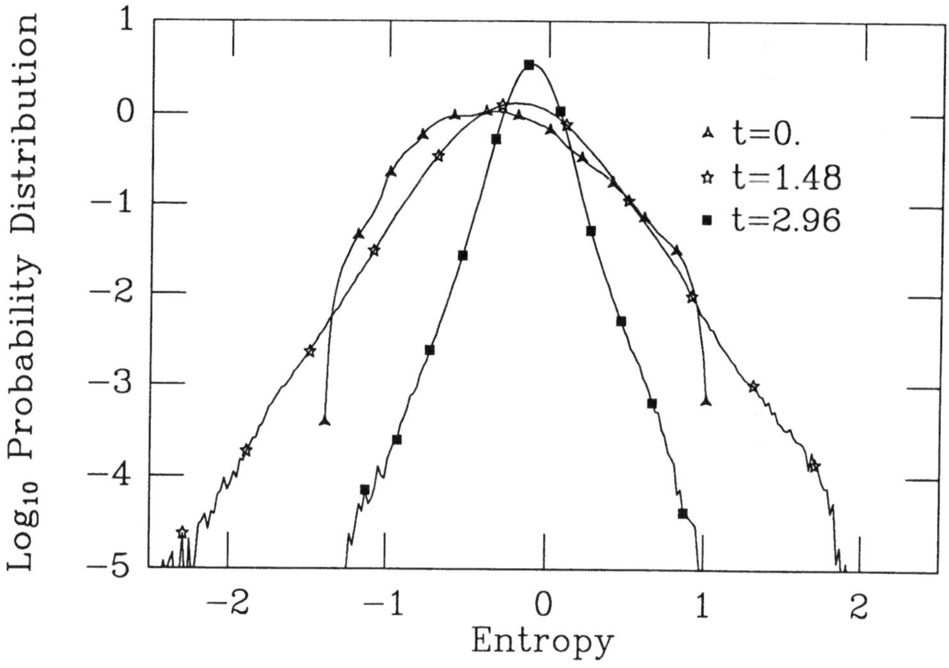

Figure 7b: Note the homogenizing of the flow with time.

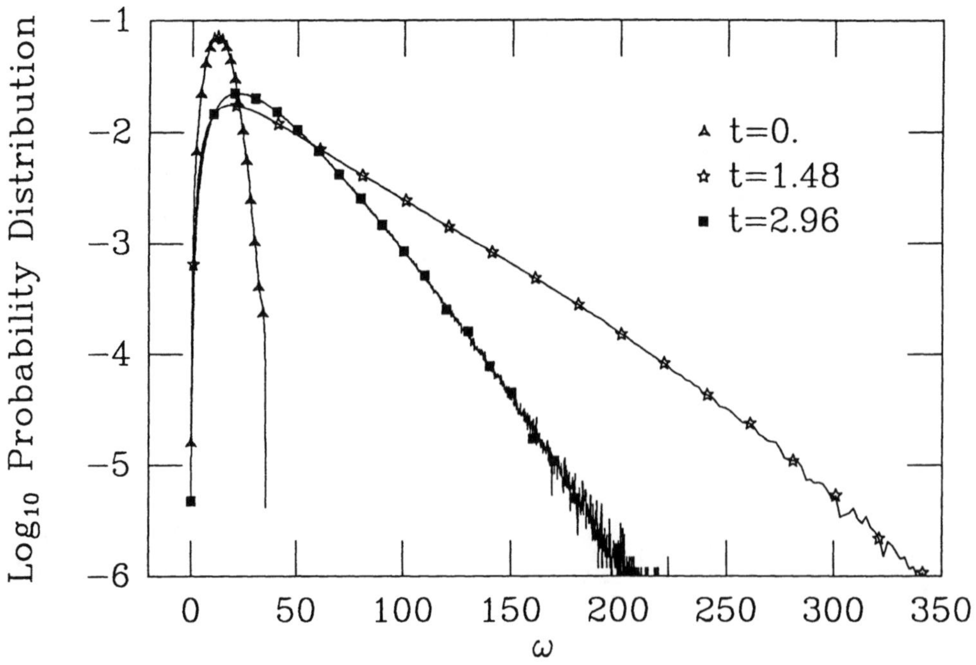

Figure 7c: Vector magnitude of vorticity; for late times, $\mathcal{P}(\omega) \sim \omega^2 e^{-\omega}$.

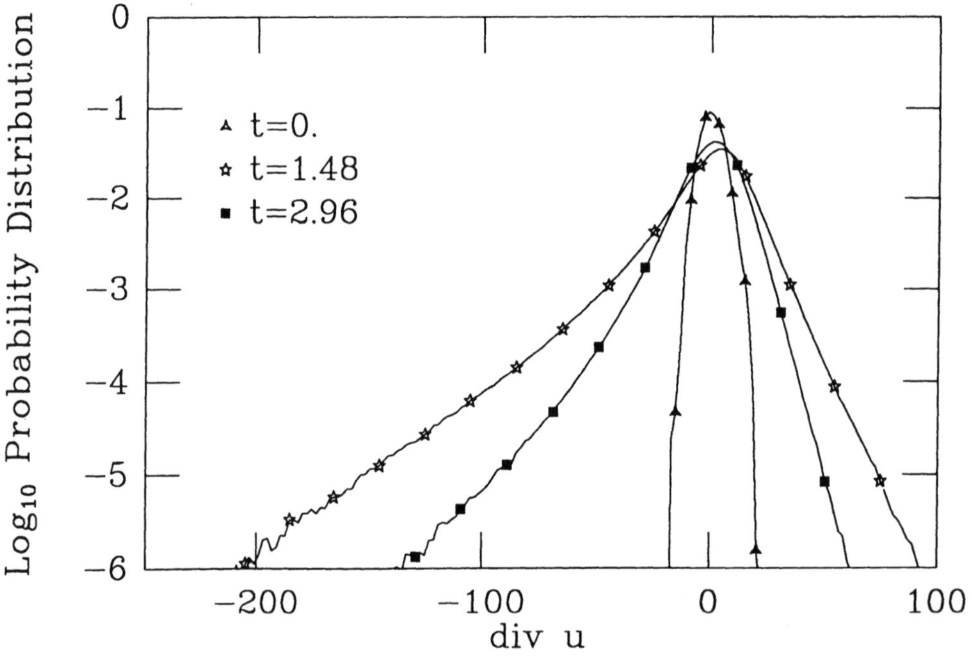

Figure 7d: Note the skewed distribution towards negative values of the velocity divergence.

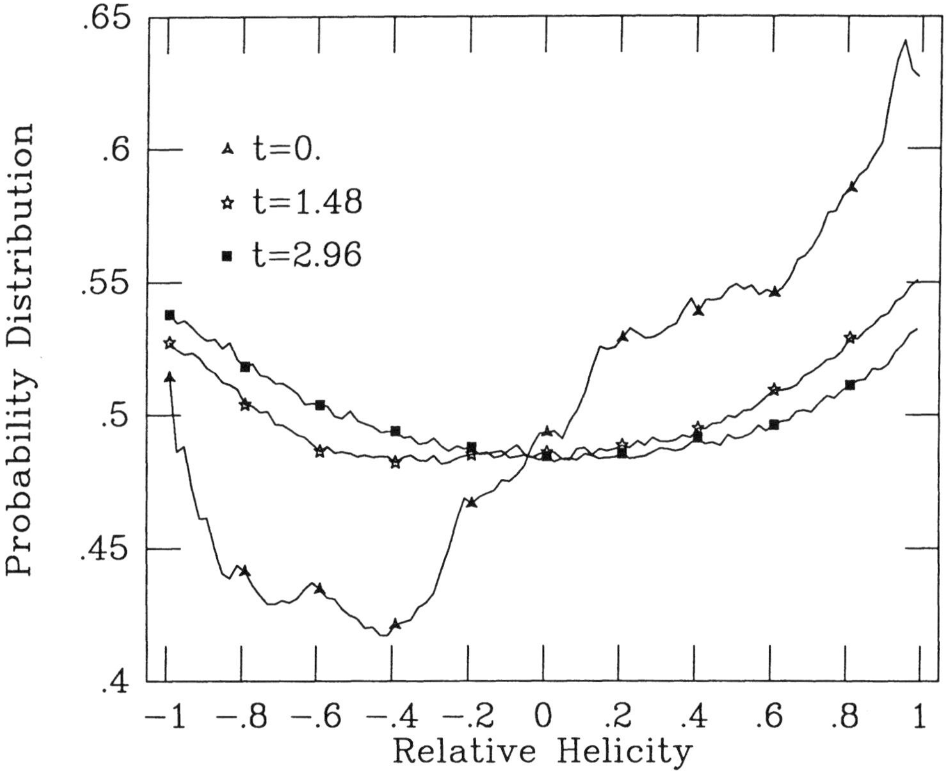

Figure 7e: Note the uniformity of the distribution.

Figure 7: Probability density functions at times 0 (triangles), 1.48 (open stars) in the supersonic phase, and 2.96 (solid squares) at the end of the post–supersonic phase.

of vorticity production, to the production stemming from pure compression; this work is now in progress.

In Figure 7, we show the evolution of the probability distribution functions of several variables: (7a) density; (7b) entropy; (7c) vorticity; (7d) divergence of velocity; and (7e) relative helicity, as defined in (2.6).

These probability distribution functions are simply histograms on the full 256^3 grid, normalized to have a unit integral on the interval of definition. They are shown at times $t = 0$ (triangles), $t = 1.48$ (stars) at which time the system is well into the quasi-supersonic phase, and $t = 2.96$ (squares) near the end of the simulation.

In the density *pdf* one can clearly see the augmentation of the density contrast $\Delta\rho$ in the supersonic phase and its narrowing later in time as the flow becomes subsonic.

As time evolves the *pdf* of the entropy defined as $S = log_e(P/\rho^\gamma)$ has a wider peak (by a factor 3) at time $t = 0.48$ (not shown here) than at $t = 2.96$. Similarly, the entropy contrast S_{max}/S_{min} is ~ 6 at $t = 0.48$, and is ~ 2.5 at $t = 2.96$. This narrowing of the entropy distribution indicates a slow evolution towards mixing and homogenizing of the flow. Fluid mixing is expected in our system, but homogenization of entropy can only happen through diffusion, which in our case arises from numerical errors: PPM cannot follow an entropy structure which is less than one zone wide. The slow rate at which the entropy homogenizes is a measure of how little numerical diffusion there is in this simulation. Even though the flow is subsonic at late times, there are still entropy fluctuations due to shocks set up by the high Mach number initial conditions. Entropy fluctuations, with their associated density fluctuations, make our flow (even at late times) violate the assumption of nearly uniform density which is typically made in analytical studies of low Mach number flow. Thus, the arguments leading to steepening of the classical $k^{-5/3}$ Kolmogorov spectrum through coupling of vortices to acoustic waves (Moissev et al., 1981) apparently do not apply here.

A wide distribution of vorticity is generated during the first half of the simulation. At $t = 1.48$ the average vorticity is 47.1 and the *rms* vorticity is 60.0, as compared to their initial values of 13.0 and 14.1 respectively. Vorticity decays in the second half of the simulation: by $t = 2.96$ the average and *rms* vorticities are 35.9 and 42.2 respectively. At this final time, the vector vorticity is exponentially distributed, that is the distribution of the magnitude of ω is given by $\mathcal{P}(\omega) \sim \omega^2 e^{-\omega}$.

The probability distribution function for $\nabla \cdot \mathbf{u}$, another small–scale variable, shows a marked development of high values (in magnitude), the distribution for the divergence being skewed on the negative – compressional – side. This is related to a volume factor, since shocks are intermittent, *ie* more concentrated in space than rarefaction waves, and since the total divergence of the velocity for a flow with periodic boundary conditions must sum to zero.

Finally, the relative helicity $\tilde{r}$ defined in (2.6) remains small ($\sim 3\%$) throughout the entire evolution of the flow. This shows that there is no generation of helicity in the flow, except possibly where the compression is strong. The *pdf* of $\tilde{r}$ shown in (7e) is flat, indicative of a uniform distribution. Histograms conditioned by the amplitude of the vorticity become flatter for higher values of the vorticity.

4. CONCLUSION

Shock waves and shock intersections play an important role in the transfer of energy from long to short wavelengths for both the compressional and solenoidal components of the velocity field. Weak shocks survive for many τ_{ac}, where τ_{ac} is the acoustic time of the energy containing modes in the initial state. In the context of our decay problems, vorticity is concentrated in long filaments and not sheets which tend to break up into shorter filaments within a few τ_{ac}. Local fluctuations in the relative helicity are strong, even though the average relative helicity may be very small.

Strong shocks in the early phases of our models produce entropy fluctuations. In 2-D we clearly see that entropy fluctuations are stretched into filaments by the solenoidal component of the velocity field. The inverse relation between entropy and density at constant pressure in our γ–law models implies that filaments in entropy are mirrored in the density field. Entropy fluctuations persist in our models because we impose no thermal conductivity. A very low thermal conductivity, which is expected in large astrophysical systems, should allow entropy fluctuations to be long lived and to develop into filaments in much the same manner as they do in our models. This could explain the observed filamentary nature of molecular clouds.

The compressional components of the velocity power spectra vary as $E^C(k) \sim k^{-2}$ in their inertial ranges for times $t > 0.25\tau_{ac}$, which is due to the presence of shocks.

For $t > 2.4\tau_{ac}$ the velocity power spectra settle into a self–similar exponential decay with time. For these decay problems, the solenoidal components of the velocity power spectra comprise $\sim 90\%$ of the total energy. Hence the spectral index of the solenoidal component $E^S(k)$ is close to that of the total $E^v(k)$. We find $E^s(k) \sim k^{-1}$, markedly flatter than the Kolmogorov law. This flat spectrum is indicative of an accumulation of small–scale motions in the form of small vortex filaments.

ACKNOWLEDGEMENTS

The computations presented in the figures here have been performed on the Cray–2 of the Minnesota Supercomputer Institute. Computations were also performed at the Centre de Calcul Vectoriel pour la Recherche (CCVR). This work was supported at the University of Minnesota by grants DE-FG02-87ER25035 from the Office of Energy Research of the Department of Energy, AST-8611404 from the National Science Foundation, and by equipment grants from Sun Microsystems, Gould Electronics, and the Air Force Office of Scientific Research (AFOSR-86-0239). Partial support for this work has also come from the Army High Performance Computing Research Center (AHPCRC) at the University of Minnesota. At Nice, this work was supported under DRET contract 500–276, the GdR–CNRS "Mécanique des Fluides" and grants from the Observatoire de la Côte d'Azur. Some of the earlier Navier-Stokes runs reviewed here have been performed in collaboration with T. Passot, and we have benefited from many useful discussions with him. We are also most grateful to T. Varghese and R. Schmidt for their assistance in carrying out the PPM computations reported here.

REFERENCES

B.J. Bayly, D.C. Levermore and T. Passot; 1990, to appear, *Phys. Fluids. A.*

S. Chandrasekhar; 1951, *Proc. Roy. Soc.* **A210**, 18; *Proc. Roy. Soc.* **A220**, 26.

Chollet J.P. and Lesieur M.; 1981, *J. Atmos. Sci.* **38**, 2747.

P. Colella and P. R. Woodward; 1984, *J. Comp. Phys.* **54**, 174.

G. Erlebacher, M.Y. Hussaini, C.G. Speziale and T.A. Tang; 1987, ICASE Nb 87–20.

G. Erlebacher, M.Y. Hussaini, H.O. Kreiss and S. Sarkar; 1990, to appear, Theoretical Computational Fluid Dynamics.

E. Falgarone; 1990, in *Structure and Dynamics of the Interstellar Medium*, IAU Symposium 120, G. Tenorio–Tagle, M. Moles & S. Melnick, Springer.

E. Falgarone and T.G. Philipps; 1990, to appear, *Astrophys. J.*

W.J. Feireisen, W.C. Reynolds and J.B. Ferziger; 1981, Report TF-13 (unpublished) Thesis, Stanford University.

H. Ha Minh and D. Vandromme; 1986, *DRET–ONERA Colloquium, Poitiers, Ecole Nationale Supérieure de Mécanique et Aérotechnique.*

C. Heiles; 1987, "Physical Processes in Interstellar Clouds", Eds. G.E. Morfill and M. Scholer, Dordretch, Reidel, 429.

R.N. Henriksen; 1990, IAU Symposium **147**, *op cit.*

S. Kida and S.A. Orszag; 1990, "Energy and Spectral Dynamics in Forced Compressible Turbulence" Preprint Kyoto University.

R.H. Kraichnan; 1953, *J. Acoust. Soc. of America*, **25**, 1096.

R.H. Kraichnan; 1990, *Phys. Rev. Lett.*, **65**, 575.

R.H. Kraichnan, and D. Montgomery; 1979, *Rep. Progress Phys.* **43**, 547.

C.J. Lada and F.H. Shu; 1990, *Science* **248**, 564.

S. Lee, S.K. Lele and P. Moin, 1991, *Phys. Fluids A*, **3**, 657.

D. C. Leslie, 1973; *Developments in the Theory of Turbulence*, Oxford Clarendon Press.

S.S. Moiseev, V.I. Petviashvili, A.V. Tur and V.V. Yanovskii; 1981, *Physica* **2D**, 218.

T. Passot and A. Pouquet; 1987, *J. Fluid Mech.* **181**, 441.

T. Passot and A. Pouquet; 1988, *J. Comp. Phys.* **75**, 300.

T. Passot, A. Pouquet and P. R. Woodward; 1988, *Astron. Astrophys.* **197**, 228.

T. Passot and A. Pouquet; 1991, to appear, *J. of Mechanics B/ Fluids.*

D. H. Porter and P. R. Woodward; 1991, to be submitted to the *Astrophysical Journal Supplement*, preprint available through the Univ. of Minnesota AHPCRC.

D. H. Porter, P. R. Woodward, W. Yang and Qi Mei; 1990, in *Nonlinear Astrophysical Fluid Dynamics*, R. Buchler Ed., Annals of the New York Academy of SciencesR **617**, 234.

D.H.Porter, P.R.Woodward, and Q.Mei, 1991a, *Video J. Eng. Res.* Vol. 1, no. 1, (in press).

D. H. Porter, A. Pouquet, and P. R. Woodward; 1991b, *Phys. Rev. Lett.*, submitted.

D. H. Porter, A. Pouquet, and P. R. Woodward; 1991c, in preparation.

D.H.Porter, A. Pouquet, P.R.Woodward, and H.Yang, 1991d, in preparation.

A. Pouquet, T. Passot and J. Léorat; 1990, IAU Symposium **147**, *Fragmentation in Molecular Clouds*, Grenoble, June 1990, F. Boulanger, G. Duvert & E. Falgarone Eds., Kluwer.

J. Scalo; 1990, Preprint Nb 95, Mc Donald Observatory, University of Texas.

J. Smagorinsky; 1963, *Mon. Weather Rev.* **91**, 99.

P. R. Woodward and P. Colella; 1984, *J. Comp. Phys.* **54**, 115.

P. R. Woodward; 1986, "Numerical Methods for Astrophysicists", in *Astrophysical Radiation Hydrodynamics*, eds. K.-H. Winkler and M. L. Norman, Reidel, 1986.

P. R. Woodward; 1990, "Numerical Methods for Unsteady Compressible Flow", to appear in *Numerical Methods in Astrophysics*, ed. P. R. Woodward, Academic Press, manuscript in preparation.

T.A. Zang, R.B. Dahlburg and J.P. Dahlburg, 1991 "Direct and Large-eddy Simulations of three-dimensional Compressible Navier-Stokes Turbulence", Preprint, NASA Langley;

LARGE-SCALE INSTABILITIES AND INVERSE CASCADES IN ORDINARY AND MHD FLOWS AT LOW REYNOLDS NUMBERS

B. Galanti and P.-L. Sulem*

CNRS, URA 1362, Observatoire de la Côte d'Azur
BP 139, 06003 Nice-Cedex, France

Abstract

The anisotropic kinetic alpha effect (AKA) and the (magnetic) alpha effect refer to large-scale instabilities which develop in low Reynolds number flows and lead to an exponential growth of a weak large-scale velocity or magnetic field respectively. When the nonlinearities become important, an inverse cascade is observed, yielding the formation of structures at larger and larger scales, up to the limits of the system.

1. Introduction

A celebrated example of large-scale instability in magnetohydrodynamics is the alpha effect [1]-[3]: a small-scale flow acting on a large-scale magnetic field generates a mean electromotive force which is a function of the large-scale magnetic field itself. When the small-scale flow is helical or otherwise lacks parity invariance [4], this effect may lead to a dynamo effect, i.e. the growth of a weak large-scale magnetic field.

An analogous effect, the AKA effect, was recently found in ordinary hydrodynamics when a large-scale velocity field is superposed on a small-scale anisotropic flow lacking parity invariance [5]-[7]. It requires that Galilean invariance be broken, for example by the body force which maintains the small-scale basic flow. In this case, due to the advection of the small-scale flow by the large-scale motion, the Reynolds stresses depend of the large-scale velocity itself. This again may lead to an instability where a weak large-scale velocity field is amplified exponentially.

For both alpha and AKA effects in the linear approximation, the destabilizing terms in the induction and momentum equations for the large-scale fields are proportional to first derivatives of the magnetic and velocity fields respectively. Consequently, these terms can dominate viscous or Ohmic dissipation, whatever small the Reynolds numbers are, provided the system be sufficiently extended. This contrasts with negative eddy viscosity instabilities for which the Reynolds numbers should exceed finite critical values, thus making explicit computations significantly more delicate [8]. Examples are provided by time-independent one-dimensional parallel periodic flows, like the Kolmogorov flow [9]-[11].

When as a result of the AKA or the alpha instability, a large-scale magnetic field or a large-scale secondary flow develops, nonlinear couplings become relevant. They

---

* Also at School of Mathematical Sciences, Tel-Aviv University, Israel

may either consist of direct interactions between the large-scale fields or result from the feed-back of the large scales on the small scales. The aim of this paper is to review the similarities and the differences of the various nonlinear regimes induced by these instabilities, and in particular the inverse cascades which develop in extended systems. Special attention is devoted to the dynamo instability when the initial (weak) large-scale magnetic field is genuinely three-dimensional. New results are presented in this case. For the sake of consistency, changes of notation have been made relatively to papers quoted in references.

2. The anisotropic kinetic alpha effect (AKA)

Let us consider a low Reynolds number basic flow stirred by a space and time periodic body force $\mathbf{f}(\mathbf{x}, t)$. A large-scale velocity $\mathbf{v}$ superposed to this flow satisfies

$$\begin{aligned} &\partial_t v_i + \partial_j(v_i v_j + r_{ij}) = -\partial_i p + \nu \nabla^2 v_i \\ &\partial_i v_i = 0, \end{aligned} \tag{2.1}$$

where the (periodicity averaged) Reynolds stresses r_{ij} read [7]

$$r_{ij} = \sum_{\mathbf{k},\, \omega} \frac{-1}{\nu^2 k^4 + (\mathbf{k}\cdot\mathbf{v} - \omega)^2} \hat{f}_i^*(\mathbf{k}, \omega) \hat{f}_j(\mathbf{k}, \omega) \, . \tag{2.2}$$

Here $\hat{\mathbf{f}}(\mathbf{k}, \omega)$ is the Fourier transform of the body force. In eq.(2.1), the space and time variables refer to variations on scales larger than those of the body force, by factors R^{-2} and R^{-4} respectively.

When the issue is the stability of the basic flow to weak large-scale perturbations, the Reynolds stresses can be linearized in the form $\partial_j r_{ij} = \alpha_{ijl} \partial_j v_l$ with $\alpha_{ijl} = \left[\frac{\delta r_{ij}}{\delta v_l}\right]_{\mathbf{v}=0}$. It immediately follows that a linear AKA instability cannot occur if the basic flow is isotropic or parity invariant. Equation (2.2) shows that it is also the case if the body force is time-independent. In the absence of AKA effect, the Reynolds stresses are to leading order proportional to the large-scale velocity gradient and induce an eddy viscosity in the momentum equation for the large-scale flow [8]. Furthermore, if the basic flow is two-dimensional or axisymmetric, the AKA effect is purely undulatory and no AKA instability occurs.

A simple example of body force leading to a linear AKA instability is

$$f_1 = f\cos(k_0 y + \nu k_0^2 t) \quad , \quad f_2 = f\cos(k_0 x - \nu k_0^2 t) \quad , \quad f_3 = f_1 + f_2 \, . \tag{2.3}$$

We write $f = \nu V_0 k_0^2$, where V_0 is a typical velocity from which we define the forcing Reynolds number $R = V_0/\nu k_0$ characteristic of the basic flow. If L denotes the largest available scale of the system, we rescale $\mathbf{x} = L\mathbf{X}$, $t = L^2 \nu^{-1} T$, $\mathbf{V} = \mathbf{v}/V_0$. The large-scale secondary flow resulting from the AKA instability then obeys

$$\begin{aligned} &\partial_T V_i + \frac{\alpha}{R} \partial_{X_j}(V_i V_j + \mathcal{R}_{ij}) = -\partial_{X_i} P + \Delta V_i \\ &\partial_{X_i} V_i = 0 \, , \end{aligned} \tag{2.4}$$

with

$$\mathcal{R}_{11} = \mathcal{R}_{13} = \frac{1}{1 + RV_2 + \frac{1}{2}R^2V_2^2} \quad ; \quad \mathcal{R}_{12} = 0$$
$$\mathcal{R}_{22} = \mathcal{R}_{23} = \frac{1}{1 - RV_1 + \frac{1}{2}R^2V_1^2} \quad ; \quad \mathcal{R}_{33} = \mathcal{R}_{11} + \mathcal{R}_{22}. \tag{2.5}$$

In eq.(2.4), we assume 2π-periodicity in the $\mathbf{X}$-variable. Note that although the forcing Reynolds number R is taken small, the parameter $\alpha = R^2 Lk_0$ can be made arbitrarily large by increasing the aspect ratio Lk_0.

2.a. Layered large-scale secondary flows

Eqs.(2.4)-(2.5) admit solutions depending on the X_3-coordinate only. We refer to these flows as "layered flows". Due to the incompressibility condition, self-advection in eq.(2.4) drops out and V_3 vanishes. The nonlinearities thus reduce to the feed-back of the large-scale velocity on the small-scale flow. The Reynolds number can then be eliminated from the resulting equations by measuring the large-scale velocity in units of R^{-1}, and α remains the only control parameter. For $\alpha < 1$, the zero solution is stable. As α crosses the value 1, the modes of wavenumber $K = 1$ become linearly unstable, and a subcritical bifurcation occurs leading to a steady solution with a finite amplitude [12]. The dynamics for larger values of α (corresponding to extended systems where a significant range of scales are linearly unstable), was addressed numerically in [6] and [7]. At early times, an initially weak large-scale velocity field is dominated by the growth of the linearly most unstable modes. When the nonlinearities become important, the maximum energy migrates to larger and larger scales, up to the limits prescribed by the size of the system. In physical space, this corresponds to a reduction of the number of structures in the velocity field by merger and destruction. This phenomenon of "inverse cascade" is illustrated in Fig.1 which displays the time-evolution of the (rescaled) energy modes $E_K = \frac{1}{2}|R\mathbf{V}_K|^2$ with $K = 1, ..., 6$ for $\alpha = 8$. In this case, the linearly most unstable mode corresponds to a wavenumber $K = 4$. A noticeable property is that the

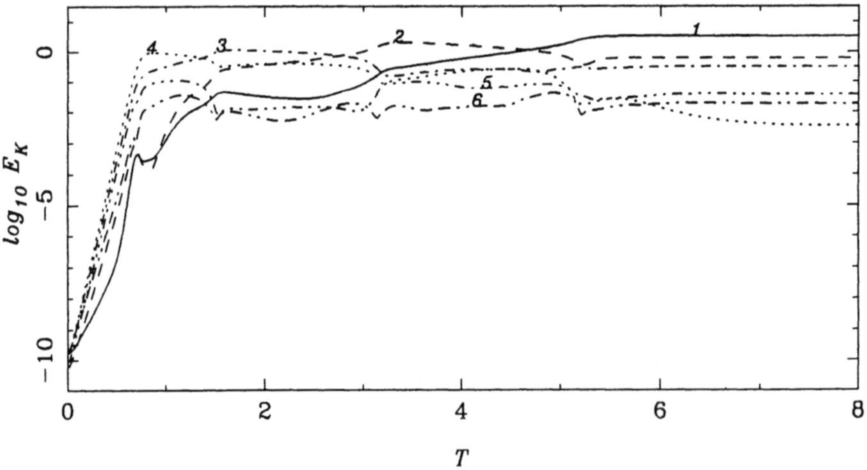

Fig. 1 : Evolution of energy modes (see labels on the curves) for the layered secondary flow generated by the AKA instability with $\alpha = 8$.

velocity of the resulting large-scale secondary flow saturates at a level of order $O(R^{-1})$, thus significantly higher than the amplitude of the basic flow.

An analysis of the inverse cascade is given in [7] where a dynamic mechanism is presented. For layered flows, eqs.(2.4)-(2.5) admit a sequence of steady cellular solutions, unstable to large-scale perturbations. During the cascade, the solution of the evolution problem is successively attracted by steady cellular solutions of larger and larger scales, until the limits prescribed by the size of the system. Note that a similar inverse cascade occurs for the Kolmogorov flow [13]. Although driven by a negative viscosity, it also appears to be the consequence of instabilities of steady cellular solutions [9].

2.b. Three-dimensional large-scale secondary flows

Numerical simulations of eqs(2.4)-(2.5) show that when the basic flow is subject to a large-scale perturbation depending on the three space coordinates, the velocity of the large-scale secondary flow resulting from the AKA effect saturates at a level of order $O(R)$ [7]. This indicates that the nonlinear dynamics is dominated by the self-advection of the secondary flow rather than the feed-back to the small-scale flow. The Reynolds stresses may be linearized and the Reynolds number can again be eliminated by measuring now the large-scale velocity field in units of R. In this regime, the large-scale secondary flow is thus significantly weaker than the basic flow. Note that this is also the case for the large-scale velocity resulting from the negative viscosity instability of the Kolmogorov flow [9], [10].

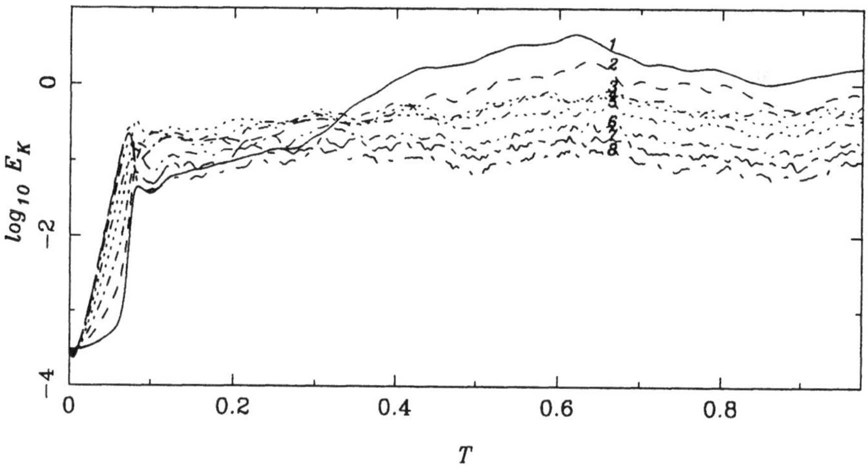

Fig. 2 : Evolution of shell averaged energy modes (see label on the curves) for the three-dimensional secondary flow generated by AKA instability with $\alpha = 16$.

Figure 2 displays the time-evolution of the (rescaled) energy modes $E_K = \frac{1}{2}\Sigma_{K'\in C_K}|R^{-1}\mathbf{V}_{K'}|^2$ in spherical shells $C_K = \{K - \frac{1}{2} \leq K' < K + \frac{1}{2}\}$, around the wavenumbers $K = 1, ..., 8$ for $\alpha = 16$. A more usual representation of the inverse cascade is seen in Fig.3 where the energy spectrum is plotted at increasing times. We observe that after the early-time linear phase during which the linearly most unstable shell $K = 8$ dominates, an inverse cascade develops like in the case of layered large-scale

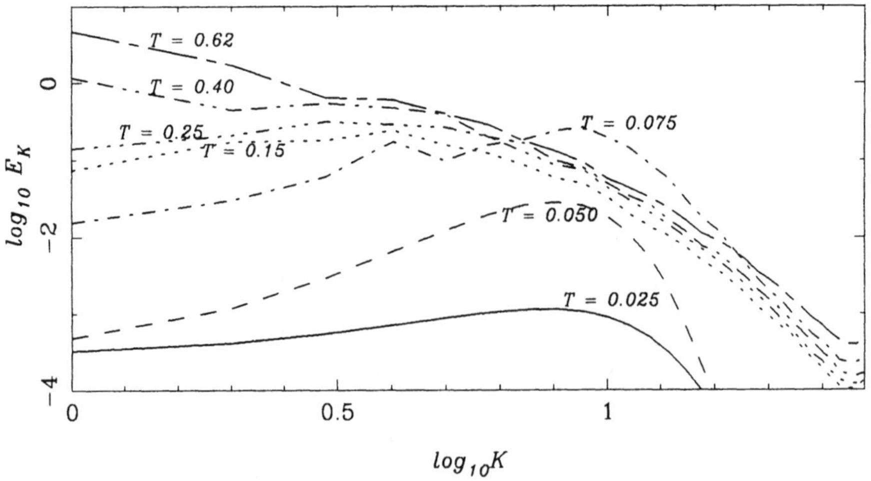

Fig. 3 : Inverse energy cascade for the three-dimensional secondary flow generated by AKA instability with $\alpha = 16$.

flows, although the nonlinear processes are different. Visualizations in physical space show the formation of structures correlated on larger and larger scales [7]. The flow is three-dimensional and displays a moderate anisotropy.

3. The alpha effect for electrically conducting fluids

Consider now an electrically conducting fluid driven at a scale $1/k_0$ by a prescribed solenoidal body force $\mathbf{f}(\mathbf{x}, t)$, periodic in space and time (or random, homogeneous and stationary). This force is now assumed to be helical. As previously, the Reynolds number R is assumed to be small and the Prandtl number is chosen of order unity. In the presence of a large-scale magnetic field $\mathbf{b}$ or velocity $\mathbf{v}$, the basic flow is modified at small scales, and this may lead to a mean electromotive force (e.m.f.) $\boldsymbol{e}(\mathbf{v}, \mathbf{b})$ and/or a mean Reynolds stresses $\mathbf{r}(\mathbf{v}, \mathbf{b})$. Note that the dependence on $\mathbf{v}$, arises because Galilean invariance is generally broken by the body force. The equations for the large-scale fields then read:

$$\begin{aligned} &\partial_t v_i + \partial_j (v_i v_j - b_i b_j + r_{ij}) = -\partial_i p + \nu \nabla^2 v_i \\ &\partial_t \mathbf{b} = \nabla \times (\mathbf{v} \times \mathbf{b} + \boldsymbol{e}) + \eta \nabla^2 \mathbf{b} \\ &\partial_i v_i = 0 \qquad ; \qquad \partial_i b_i = 0 \,. \end{aligned} \tag{3.1}$$

In the limit of small Reynolds numbers, we have [16]:

$$\begin{aligned} r_{ij} &= \sum_{\mathbf{k},\, \omega} \frac{(\mathbf{k}\cdot\mathbf{b})^2 - (\mathbf{k}\cdot\mathbf{v} - \omega)^2 - \eta^2 k^4}{|Q|^2} \hat{f}_i^*(\mathbf{k}, \omega) \hat{f}_j(\mathbf{k}, \omega) \\ e_i &= \sum_{\mathbf{k},\, \omega} \frac{-\eta k_i (\mathbf{k}\cdot\mathbf{b})}{|Q|^2} h^f(\mathbf{k}, \omega) \,, \end{aligned} \tag{3.2}$$

with

$$Q = (\nu k^2 + i(\mathbf{k}\cdot\mathbf{v} - \omega))(\eta k^2 + i(\mathbf{k}\cdot\mathbf{v} - \omega)) + (\mathbf{k}\cdot\mathbf{b})^2.$$

Here $\hat{\mathbf{f}}(\mathbf{k},\omega)$ is again the Fourier transform of the body force, while $h^f(\mathbf{k},\omega) = \hat{\mathbf{f}}^*(\mathbf{k},\omega) \cdot i\mathbf{k} \times \hat{\mathbf{f}}(\mathbf{k},\omega)$ denotes the helicity in each Fourier mode of the body force. When the forcing is assumed to be random, $\hat{f}_i^* \hat{f}_j$ and h^f in eqs. (3.2) are replaced by their statistical average. Equations (2.1)-(2.2) for the AKA effect are recovered in the absence of magnetic field. At the opposite, when the small-scale dynamics is insensitive to a large-scale velocity field (Galilean invariance), the Reynolds stresses and mean e.m.f. depend on $\mathbf{b}$ but not on $\mathbf{v}$. The corresponding equations were first considered in [14] in the case of an homogeneous isotropic steady basic flow.

When in the limit of weak large-scale fields, the mean Reynolds stresses $\mathbf{r}$ and mean e.m.f. $\boldsymbol{e}$ are linearized, we may obtain an AKA instability if the basic flow is anisotropic and lacks parity-invariance and a dynamo instability at comparable scales if it possesses helicity. Furthermore, there are two different mechanisms for the saturation of these instabilities. The mean fields $\mathbf{v}$ and $\mathbf{b}$ may interact through the quadratic terms in eqs. (3.1), and such direct couplings may lead to saturation. However these terms may be zero, for example, if the large-scale fields depend on only one space coordinate [15]-[17]. In this case, saturation is only due to the nonlinear dependence of the mean e.m.f. and Reynolds stresses on the large-scale fields.

3.a. Layered large-scale magnetic fields

In this section, we restrict our attention to situations where there is no large-scale velocity field and where the large-scale magnetic field depends on only one space coordinate (for example the Z-coordinate). For simplicity we also assume a unit magnetic Prandtl number. After changes of variables analogous to those discussed in the case of the AKA effect, the equations governing the evolution of a large-scale magnetic field lying in a domain of characteristic length-scale L read [16]:

$$\begin{aligned} \partial_T B_1 &= -\alpha \partial_Z \mathcal{E}_2(B_1, B_2) + \partial_Z^2 B_1 \\ \partial_T B_2 &= \alpha \partial_Z \mathcal{E}_1(B_1, B_2) + \partial_Z^2 B_2 \\ B_3 &= 0 \; . \end{aligned} \tag{3.3}$$

In eqs.(3.3), $\alpha = R^2 L k_0$ and the magnetic field $\mathbf{B}$ (measured in units of R^{-1}) is taken 2π-periodic in Z. The normalized e.m.f. is given by:

$$\mathcal{E}_i = -\sum_{\mathbf{k},\,\omega} \frac{k_i(\mathbf{k}\cdot\mathbf{B}) h^f(\mathbf{k},\omega)}{\left|(\mathbf{k}^2 - i\omega)^2 + (\mathbf{k}\cdot\mathbf{B})^2\right|^2} \; . \tag{3.4}$$

A simple steady helical body force leading to an alpha effect instability consists of the sum of two Beltrami waves:

$$\mathbf{f}(\mathbf{x},t) = (f\sin(k_0 y), f\cos(k_0 x), f[\sin(k_0 x) + \cos(k_0 y)]) \; . \tag{3.5}$$

In this case,

$$\mathcal{E}_1 = \frac{B_2}{(B_2^2+1)^2} \quad , \quad \mathcal{E}_2 = \frac{B_1}{(B_1^2+1)^2} \; . \tag{3.6}$$

The Reynolds stresses vanish and any large-scale velocity perturbation decays by viscous dissipation. In contrast, a large-scale magnetic perturbation is exponentially amplified.

When for large α, eqs.(3.3),(3.6) are integrated numerically starting with weak initial conditions, an evolution qualitatively similar to that of the AKA effect is observed. At early times, the magnetic field is dominated by the growth of the linearly most unstable modes. At larger amplitudes, the nonlinearities become important and magnetic energy is transferred to successively larger and larger scales in an inverse cascade.

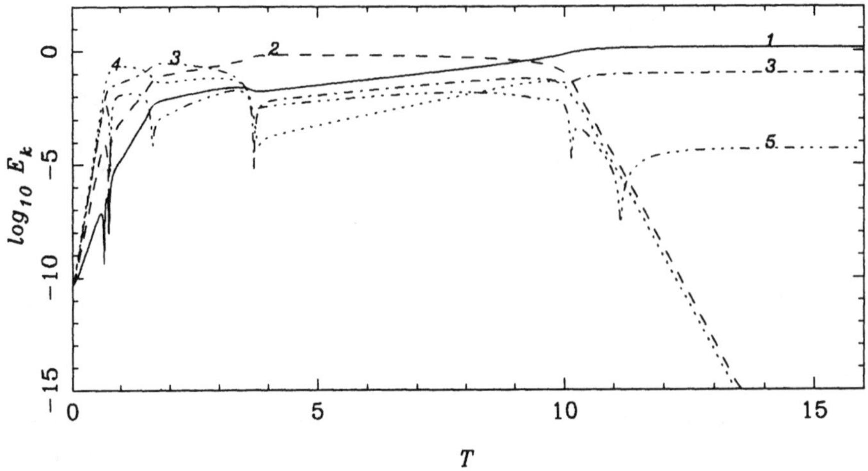

Fig. 4 : Evolution of energy modes (see label on the curves) of the layered magnetic field generated by alpha effect for the body force (3.5) with $\alpha = 8$.

This phenomenon is illustrated in Fig. 4 for the body force (3.5) with $\alpha = 8$. The (rescaled) energy E_K of the large-scale magnetic modes is plotted against time for $K = 1, ..., 5$. A simulation for a larger number of unstable modes ($\alpha = 20$) is presented in [18]. Like in the case of the AKA effect, during the inverse cascade the system is attracted by a sequence of steady cellular solutions with the form of hyperbolic fixed points, stable to perturbations of smaller scales and unstable to perturbations of larger scales. Ultimately, the system stabilizes on the steady cellular solution corresponding to the largest available scale. For this solution, the Fourier modes of even wavenumber vanish, a property reflecting a symmetry $\mathbf{B}(Z+\pi) = -\mathbf{B}(Z)$ of the solution in physical space.

When the body force is random, homogeneous and isotropic, the mean e.m.f. has the form:

$$\boldsymbol{\mathcal{E}} = g(|\mathbf{B}|)\mathbf{B} \ , \tag{3.7}$$

where g is a scalar function depending on specific hypotheses about the forcing. Phenomenological models where g is given *a priori* are considered in [15],[17]. A specially interesting model is that introduced by Kraichnan [19],[20], in which the body force injects fluid helicity into the system at a constant rate, an idealization commonly used in turbulence theory. Kraichnan achieves this by taking a body force which is a white noise in time; however a constant helicity injection rate may also be prescribed within the framework of multiple-scale dynamics by assuming that the body force is advected by the large-scale flow in order to preserve Galilean invariance. In this case [16]:

$$g(|\mathbf{B}|) = (1 - |\mathbf{B}|^{-1} \arctan(|\mathbf{B}|))\frac{1}{|\mathbf{B}|^2} \; . \tag{3.8}$$

The mean e.m.f. given in [19] and [20] is recovered in the limit of strong magnetic field.

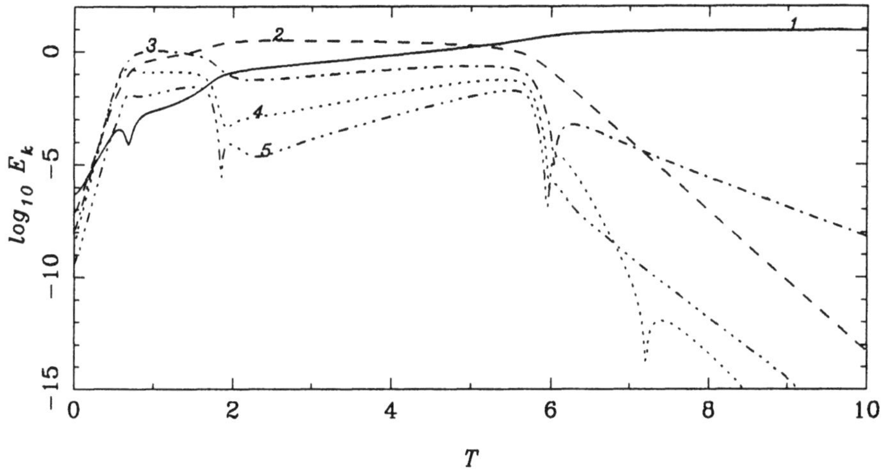

Fig. 5 : Same as Fig. 4 in the case of an isotropic forcing corresponding to a prescribed helicity injection rate with $\alpha = 8$.

Figure 5 shows the dynamics which develops with $\alpha = 8$. We again observe an inverse cascade for the magnetic field. It is easily checked that the system admits steady cellular solutions of the form of Beltrami waves associated to different wavenumber. As a result of the inverse cascade, the system eventually evolves to a Beltrami wave of unit wavenumber.

These observations are not specific of a constant helicity injection rate. A very similar evolution is observed with a prescribed isotropic helical body force assumed to be advected by the flow in order to preserve Galilean invariance [16].

3.b. Three-dimensional large-scale magnetic fields

The body forces used in Section 3.a have also been considered for initial large-scale perturbations depending on the three space coordinates. In this case, the Reynolds stresses do not vanish. When the basic flow is stirred by a random homogeneous isotropic body force associated to a prescribed helicity injection rate, the equations for the large-scale velocity and magnetic field read

$$\begin{aligned} &\partial_T V_i + \alpha \partial_{X_j}\left(R^{-2}(V_i V_j - B_i B_j) + \mathcal{R}_{ij}\right) = -\partial_{X_i} P + \Delta V_i \\ &\partial_T \mathbf{B} = \alpha \nabla \times \left(R^{-2} \mathbf{V} \times \mathbf{B} + \mathcal{E}\right) + \Delta \mathbf{B} \\ &\partial_{X_i} V_i = 0 \quad ; \quad \partial_{X_i} B_i = 0 \; , \end{aligned} \tag{3.9}$$

with

$$\mathcal{R}_{ij} = 3\{3 - (1 + 3|B|^{-2})|B| \arctan(|B|)\} \frac{B_i B_j}{|B|^4}$$
$$\mathcal{E}_i = 3(1 - |B|^{-1} \arctan(|B|)) \frac{B_i}{|B|^2} , \quad (3.10)$$

where velocity and magnetic field are again measured in units of R^{-1}. Figures 6 and 7 display the result of the numerical integration of eq.(3.9) for $R^2 = 1/10$ and $\alpha = 8$. The initial conditions consist in weak three-dimensional kinetic and magnetic noises.

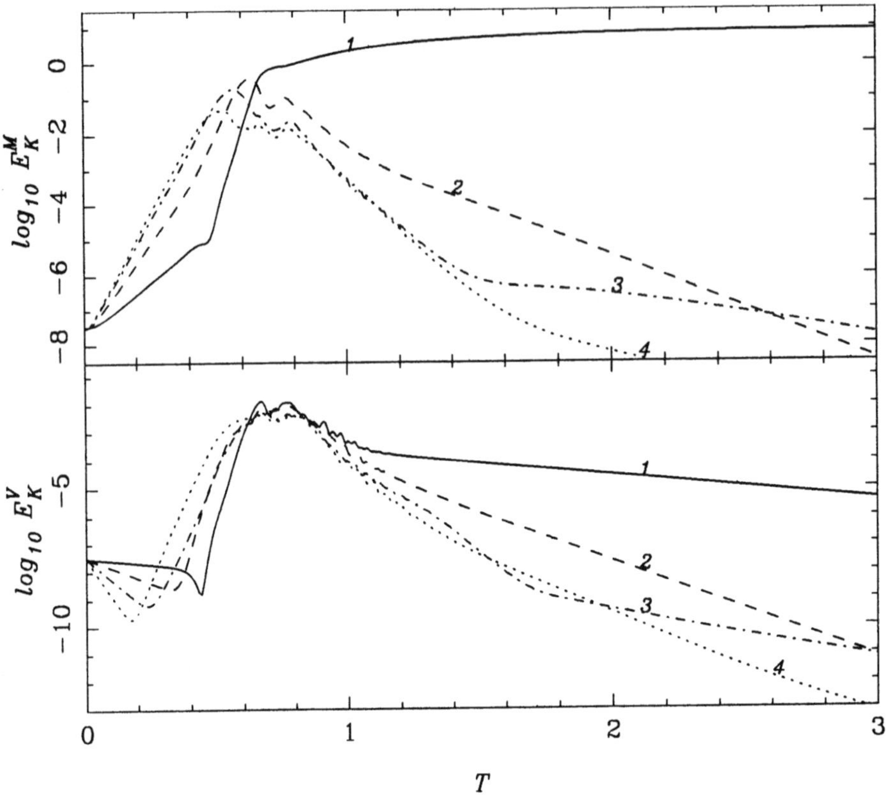

Fig. 6 : Evolution of shell averaged magnetic and kinetic energy modes of initially three-dimensional large-scale perturbations when the small-scale flow is stirred by the same forcing as in Fig. 5 with $\alpha = 8$.

In Figure 6, the shell-averaged magnetic and kinetic energy spectra are shown versus time. In Fig.7 the energy of a few individual magnetic or kinetic Fourier modes belonging to the shells $K = 1, 2$ and 3 are plotted versus time. At early time, magnetic modes depending on the three coordinates are amplified exponentially as a result of the linear alpha instability. For the considered value of α, the linearly most unstable modes correspond to a wavenumber $K = 4$. When the magnetic field starts growing, the velocity field is also amplified under the effect of Lorentz force and Reynolds stresses. A rapid magnetic inverse cascade is visible in Fig.6 after the linear phase: magnetic energy modes corresponding to larger and larger scales become dominant successively. Figs.7 show that during this process, velocity and magnetic field depend on the three space coordinates. Nevertheless, when as a result of the inverse cascade, the magnetic energy shell $K = 1$ becomes dominant, the magnetic energy in the other shells falls down rapidly. The velocity also decays and the magnetic field eventually reduces to a pure Beltrami wave of a unit wavenumber. The system thus evolves to the same solution as in the case of an initially layered magnetic field, except that the wavevector may now be directed along any of the coordinate axes with an equal probability.

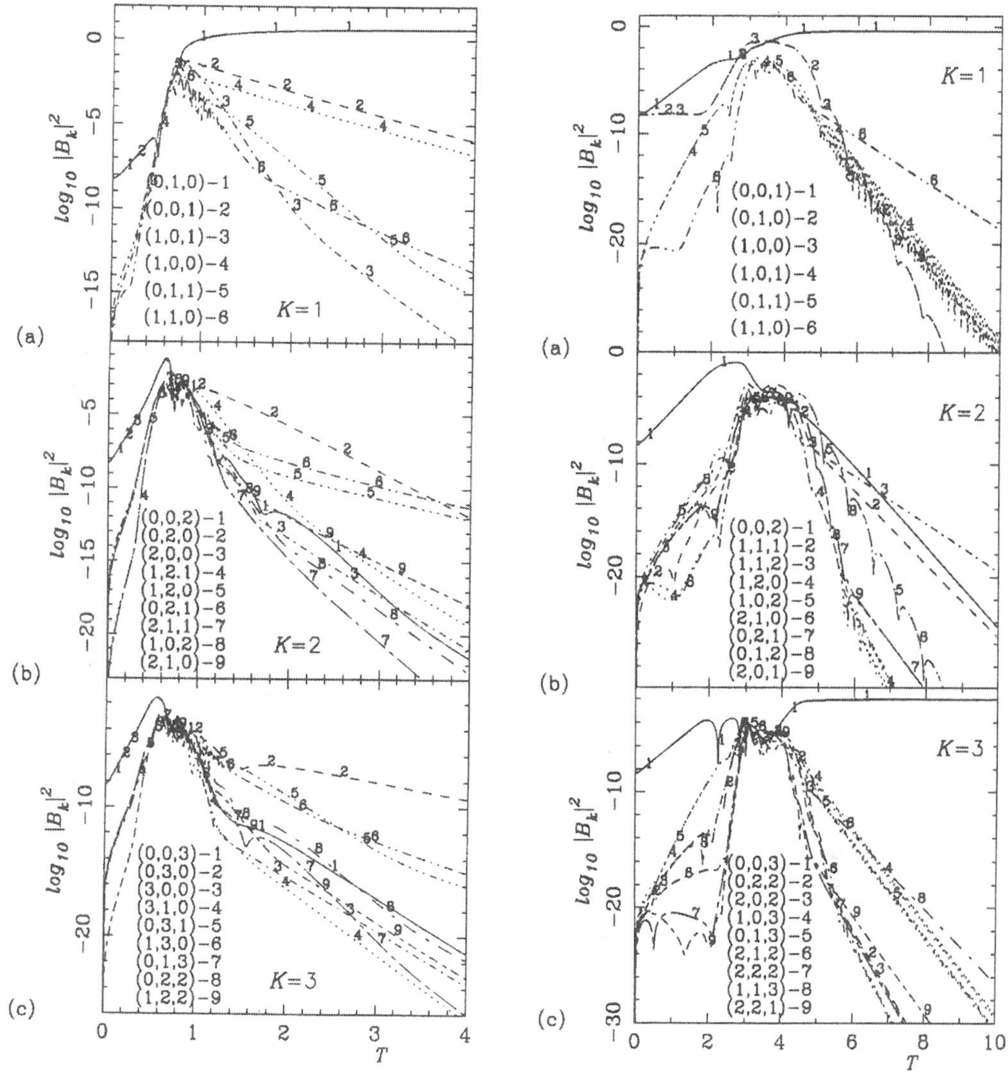

Fig. 7 : Evolution a few individual magnetic and kinetic modes of the shells $K = 1, 2, 3$ for the conditions of Fig. 6.

The three-dimensional transient nevertheless depends on the Reynolds number. A numerical integration with the same α but $R^2 = 1/25$ shows that the maximum of the eventually damped modes has a lower amplitude and occurs at an earlier time.

Relaxation to a layered solution is also observed with a prescribed isotropic helical body force, assumed to be advected by the large-scale flow in order to preserve Galilean invariance (see section 3.a).

In the case of the deterministic body force (3.5), an initially three-dimensional large-scale perturbation evolves according to eq.(3.9) with

$$
\begin{aligned}
\mathcal{R}_{11} = \mathcal{R}_{13} &= \frac{1}{2}\frac{V_2^2 - B_2^2 - 1}{(B_2^2 - V_2^2 + 1)^2 + 4V_2^2} \quad ; \quad \mathcal{R}_{12} = 0 \\
\mathcal{R}_{22} = \mathcal{R}_{23} &= \frac{1}{2}\frac{V_1^2 - B_1^2 - 1}{(B_1^2 - V_1^2 + 1)^2 + 4V_1^2} \quad ; \quad \mathcal{R}_{33} = \mathcal{R}_{11} + \mathcal{R}_{22} \\
\mathcal{E}_1 &= \frac{B_1}{(B_1^2 - V_1^2 + 1)^2 + 4V_1^2} \quad ; \quad \mathcal{E}_2 = \frac{B_2}{(B_2^2 - V_2^2 + 1)^2 + 4V_2^2} \\
\mathcal{E}_3 &= 0,
\end{aligned}
\tag{3.11}
$$

where velocity and magnetic field are again measured in units of R^{-1}.

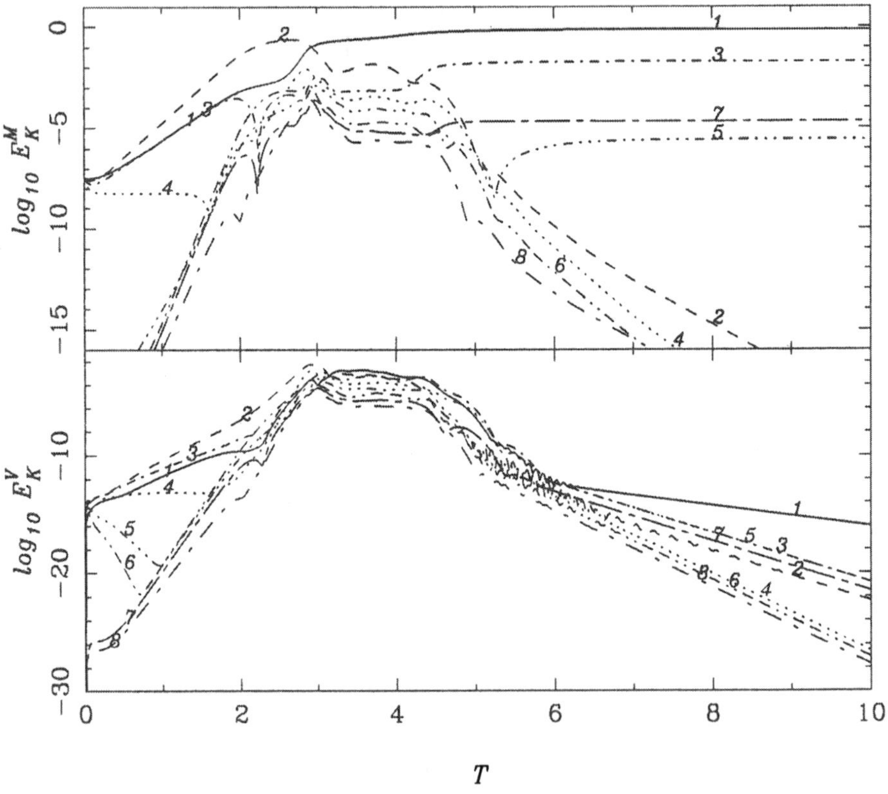

Fig. 8 : Same as Fig. 6 for body force (3.5) with $\alpha = 4$.

Figure 8 displays the magnetic and kinetic energy modes versus time for $\alpha = 4$ and $R^2 = 1/10$. The details of the spectral distribution is presented in Fig.9 which shows the evolution of the energy of a few individual modes in the first three shells. It appears that after a three-dimensional transient which takes place just after the linear phase, the velocity tends to zero and the magnetic field relaxes to the symmetric layered steady state obtained in section 3.a.

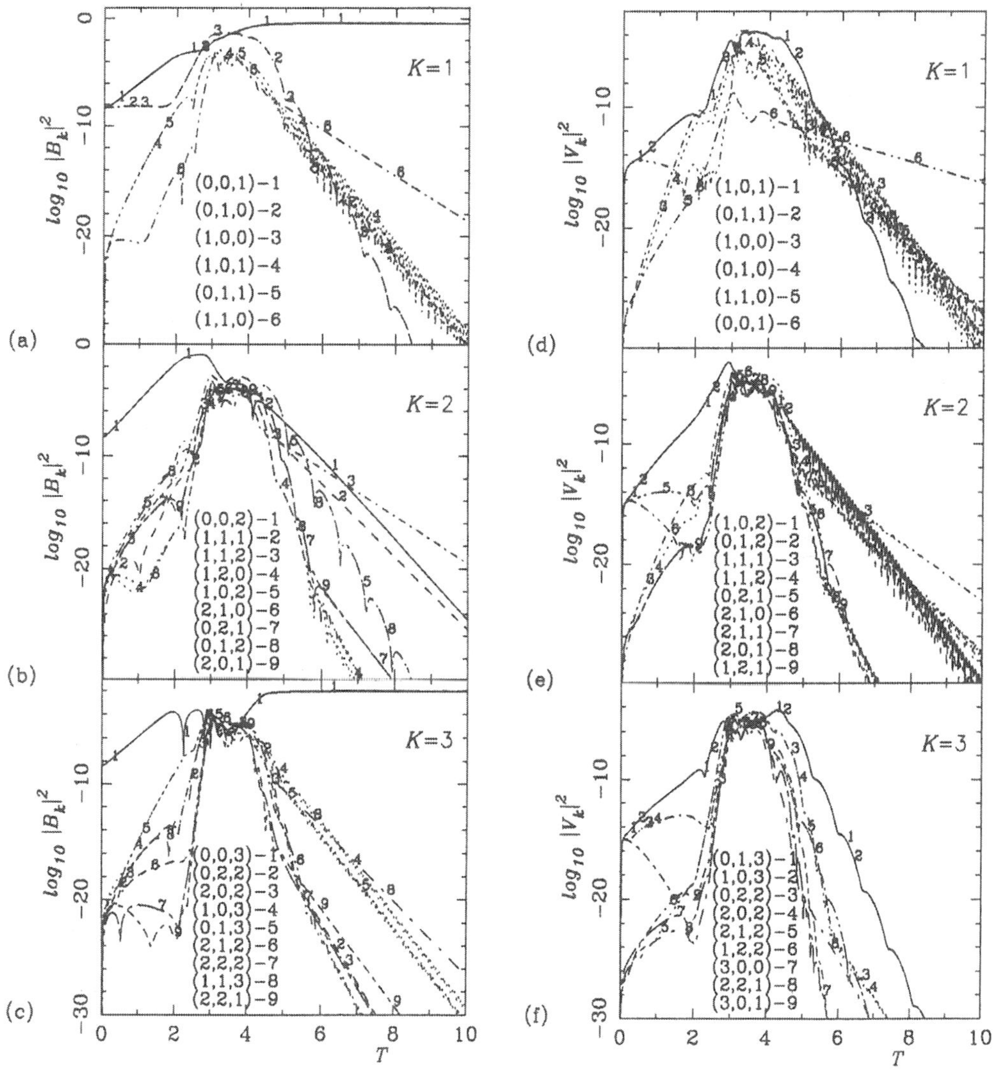

Fig. 9 : Same as figure 7 for body force (3.5) with $\alpha = 4$.

A more complex evolution is seen with $\alpha = 8$ (Figs. 10 and 11). Just after the linear phase, we observe the dominance of the modes $(1,0,0)$ and $(3,0,0)$. This regime

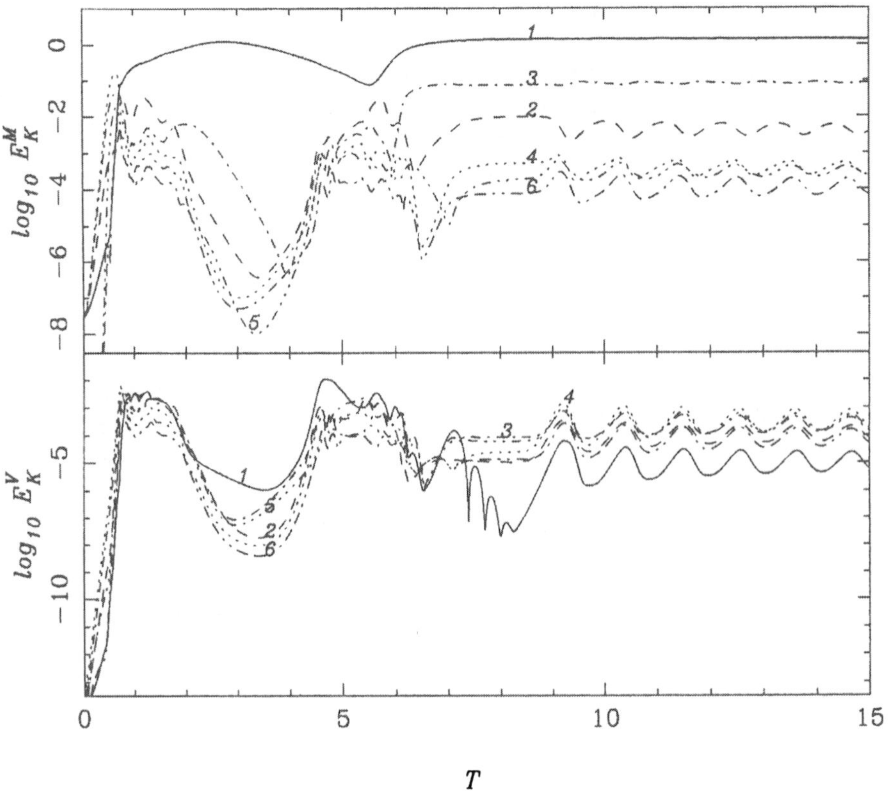

Fig. 10 : Same as figure 6 for body force (3.5) with $\alpha = 8$.

is in fact a transient and around $T \approx 5$, these modes rapidly decay. The system approaches the steady symmetric solution depending only on the Z-variable discussed in Section 3.a. Nevertheless, three-dimensional velocity and magnetic field noises displaying slowly damped regular oscillations in time, are seen to survive at a very relatively low level. This contrasts with the regime obtained with $\alpha = 4$ and also with the case of isotropic body forces where three-dimensional modes rapidly tend to zero. Integration of the eqs.(3.9)-(3.10) with the same value of α but $R^2 = 1/25$ shows the same qualitative behavior but indicates that the three-dimensional components are reduced when the Reynolds number is decreased. Their precise dynamics is probably outside the scope of eq.(3.9) which is restricted to leading order in Reynolds number. When keeping $R^2 = 1/10$, eqs.(3.9)-(3.10) are integrated with $\alpha = 16$, we obtain a similar regime but the transient during which the solution is not dominated by the Z-modes, is significantly reduced.

It is noticeable that with all examples we have considered we never obtained a regime where a significant three-dimensional velocity field develops and by interacting with the magnetic field, leads to its saturation. Such a dynamics was studied in [23],[24] (see also [25]) for a large-scale velocity field resulting from a balance between the Lorentz force and the Coriolis force for a rotating fluid in a bounded domain.

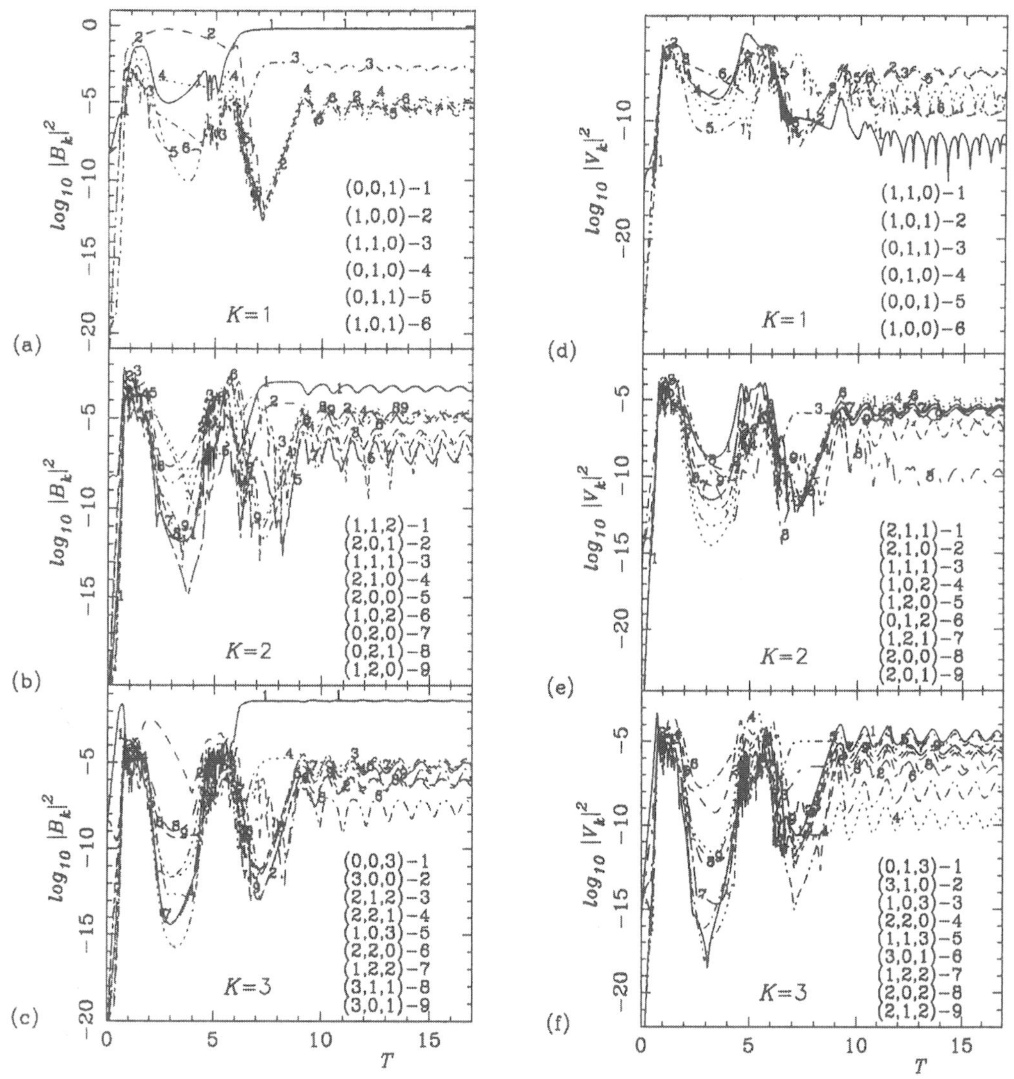

Fig. 11 : Same as figure 7 for body force (3.5) with $\alpha = 8$.

4. Coexisting alpha and AKA effects in conducting fluids

A simple example of body force which when stirring a conducting fluid, leads to both a linear AKA instability and a linear alpha instability reads:

$$\begin{aligned} f_1 &= f\cos(k_0 y + \nu k_0^2 t) \quad ; \quad f_2 = f\cos(k_0 x - \nu k_0^2 t); \\ f_3 &= (f_1 + f_2) + f\{\sin(k_0 x - \nu k_0^2 t) - \sin(k_0 y + \nu k_0^2 t)\}. \end{aligned} \tag{4.1}$$

4.a. Layered large-scale flows and magnetic fields

Using the same units as in Section 3, eqs. (2.3) and (2.5) become [(1)]

$$
\begin{aligned}
\partial_T B_1 &= -4\alpha\partial_Z \frac{B_2}{(B_2^2-(V_2+1)^2+1)^2+4(V_2+1)^2} + \partial_Z^2 B_1\\
\partial_T B_2 &= 4\alpha\partial_Z \frac{B_1}{(B_1^2-(V_1-1)^2+1)^2+4(V_1-1)^2} + \partial_Z^2 B_2\\
\partial_T V_1 &= 2\alpha\partial_Z \frac{B_2^2-(V_2+1)^2-1}{(B_2^2-(V_2+1)^2+1)^2+4(V_2+1)^2} + \partial_Z^2 V_1\\
\partial_T V_2 &= 2\alpha\partial_Z \frac{B_1^2-(V_1-1)^2-1}{(B_1^2-(V_1-1)^2+1)^2+4(V_1-1)^2} + \partial_Z^2 V_2.
\end{aligned}
\tag{4.2}
$$

when the large-scale velocity $\mathbf{V}$ and magnetic field $\mathbf{B}$ are again assumed 2π-periodic in Z. As previously the parameter α controls the number of linearly unstable and velocity field modes in the periodic system.

Equations (4.2) were integrated in [18] for increasing values of α. An inverse cascade is always observed at early times. However, whereas in the absence of coupling with a velocity field, the magnetic field eventually relaxes to a steady state (Section 3.a), a variety of dynamical regimes (steady, periodic or chaotic) are obtained in the present case.

Figure 12 displays the time evolution of magnetic and kinetic energy modes, $E_K^M = \frac{1}{2}|\mathbf{B}_K|^2$ and $E_K^V = \frac{1}{2}|\mathbf{V}_K|^2$ for various values of α corresponding to steady, periodic and chaotic regimes. For $\alpha = 4$ (Fig. 12a), a steady regime is obtained for which the even modes of the magnetic field and the odd modes of the velocity vanish. In physical space, the magnetic field thus displays a symmetry $\mathbf{B}(Z+\pi) = -\mathbf{B}(Z)$ and the velocity is π-periodic. For $\alpha = 10$ (Fig. 12b), a periodic regime displaying simple oscillations is obtained. In physical space, velocity and magnetic fields show the same symmetries as the steady solution. Furthermore, a transient of several time units is visible between the linear phase and the onset of the periodic oscillations. This transient reflects the existence of the steady state discussed above, and becomes longer as α is closer to the bifurcation value which is slightly in excess of 7.

The regime shown in Fig.12c for $\alpha = 30.5$ is also strictly periodic, but significantly more complex: the solution displays sudden transitions between time intervals of quiescent evolution and bursts of violent oscillations. For larger values of α, the solution loses its periodicity in time, and a transition to chaos is observed, although intervals of quiescent evolution still survive (Fig.12d). Note that oscillatory dynamos have also been discussed in the context of simple nonlinear models for solar dynamos in the presence of differential rotation (omega effect) ([26] and references therein).

---

(1) Eq. (4.2) should replace eq.(3.2) of ref.[18] and eq.(9) of ref.[21] where a few coefficients were misprinted.

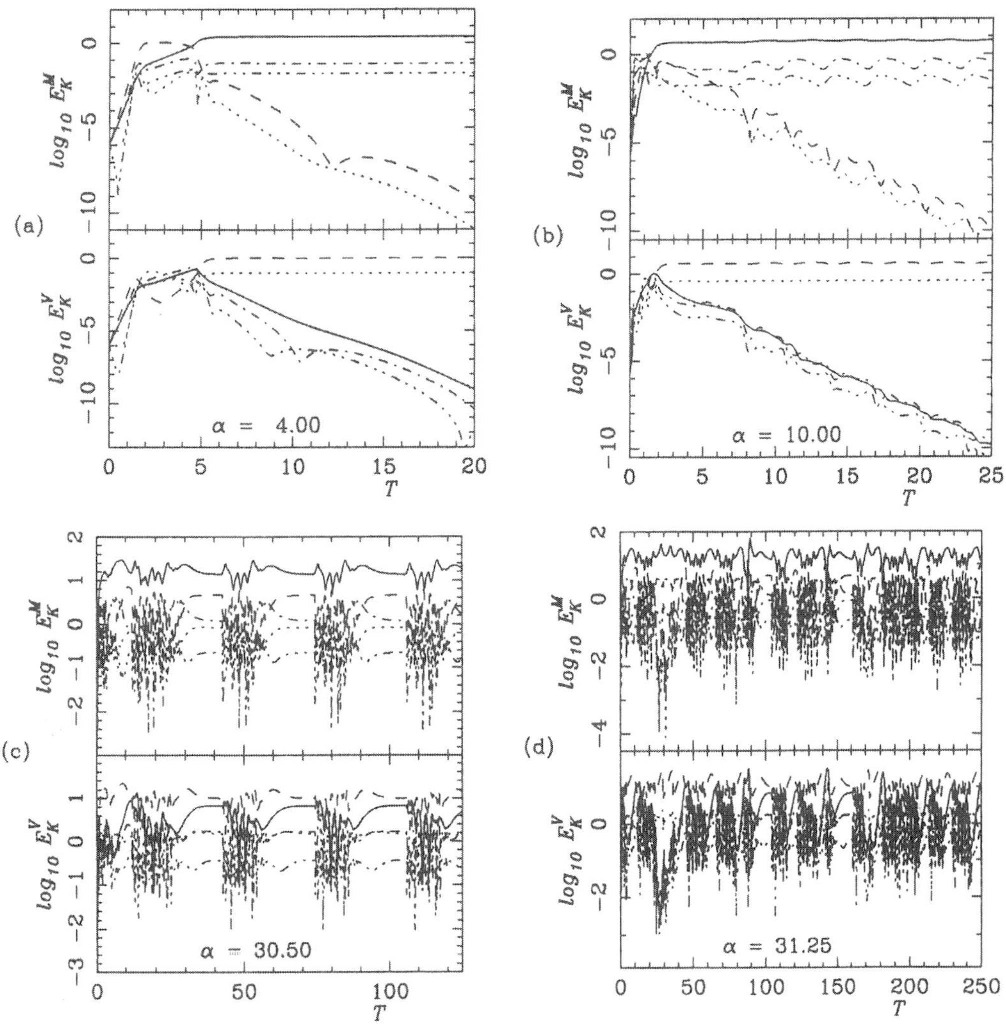

Fig. 12 : Evolution of magnetic and kinetic energy modes E_K^M and E_K^V for the body force (4.1) and various values of α in the case of layered solutions; K=1 (—), K=2 (- - -), K=3 (-.-.-), K=4 (....), K=5 (-....-).

4.b. Three-dimensional large-scale flows and magnetic fields

When for the same body force (4.1) as in section 4.a, the large-scale perturbations are three-dimensional, the large-scale dynamics is governed by eqs.(3.9) with

$$
\begin{aligned}
\mathcal{R}_{11} = \mathcal{R}_{13} &= 2\frac{(V_2+1)^2 - B_2^2 + 1}{(B_2^2-(V_2+1)^2+1)^2+4(V_2+1)^2} \quad ; \quad \mathcal{R}_{12}=0 \\
\mathcal{R}_{22} = \mathcal{R}_{23} &= 2\frac{(V_1-1)^2 - B_1^2 + 1}{(B_1^2-(V_1-1)^2+1)^2+4(V_1-1)^2} \quad ; \quad \mathcal{R}_{33}=\mathcal{R}_{11}+\mathcal{R}_{22} \\
\mathcal{E}_1 &= 4\frac{B_1}{(B_1^2-(V_1-1)^2+1)^2+4(V_1-1)^2} \\
\mathcal{E}_2 &= 4\frac{B_2}{(B_2^2-(V_2+1)^2+1)^2+4(V_2+1)^2} \\
\mathcal{E}_3 &= 0.
\end{aligned}
\tag{4.3}
$$

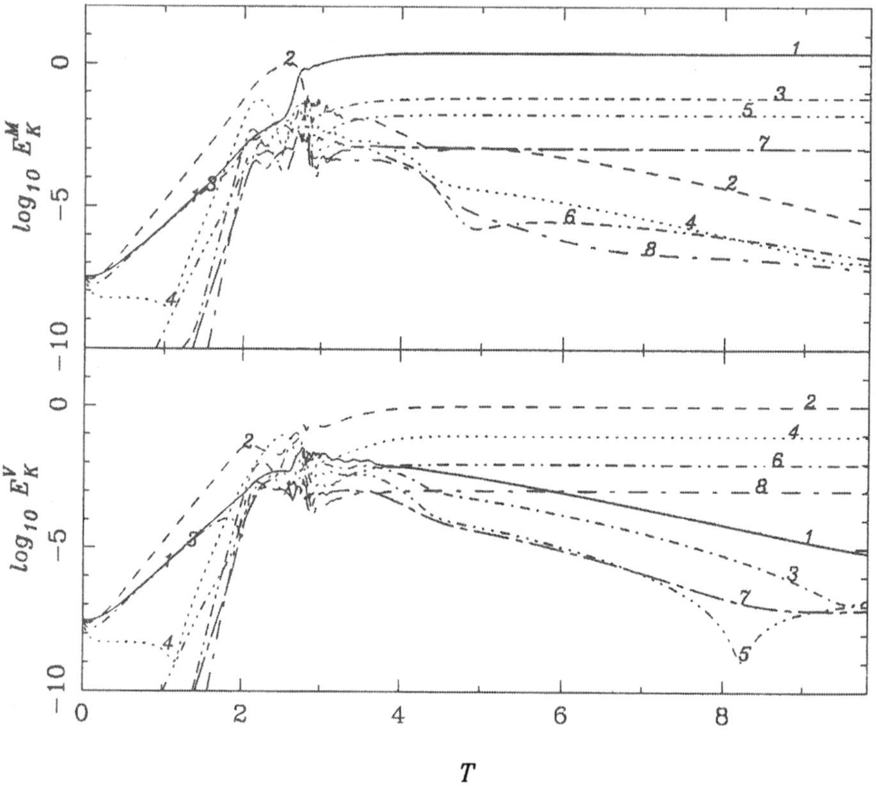

Fig. 13 : same as Fig. 6 for body force (4.1) with $\alpha = 4$.

This system was first integrated for $\alpha = 4$ and $R^2 = 1/10$. We observe in Figs. 13 and 14 that after a three-dimensional transient, the velocity and the magnetic field relax to the steady layered solution discussed in Section 4.a. The magnetic field displays the symmetry $\mathbf{B}(Z+\pi) = -\mathbf{B}(Z)$ while the velocity is π-periodic.

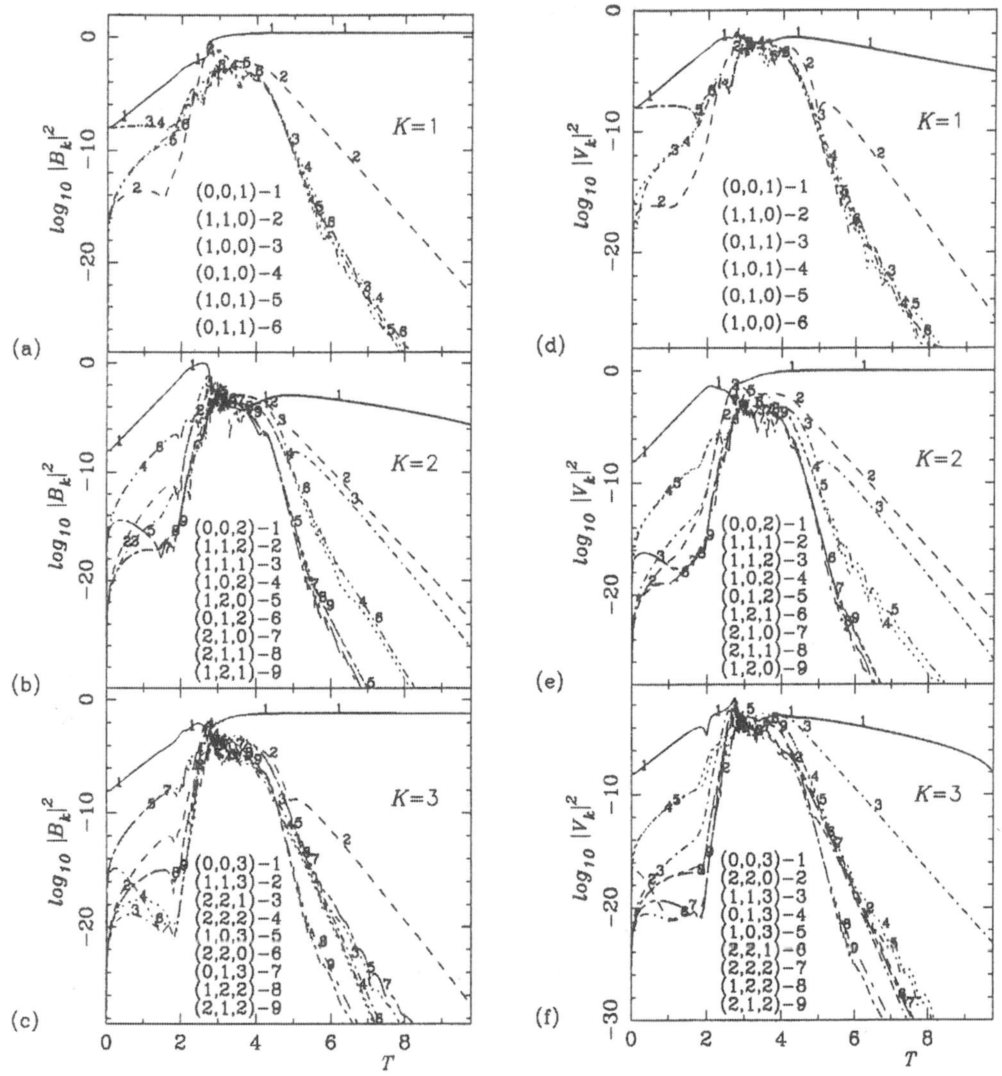

Fig. 14 : Same as Fig. 7 for body force (4.1) with $\alpha = 4$.

The problem was also considered with $\alpha = 10$ and the same Reynolds as above (Figs. 15 and 16). We observe that at the end of the integration, the solution approaches the layered solution described in section 4.a. A low three-dimensional background is nevertheless visible. At this time, both solutions are in the quasi-steady phase which develops before the appearence of the periodic oscillations and the onset of the symmetries seen at longer times in the case of the layered solution. Note that the rapid oscillations visible in Fig. 15, are numerical. Indeed, even when the solution is dominantly one-dimensional, the problem is stiff because the large coefficients multiplying direct couplings such as advection and Lorentz force.

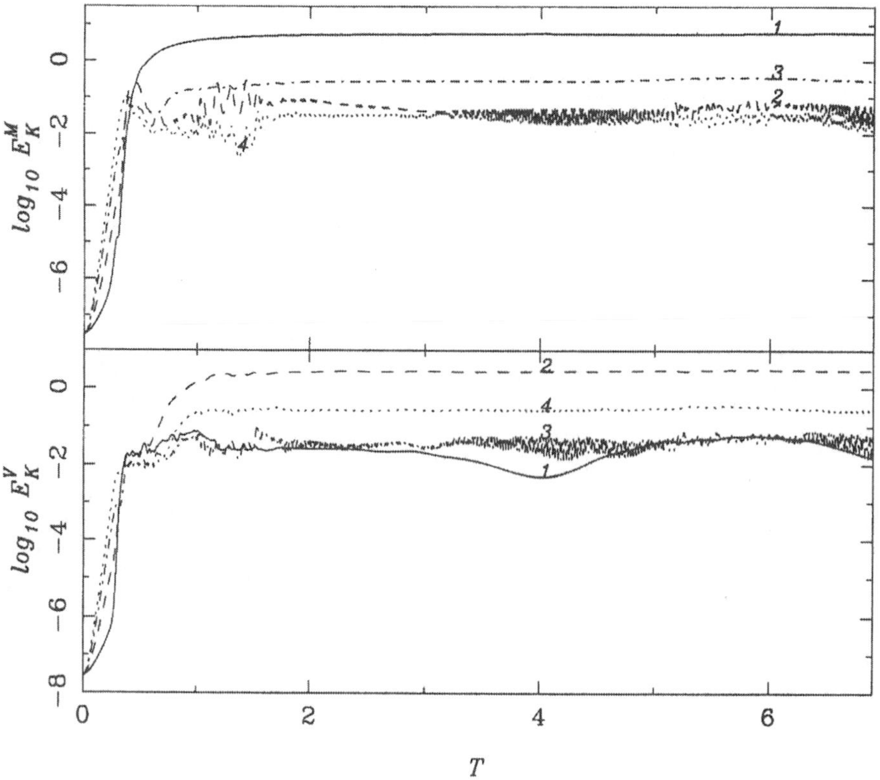

Fig. 15 : same as Fig. 6 for body force (4.1) with $\alpha = 10$.

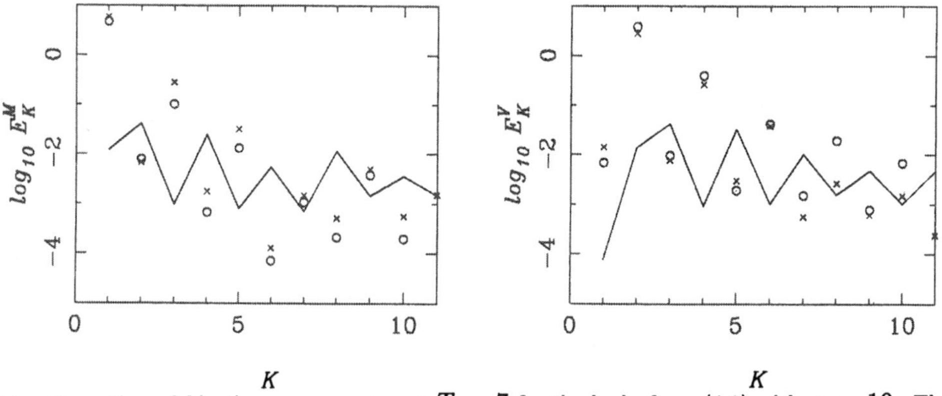

Fig. 16 : Magnetic and kinetic energy spectra at $T = 7$ for the body force (4.1) with $\alpha = 10$. The symbols $\times$ indicate the energy level in magnetic modes (0,0,$2n+1$) and in kinetic modes (0,0,$2n$). The solid line gives the cumulated energy in all other modes. For comparison, the symbols o show the magnetic and kinetic energy modes for the layered solution during the quasi-steady regime which occurs before the onset of periodic oscillations.

To conclude this section, we stress that for initially three-dimensional large-scale perturbation, the velocity of the secondary flow is significantly larger in the case of coexisting alpha and AKA instabilities than in the case of pure AKA effect.

5. Summary

In contrast with instabilities driven by a negative eddy viscosity, both AKA and alpha effects can develop in low Reynolds number flows, provided the system is sufficiently extended. It follows that the large-scale dynamics can be computed for a large class of body forces stirring the basic flow. Although the AKA instability refers to the amplifications of a seed large-scale velocity, while the alpha instability produces a dynamo, these phenomena display similar properties in the case where the large-scale perturbations depend on a unique space coordinate. Due the existence of a sequence of cellular steady solutions which are unstable to large-scale perturbations, the nonlinear dynamics is dominated by an inverse cascade leading to the formation of strongly energetic structures of larger and larger scales, up to the limits of the system. Furthermore, when both instabilities are present, their coupling can lead to various time-dependent regimes where velocity and magnetic field can display periodic or chaotic oscillations.

The nonlinear dynamics of the AKA instability is however significantly modified when the large-scale initial perturbation is genuinely three-dimensional. Self-advection of the large-scale flow is the dominant nonlinearity and it limits the growth of the secondary flow to a relatively low level. An inverse cascade nevertheless develops, leading to a three-dimensional flow dominated by the largest scales.

The situation is different for the alpha dynamo. The Lorentz force does not produce a significant large-scale flow and one-dimensional solutions are generally dynamically stable. This observation provides a basis to various alpha dynamo models, often constructed in an astrophysical context, where the nonlinear saturation is mainly due to the decay of the effective alpha coefficient with the intensity of the large-scale magnetic field [15], [17].

Acknowledgments

The three-dimensional numerical simulations were done on the CRAY2 of the Centre de Calcul Vectoriel pour la Recherche (Palaiseau). The work of B.G. was supported by a grant from the Direction des Recherches Etudes et Techniques (901515/A000).

References

[1] E.N. Parker, Astrophys. J. **122** (1955) 293.
[2] M. Steenbeck, F. Krause and K.-H Rädler, Z. Naturforsch **21A** (1966) 369.
[3] H.K. Moffatt, *Magnetic Field Generation in Electrically Conducting Fluids*, Cambridge University Press (1978).
[4] A.D. Gilbert, U. Frisch and A. Pouquet, Geophys. Astrophys. Fluid Dyn. **42** (1988) 151.
[5] U. Frisch, Z.-S. She and P.-L. Sulem, Physica **28D** (1987) 382.
[6] P.-L Sulem, Z.-S. She, H. Scholl and U. Frisch, J. Fluid Mech. **205** (1989) 341.
[7] B. Galanti and P.-L. Sulem, Phys. Fluids **A 3** (7) (1991) 1778.
[8] B. Dubrulle and U. Frisch, Phys. Rev. A **43** (1991) 5355.
[9] A.A. Nepomniashchii, Prikl. Math. Mech. **40** (1976) 886; J. Applied Math. Mech. **40** (1976) 836.

[10] G.I. Sivashinsky, Physica **17D** (1985) 243.
[11] Z.-S. She, Prog. Astronaut. Aeronaut. **112** (1988) 374.
[12] U. Frisch, Z.-S. She and P.-L. Sulem, Prog. Astronaut. Aeronaut. **112** (1988) 262.
[13] Z.-S. She, Phys. Lett. A. **124** (1987) 161.
[14] G. Rüdiger, Astron. Nachr. **295** (1974) 275.
[15] F. Krause and K.-H Rädler, *Mean-field magnetohydrodynamics and dynamo theory*, Pergamon Press, 1980.
[16] A.D. Gilbert and P.-L. Sulem, Geophys. Astro. Fluid Dyn., **5** (1990) 243.
[17] Ya.B. Zeldovich, A.A. Ruzmaikin and D.D. Sokoloff, *Magnetic fields in astrophysics*, Gordon and Breach, 1983.
[18] B. Galanti, P.-L. Sulem and A.D. Gilbert, Physica **47D** (1991) 416.
[19] R.H. Kraichnan, Phys. Rev. Lett. **42** (1979) 1677.
[20] M. Meneguzzi, U. Frisch and A. Pouquet, Phys. Rev. Lett. **47** (1981) 1060.
[21] B. Galanti, A.D Gilbert and P.-L. Sulem, in *Topological Fluid Mechanics*, H.K. Moffatt and A. Tsinober eds. p 138, Cambridge University Press (1990).
[23] W.V.R. Malkus and M.R.E. Proctor, J. Fluid Mech. **67** (1975) 417.
[24] M.R.E. Proctor, J. Fluid Mech. **80** (1977) 769.
[25] A. Brandenburg and I. Tuominen Geophys. Astrophys. Fluid Dyn. **49** (1989) 129.
[26] F. Krause and R. Meinel, Geophys. Astrophys. Fluid Dyn. **43** (1988) 95.

FROM QUANTUM FIELDS TO FRACTAL STRUCTURES: INTERMITTENCY IN PARTICLE PHYSICS

by

R. *Peschanski**

CERN, Theory Division

abstract

Some features and theoretical interpretations of the intermittency phenomenon observed in high-energy multi-particle production are recalled. One develops on the various connections found with fractal structuration of fluctuations in turbulence, spin-glass physics and aggregation phenomena described by the non-linear Smoluchowski equation. This may lead to a new approach to quantum field properties.

Invited talk at the Workshop on
"Large-scale structures in non-linear physics"
La Citadelle, Villefranche-sur-mer, France, January 13-18, 1991

- -

**Adress*: Theory Division, CERN, CH-1211 Geneva 23, Switzerland.
On leave of absence till September 1991 from:
Service de Physique Théorique de Saclay
91191 Gif-sur-Yvette Cedex, France.

1. Introduction: The intermittency phenomenon in particle physics

Intermittency has been first invoked in Particle Physics in the study[1]of *dynamical* fluctuations observed in the particle density distribution in small intervals (bins) in relativistic phase-space variables. These fluctuations have been called *dynamical* by contrast with the purely *statistical* ones due to the limited (often small) number of particles registered in small bins. Indeed, current high-energy collisions produce dozens to hundreds of particles per event, which is large considered by particle physics standards but is rather few for applying statistical concepts, such as intermittency, without care. Hence, the first step of ref. [1] was to propose a method for distinguishing *dynamical* fluctuations from the *statistical* "noise". Assuming a simple Poissonian noise, or Bernouilli one if the total multiplicity of events is constrained, one can write a formula for the normalised factorial moments of the multiplicity distribution, namely:

$$\langle F_q \rangle \equiv \frac{\langle k_m\,(k_m - 1)\ldots(k_m - q + 1)\rangle}{\langle k_m \rangle^q} = \frac{\langle \rho_m^q \rangle}{\langle \rho_m \rangle^q}, \tag{1}$$

where k_m is the observed number of particles in the bin [m] and the "density" ρ_m , with its "ordinary" moments, would correspond to the absence of statistical bias. This is nothing other than a simple application of the difference between frequency and probability weight in Statistics. The assumption about the noise has been surprisingly wel adapted to the study of various reactions[2], such as $e^+ - e^-$annihilations into hadrons and those involving incident hadrons and/or nuclei. In all cases, realistic models except for eventual dynamical fluctuations were found consistent with the noise assumption.

The main yet unexpected outcome of these studies is the power-law behaviour of moments with the binsize δ in the suitable variables, namely the *rapidity* in the standard case displayed on the Figure 1. *Rapidity*[3]is the variable analoguous to the velocity in a relativistic (Lorentzian) frame of reference oriented along the preferred production axis. It is related to the angular distribution of particles with respect to the axis in the total center-of-mass frame. One finds:

$$\langle F_q \rangle \sim \delta^{-f_q} \tag{2}$$

where f_q is constant in some binsize range. In fact the inverse binsize δ^{-1} is a measure of the chosen experimental resolution. Note that other variables, such as the azimuthal angle with respect to the axis, has revealed the same behaviour, and it is even more pronounced, in a two or three-dimensional analysis[4] combining these variables. As we shall develop further on, formula (2) was suggestive of a fractal behaviour of fluctuations similar to fluid turbulence and related to intermittency, i.e. the property of dynamical fluctuations with a hierarchy of scales, showing a fractal dependence on the experimental resolution.

In the next section **2**, we show how the quantum field theory of particle interactions is put into question by the intermittency phenomenon, impredictably. In section **3**, one introduces the random cascading models by analogy with fluid turbulence, but with some clear difference due to relativistic kinematics. Section **4** discusses the connections with phase transitions and also spin-glass systems using the fruitful example of Statistical Mechanics. Section **5** is devoted to the unexpected relations with fractal growth, aggregation processes, and the non-linear Smoluchowski equation. Conclusions and outlook are given in section **6**.

It must be clear to the reader that the following sections mainly reflect our personal views, much influenced of course by the numerous friends and colleagues with whom the author shares the passion for the present subject, and whose names can be found in part in the references list. It is a pleasure to have been invited to talk at the workshop since it allows one to be less rigorous but perhaps more intuitive than for a registered paper. I also took this opportunity to add new pieces of material concerning aggregation phenomena which seem to fit well the subject of the workshop. Many thanks for the organizers of this nice interdisciplinary meeting.

2. Intermittency and the quantum field theory of particles.

The experimentally observed behaviour compatible with (2) has been largely commented in the theoretical literature[5]. So-called "conventional" mechanisms have been proposed to take into account the behaviour (2) while using already known models. Depending on the reaction, short-range and/or long-range rapidity correlations, a hierarchy of resonance decays, and the Bose-Einstein enhancement effect for identical pions, have been invoked[5]. However the higher-dimensional analysis[4] is difficult to explain in this context and the "universal" presence of the phenomenon is difficult to understand, since, "conventionnally", different mechanisms are to be introduced for different reactions. However a "conventional" model is not completely excluded but, as we shall see now, this is another way to formulate our ignorance of quantum fields at strong coupling.

Particle interactions are expected to be well-described by quantum field theories. Indeed the so-called "standard model" of fundamental interactions has met considerable success in the unified description of electro-magnetic interactions. In the domain of strong interactions, the success is also remarkable with the important restriction that the theory remains incomplete: at short time-distances the interactions between quarks and gluons _ the fundamental building blocks of matter or *partons* _ is well understood; However at longer distances the transformation of quarks and gluons into the observed hadrons (pions, nucleons, resonances etc..) is rather problematic. Technically, the difficulty is in treating the strong coupling regime of Quantum-Chromo-Dynamics (QCD), the interaction theory of quarks and gluons. Most probably, the difficulty is basic and related to the problem of *confinement*, that is the formulation of a theory where the objects existing at short distances are not the asymptotic particles, the hadrons, which posess a complex composite structure in terms of partons.

The relation of the intermittency phenomenon with QCD has naturally been a constant subject of interest. Even if the only strict theoretical understanding of strong interactions is given by the week coupling limit of QCD, that is the short-distance behaviour, it is interesting to adopt first this framework. Two ways have been investigated; One is to look for an eventual intermittent behaviour of the interaction of quarks and gluons; Another one is to add to the parton-interaction stage a phenomenological description of a second "hadronisation" stage and compare the results of the simulation to the experimental data on factorial models. Both ways have led to interesting results showing better the limitation between the known and the unknown in QCD.

Interestingly enough, it has been known since a long time[6] that QCD admits a hierarchical solution at short-distances, under the form of a parton cascading structure[6], at least for quark and glu jets produced e.g. in e^+e^-reactions. However, while the fractal character of this process was noticed[7] early, its intermittency properties have only recently been numerically proven[8]. On a more phenomenological ground, a reasonably good description of the observed factorial moments has been obtained for Monte-Carlo simulations based on the Lund Model for LEP data[9] on quark jets, where the large production of the famous intermediate boson $Z^0$and its decay into quark jets allows a detailed analysis. It is to be noticed that the underlying mechanism of the appropriate Lund Model is based on the QCD cascading structure. For all other reactions, for which the first quark-gluon cascading stage is not proven to exist, no satisfactory simulation including factorial moments has been found so far.

These facts, together with the unresolved problem of the confinment of quark and gluon jets, points towards the long-lasting problem of quantum field theory at strong coupling. Indeed it is known from renormalization group properties of QCD that this theory is asymptotically free, that is its effective coupling is weak at small distances and becomes strong at long distances, precisely the region where hadrons are formed. Hence, the observed intermittency pattern gives a new angle of attack for the strong coupling problem.

3. Turbulence and random cascading models of particle production

Fluctuation structures leading to the power-law (2), section **2**, is not unknown in Physics. In fully-developed turbulence, as observed in fluids, moments of the eddy velocity distribution are compatible with such a behaviour up to high values of the rank q. One finds general models[10] of cascading which fulfill relation (2) with a phase-space bin δ corresponding to small volumes of the fluid. This is very different from the intermittency for particles, which is present in momentum space and not in coordinate space. Moreover, the problem of statistical noise, as mentionned in introduction, is very different. However the structure of the cascading models of refs. [10] can be adapted to the case of particles and transposed in the appropriate relativistic kinematics.

Let us introduce the specific α-models, introduced by D. Schertzer and S. Lovejoy,(see the third ref. [10]), in atmospheric turbulence and considered in ref. [1] in the context of particle physics. Following the scheme of Fig. 2, one considers a series of cascading steps n, ranging from 1 to ν, each of them corresponding to a new λ-partition of phase-space (λ is 2, for simplicity, on the Fig.). In this way, one establishes a correspondence between the value of ν and the desired binning resolution $\frac{\Delta}{\delta}$, where δ is as previously the smaller bin unit and Δ, the larger one. One has the identities:

$$\frac{\Delta}{\delta} \equiv \lambda^{\nu} \equiv M \tag{3}$$

where M is the total number of bins. Let assume that at each step n, density fluctuations may occur and are represented by random factors W for each link of the tree structure (see Fig. 2); One gets after ν steps, for the bin [m] :

$$\frac{\langle \rho_m^q \rangle}{\langle \rho_m \rangle^q} \equiv \left\langle \prod_{n=1}^{\nu} W^q \right\rangle = \{W^q\}^\nu = \left(\frac{\Delta}{\delta}\right)^{\frac{ln\{W^q\}}{ln\lambda}} \tag{4}$$

where one has used the mutual independance of the random factors W, and of their normalization conditions $\{W\} = \{1\} = 1$, the brackets{} meaning the averaging over the distribution of W's. Using expression (1) for the factorial moments one gets the required relation (2) with:

$$f_q = \frac{ln\{W^q\}}{ln\lambda} \tag{5}$$

where the exponent f_q has been called the "intermittency index of rank q" and is related in this model to the *local* probability distribution of the density fluctuations in rapidity.

In fact, random cascading models can be shown[1] to be consistent with the relativistic kinematic constraints on particle collisions. After the collision between initial particles, the produced "medium", whatever it is, is subjected to the Lorentz expansion of distances and times of interaction. When this expansion reaches the canonical correlation length of 1 fermi ($10^{-13} cm.$), the system breaks into pieces and dynamical fluctuations can be generated and persist, since they cannot be destroyed by re-interaction. However, contrary to the conventional picture in which the finally observed hadrons are all created at this length scale (see Fig. 3a), the intermittency phenomenon implies a self-similar process: the system develops further on in time and undergoes, after some expansion time, a new breaking into pieces, with new fluctuations superimposed onto the old ones, and again expands etc..., see Fig. 3b. In fact, the 1-fermi scale remains the basic length, not as the absolute scale of hadron production as in the conventional picture, but as the average scale of dynamical fluctuations and the intrinsic repetition scale of the hierarchical fluctuation pattern. On average, the intermediate system lasts 1 fermi, but from event-to-event or inside the same event, its time-life may vary considerably, generating self-similarity.

Then, as a particle collision, considering its space-time development, the process is compatible with the random geometrical structure of cascading, as schematized in *Fig.2*. However, the intermittency structure is observed in the short range of momentum space and thus cannot find any justification from fluid turbulence theory, where the intermittent structuration appears in coordinate space; One has to find the appropriate theoretical approach. Statistical Mechanics, as often in field theory, is of great help, as discussed in the next section.

4. Phase transitions, Spin-Glasses and random cascading

Whilst trying to understand the intermittency mechanism in terms of quantum field theory at strong coupling, it is natural to address the same question to spin systems, in particular, when they posess a phase transition. Indeed, it is known that the behaviour of spin systems at a second-order phase transition point, that is when the correlation length diverges, is related by a scale transformation to a quantum field theory[11]. As an application of the factorial moment method, intermittency patterns have been searched for in numerical simulations of the 2-dimensional Ising system near its (pseudo-)phase transition coupling. The idea[12] is to consider the subdivisions of the Ising lattice as the bins of Fig.2. In each "box" in the lattice corresponds a bin $[m]$, and the number K_m , cf. definition (1), is taken to be the number of spins with same orientation (a magnetic cluster).

One important point of interest of these statistical systems is that the assumptions about the "noise" and analytic predictions for the behaviour of moments can be confronted with accurate computer simulations[12]. Intermittency structuration has indeed been clearly seen and the scaling properties of a 2nd order phase transition lead to a specific prediction for the intermittency indices (5), namely

$$f_q = (q-1)\bar{D} \tag{6}$$

where $\bar{D}$ is independent of q and related[13] to the critical indices of the relevant spin Hamiltonian. However, ambiguities seem to persist concerning the last point. While one expected[13] t find $\bar{D} = \frac{1}{8}$ from the Ising Hamiltonian, the remark was made[14]that the result seems to be dependent of the definition of clusters. For connected (percolation) clusters, one would find a Potts Hamiltonian and the value[14] $\bar{D} = \frac{5}{96}$ which happens to be confirmed by recent simulations[15]. In fact, the problem remain open, not forgetting its extension to other second-order transitions or, more importantly for particle physics, to a first-order phase transition such as the one predicted for QCD at high temperature.

In fact, formula (6) can also be expressed as[16] the existence of a fixed fractal dimension $\bar{D}$ of the strucure of fluctuations. This property is specific to a phase transition at equilibrium. However, collision processes, except eventually heavy-ion collisions, are probably far from equilibrium. Thus it is thus important to consider other theoretical schemes, which may lead to fractal dimensions $\bar{D}$ depending of q, and are known as *multi-fractal* systems. This is generally the case of random cascading as presented in the previous section ($\bar{D} = \frac{f_q}{q-1}$, see formula 4.) It was found that[17] the relevant statistical systems possess a spin-glass structure, that is spin systems with quenched random interactions instead of deterministic ones as in the case of the first lattice problems considered.

One may introduce the concept of a Partition Function Z by summation of density powers over all bins of same width δ. For random cascading models of the type of Fig. 2, one finds:

$$Z(\beta) \equiv \sum_{m=1}^{M} \rho_m^q = \sum_{m=1}^{M} exp\left(-\beta \sum_{n=1}^{\nu} \varepsilon_n\right) \tag{7}$$

where the substitution $exp(-\beta\varepsilon_n) \equiv \left(\frac{W}{\lambda}\right)^q$ using the same definition of ρ_m as in formula (4), allows one to identify ϵ_n as a random energy level and $q \sim \beta$ as the inverse temperature of the spin-glass system[17]. In such a way the identification is proven with the Generalized Energy Spin Models[18]. Among the interesting consequences of this identification, one may quote a non-trivial pattern[19] of phase transitions leading to a hierarchical structure very different from the usual order-disorder transition. There exists a breaking of ergodicity at low temperature (high β), and a specific classification of the multi-fractal spectrum[19].

The interpretation of this phase transition in the context of Particle Physics is under study. However, no Lagrangian or Hamiltonian formulation exists for these systems and the explicit formulation in terms of a field theory appears to be difficult. In fact, one recently found that the link with quantum field theory could be made easier using the emergence of an underlying non-linear equation which is discussed in the next section.

5. Fractal growth, aggregates and random cascading equations

Among the unexpected connections between random cascading models of particle production and Statistical Physics, last but not least is the relation with the Physics of aggregation and gelling via the well-known Smoluchowski's formulation[20], leading to non-linear rate equations. It is well known that these equations were originally proposed for the description of the coagulation of colloids submitted to Brownian motion more than 74 years ago. However, quite recently, these equations met a revived interest in the numerous studies on the fractal growth, in particular for cluster-cluster aggregation[21].

In all these studies, the number N of clusters of a given number k of mass units is followed as a function of time. One writes a very general mean-field equation (the effect of spatial fluctuations of N_k being neglected) for the aggregation rate, namely:

$$\frac{dN_k}{dt} = \sum_{i+j=k} N_i K^{ij} N_j - \sum_i N_i K^{ik} N_k \tag{8}$$

where the *fusion weights* K^{ij} are the dynamical input. Eqn. (8) establishes the balance at each time and for each mass k between clusters aggregating to form the mass k and clusters of mass k transferred to higher mass, by aggregation.

The way the connection is made[22] with random cascading is through the derivation of a non-linear equation for the Partition Function of random cascading, see (7) and an appropriate transformation of the Smoluchowski equation. In fact one can show that the following two generating functionals verify the same equation: $\mathcal{G}$ for aggregation and $\mathcal{H}$ for the partition function Z, defined as follows:

$$\begin{aligned} \mathcal{G} &= \langle 1 - u^l \rangle_{N_l} \equiv \sum_l N_l \left(1 - u^l\right) . \frac{1}{\sum_l N_l} \\ \mathcal{H} &= \langle e^{uZ} \rangle_{\mathcal{P}(Z)} \equiv \int_0^\infty \mathcal{P}(Z) e^{uZ} dZ \end{aligned} \tag{9}$$

where $\mathcal{P}(Z)$ is the Z probability distribution, computable for random cascading models

One finds the same non-linear equation, namely:

$$\frac{d\mathcal{H}}{d\nu} = \mathcal{H} * \mathcal{H} - \mathcal{H} \tag{10}$$

where $\nu = ln \left|\sum_l N_l\right|$ can be identified with the generation number of random cascading, see section **3**, and the convolution * is defined in terms of the fusion coefficients K^{ij}[22]. In the case of "multiplicative" aggregation, with $K^{ij} = K^i K^j$, one finds a random-branching random cascading model, while the "monodisperse" aggregation case[21] with discrete time intervals gives back the random cascading models already discussed.

The identity of the equations does not mean identity of the solutions, due to possible different initial conditions. In the first investigations of this problem, it seems that the so-called "scaling" solutions of the Smoluchowski equation, which can be derived in a quite general way[23], lead to interesting solutions for the Partition Function of random cascading with asymptotic freedom properties and intermittency. It appears that the world of these non-linear equations has not revealed all its secrets!

6. Conclusions and future prospects

Intermittency in Particle Physics has revealed a quite fructuous field of research. In fact, the study of fluctuations has always led to an interesting insight on physical phenomena. The study of these fluctuations in particle physics is at a very early stage and one does not yet possess a complete picture of its properties. Much more experimental work and critical comparisons with existing models are needed; However it is already possible to guess that a better understanding of these patterns of fluctuations, combined with the already known facts about the different distributions of particles in high-energy experiments may lead to a deeper vew on subnuclear physics. Fractals and Chaos have probably something to teach us about quantum field theory and elementary particles.

On the other hand, there is a hope that the field theoretical techniques developed in particle physics and briefly described in the present talk could be useful in other domains where the problem and structures of intermittency appear. For instance the combination of the "local" study of fluctuations in phase-space with the "global" thermodynamical apparatus related to the partition functions of spin systems could be useful in many problems. As an unexpected example, the intermittency analysis in terms of factorial moments in relation with percolation models has been shown[24] to be relevant in the study of nuclear multi-fragmentation. Perhaps the long-standing respectable unsolved problem of fluid turbulence itself, could benefit from some ideas, as a return gift from particle physics which seems to have taken interest in methods inspired by turbulence!

REFERENCES

[1] A. Bialas and R. Peschanski, *Nucl. Phys. B* **273** (1986) 703, **308** (1988) 857.

[2] For a recent experimental review, see B. Buschbeck, "Moriond Conference", hadronic session, March 1991, to appear in the Proceedings of the Moriond Conference.

[3] the rapidity y of a particle is defined by: $y = \frac{1}{2} ln \left(\frac{E+P_L}{E-P_L} \right)$, where E and P_L are the energy and longitudinal momentum of the particle along the production axis. As a key property $\delta y \equiv \frac{\delta P_L}{E}$ is a Lorentz-invariant along the same axis.

[4] W. Ochs, *Phys. Lett.* **B247** (1990) 101 and preprint MPI-PAE/PTh 63/90; A. Bialas and J. Seixas, *Phys. Lett.* **B250** (1990) 161.

[5] For theoretical reviews, see for instance A. Bialas, "Intermittency 90", invited talk at the Quark Matter Conference of Menton, preprint CERN/Th 5791/90; R. Peschanski, "Intermittency in particle collisions", preprint CERN-TH 5891/90.

[6] See, for instance the book "Perturbative quantum chromodynamics", ed. A.H. Mueller (World Scientific, Singapore, 1989.)

[7] G. Veneziano, Talk given at the 3rd Workshop on "Current Problems in High Energy Particle Theory", Florence, 1979, and preprint CERN-TH.2691/79. A. Giovanninni, Talk given at the X Int. Symp. on Multi-particle Dynamics, Goa, India, 1979.

[8] M. Jedrzejczak, Phys. Lett. **B228** (1989) 259. P. Dahlqvist, B. Anderson, G. Gustafson, *Nucl. Phys.* **B328** (1989) 76.C. Chiu and R. Hwa, , *Phys. Lett.* **B326** (1990) 466.

[9] LEP data: DELPHI coll., P. Abreu et al. *Phys. Lett.* **B247** (1990) 137; OPAL coll., CERN preprint EP- /91.

[10] B. Mandelbrot, J. Fluid Mech. **62** (1974) 331. U. Frisch, P. Sulem and N. Nelkin, J. Fluid Mech. **87** (1978) 719. D. Shertzer and S. Lovejoy, *in* Turbulent shear flows 4. Selected papers from the Fourth Int. Symp. on Turbulent Shear Flows, University of Karlsruhe (1983) ed. L.J.S. Bradbury et al. (Springer, 1984) and references therein.

[11] see for instance, J.-M. Drouffe and C. Itzykson, "Statistical Field Theory", Hermann ed., 1990.

[12] J. Wosiek "Intermittency in the Ising Systems", *Acta Phys. Pol.* **B19** (1988) 863; *Nucl. Phys.* **B (proc. Suppl.) 9** (1989) 640.

[13] H. Satz, *Nucl. Phys.* **B 326** (1989) 613; B. Bambah, J. Fingberg and H. Satz, *Nucl. Phys.* **B 332** (1990) 629.

[14] R. Peschanski, communication in Léon Van Hove Festschrift, eds. A. Giovannini and W. Kittel, World Scientific 1990.

[15] S. Gupta, A. La Cock and H. Satz, Bielefeld preprint 1990.

[16] P. Lipa and B. Buschbeck, *Phys. Lett.* **B234** (1990).

[17] R. Peschanski, *Nucl. Phys. B* **235** (1990) 317.

[18] B. Derrida, J. Phys. Lett. **46** (1986) 1410; B. Derrida and E. Gardner, J. Phys. **C19** (1986) 5783; B. Derrida and H. Flyvbjerg, J. Phys. **A20** (1987) 5273; B. Derrida and H. Spohn, J. Stat. Phys. **51** (1988) 817.

[19] Ph. Brax and R. Peschanski, *Nucl. Phys.* **B.235** (1990) 317; *Nucl. Phys.* **B.353** (1991) 165.

[20] M. Smoluchowski, *Physickalische Zeit.* , XVII (1916) 557, 787; *Zeit. für Phys Chem* XCII (1917) 129.

[21] See, for instance, R. Jullien, F. Botet "Aggregation and fractal aggregates" (World Scientific, Singapore, 1987).

[22] R. Peschanski, "On a transformation property of the Smoluchowski aggregation equation", preprint CERN-TH.6044/91 and SPhT/91-043, March 1991, to be published in *Acta Physica Polonica.*

[23] Y. Gabellini, J.-L. Meunier, Nice University preprints INTh 90/19, INLN 91/4.

[24] M. Ploszajczak and A. Tucholski, *Phys. Rev. Lett.* **65** (1990) 1539.

[25] JACEE Coll., T.H.Burnett et al. *Phys. Rev. Lett.* **50** (1983) 2062.

[26] KLM Coll., R. Holynski et al. *Phys. Rev. Lett.* **62** (1989) 733.

[27] P. Lipa, unpublished UA1 note (1988).

FIGURE CAPTIONS

Figure 1 : **Intermittency: first example**
a) *Particle number distribution*
observed by the JACEE Collaboration for a 5 Tev/nucleon, Si + Ag(Br) cosmic-ray collision on an emulsion plate carried by a balloon[25]. Cosmic-rays provide the only opportunity at present to reach very large energies, and thus high multiplicity-per-event. The next generation of accelerators(LHC at CERN, SSC in Texas), will allow to reach such energies but in a reproducible and controlable way.
b) *Factorial moment of rank 5.*
The factorial moment obtained from Fig. 1a (black dots) is compared with a simulation with Gaussian statistical noise (crosses). The straightline is a typical prediction for intermittent pattern of fluctuations. The Figure is from ref [1].
c)*Intermittency at accelerator.* The factorial moments of rank 2 and 4 are displayed (black squares) and compared with a simulation with only statistical fluctuations (white squares) for an $^{16}O - emulsion$ experiment at CERN.This reaction is very similar to the previous one, but at smaller energy, with much less particle produced (around 120, compared to more than 1000). The straight lines correspond to the intermittency prediction. These results[26] were the first ones published showing that the methods of factorial moments was applicable at present accelerator energies (with moderate multiplicity-per-event), as proposed shortly before[27], and led to intermittency-like fluctuations very similar to the cosmic-ray event.

Figure 2 : **Random cascading model of intermittency**

The figure represents three stages of realization of a random cascading process.

a) *The tree structure of the model*

at each step n the branches of the tree are subdivided into λlinks. For a given resolution, i.e a given total number of steps ν, there is a one-to-one correspondence between a "box $[m]$" in the phase space and a series of integers $\{\alpha_1, \alpha_2, \cdots, \alpha_\nu\}$. In the example we have chosen $\lambda = 2$, $\nu = 4$.

b) *The "Rapidity-box" representation of the α-model*

of intermittency. For each box of this diagram, one chooses a random factor $W(\alpha)$. In each box, the sign " + " or " - " represents the enhancement (resp. damping) density factor of an $\alpha - model$ (see text). The final density ρ_m of states in the box $[m]$ is the product of the factors $W(\alpha_1)\, W(\alpha_2) \cdots, W(\alpha_\nu)$.

c) *The "event".*

The fluctuation pattern obtained after ν steps is displayed, following the random values attributed to the boxes of Fig. b).

Figure 3 : **Space-time representation of intermittent fluctuations**

This figure shows the space-time relativistic Lorentz frame in which a particle collision takes place, at least when projected on the (t, z)plane, where t is the time and z the longitudinal distance. The causal conus ($t^2 - z^2 \geq 0$, $t \geq 0$), the region where particles propagate, is dispayed in both figures, together with the causal hyperbolae, $\tau \equiv \sqrt{t^2 - z^2} = cste$, t$\geq$0, which are the curves of same proper-time, that is which correspond to simultaneity in the intrinsic frame of reference. The figures are from A. Bialas' review, see refs. [5].

a)*In-out conventional picture of hadron production*

The figure shows the conventional picture of hadron production after a high-energy collision. At the origin (O) an interacting "piece" of partonic or hadronic matter (hatched region) is created. After a proper time duration τ of order 1 fermi ($10^{-13} cm$) of relativistic expansion, the "piece" breaks into new pieces of size (on average)equal to the conventional correlation length ξ, giving rise to dynamically independent hadrons. The only scale in the problem is $\tau \sim \xi \sim 1$ in fermi units where the speed of light is unity.

b) *Random cascading in space-time*

In the (1 + 1)space-time frame, Fig. 3b shows the (random) generation of intermittent fluctuations following the geometrical structure of the α-models, but satisfying the constraints of Lorentzian relativistics kinematics. In fact, at each step of the cascade, one iterates the individual process shown in Fig. 3a. A piece of partonic or hadronic matter is locally streched by the relativistic expansion to a length larger than ξ, and breaks into λ new pieces, with each a new value of the density, due to an additional random factor W. Appearing at successive proper time values τ_n, the fluctuations lead to the superposition of structures at different rapidity scales of size $\delta y \sim \xi/\tau_n$. In the figure, one has chosen for simplicity sake, $\lambda = 2$, $\tau_n = 2^n$, and random factors $W_+$and W_- as in Fig. 2.

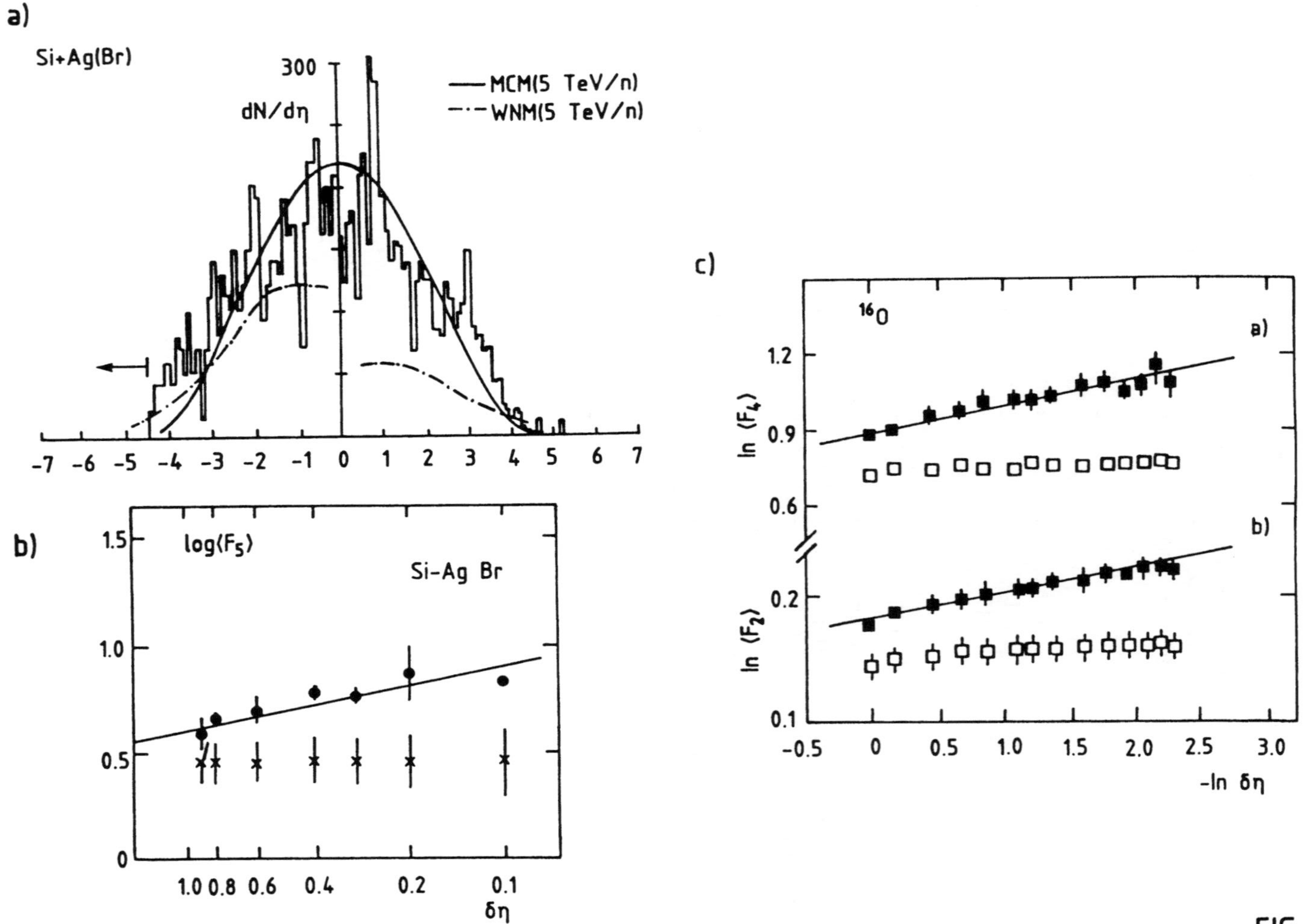
a)
Si+Ag(Br)
300
dN/dη
—MCM(5 TeV/n)
-·-WNM(5 TeV/n)
-7 -6 -5 -4 -3 -2 -1 0 1 2 3 4 5 6 7
b)
log⟨F5⟩
Si-Ag Br
1.5
1.0
0.5
0
1.0 0.8 0.6 0.4 0.2 0.1
δη
c)
16O
a)
b)
ln ⟨F4⟩
ln ⟨F2⟩
1.2
0.9
0.6
0.2
0.1
-0.5 0 0.5 1.0 1.5 2.0 2.5 3.0
-ln δη

FIG. 1

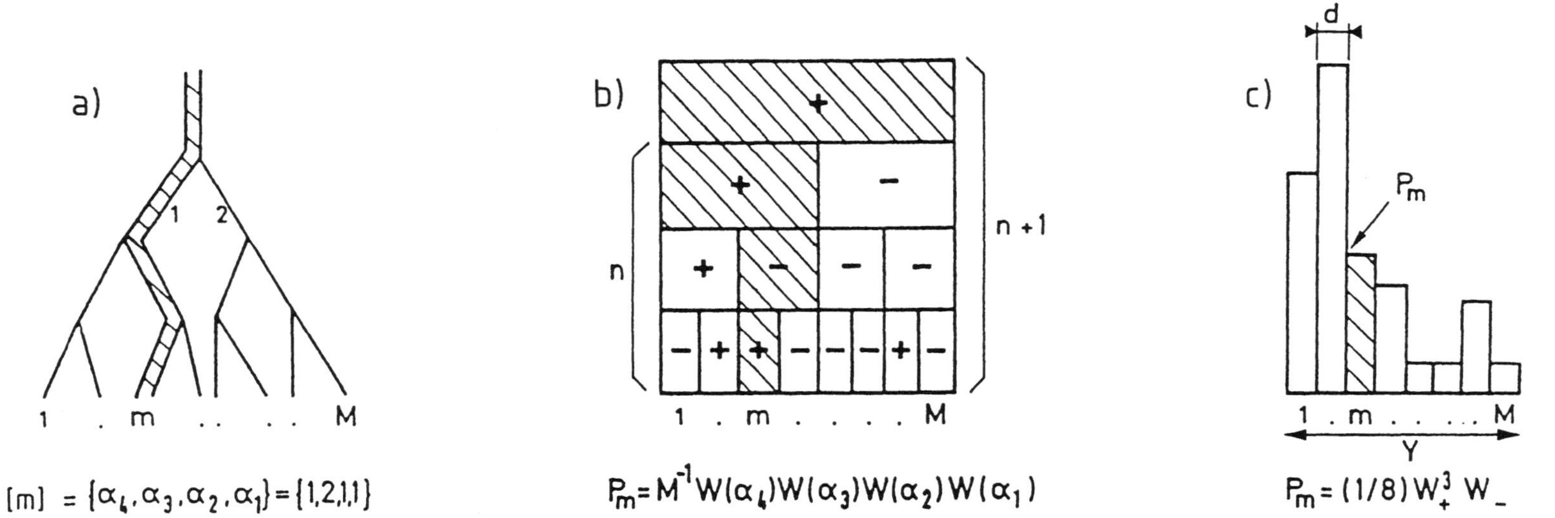
a)
1 . m . . . . M
[m] = {α4, α3, α2, α1} = {1,2,1,1}
b)
n
n + 1
1 . m . . . . M
Pm = M^-1 W(α4) W(α3) W(α2) W(α1)
c)
d
Pm
1 . m . . . . M
Y
Pm = (1/8) W+^3 W−

FIG. 2

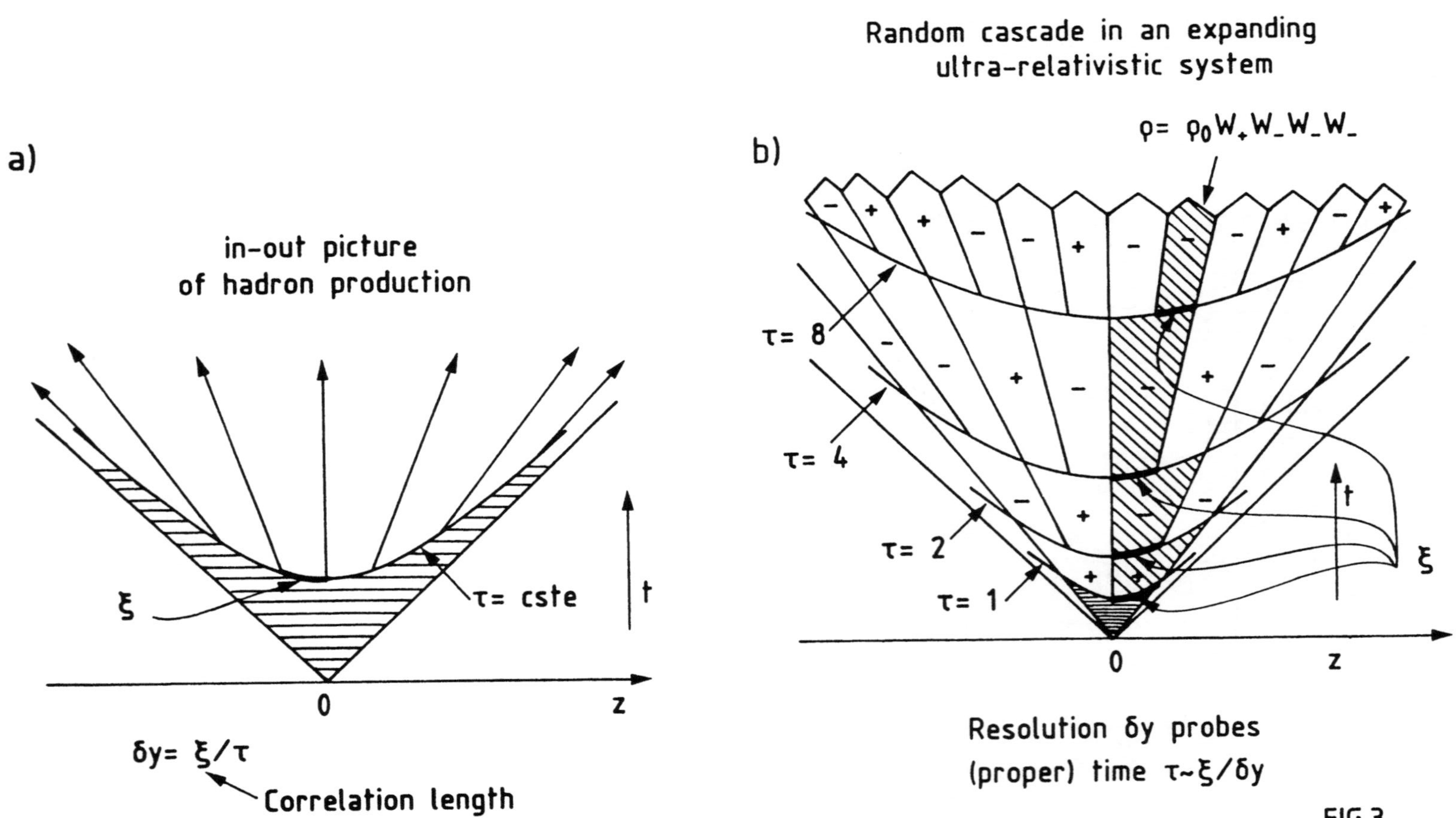
a)
in-out picture
of hadron production
ξ
τ= cste
t
0
z
δy= ξ/τ
Correlation length
b)
Random cascade in an expanding
ultra-relativistic system
ρ= ρ₀W₊W₋W₋W₋
τ= 8
τ= 4
τ= 2
τ= 1
t
ξ
0
z
Resolution δy probes
(proper) time τ~ξ/δy

FIG.3

Hot QCD
and the quark-gluon plasma

Michel Le Bellac

Institut Non Linéaire de Nice*, Université de Nice
Parc Valrose, 06034 Nice Cedex, France

Abstract

We review recent developments in the physics of the quark-gluon plasma. After giving a short account of results in lattice QCD at finite temperature, we describe briefly the Feynman rule at nonzero T. Then we discuss the spectrum of collective excitations (quasiparticles) in the quark-gluon plasma: fermionic as well as gluonic excitations. Finally we explain how a resummation method duc to Braaten and Pisarski allows to deal with some infrared problems of perturbative QCD at finite temperature. As an application we show how one solves the so-called "plasmon puzzle".

* Unité Mixte de Recherche 129 du CNRS

1. Introduction

The main motivation for being interested today in finite temperature QCD (and more generally in finite temperature quantum field theory (FTQFT)) is the strong belief that we might be able to produce a deconfined state of matter in laboratory experiments. At sufficiently high temperatures or densities, quarks and gluons are no more confined into hadrons, and constitute what has been called a "quark-gluon plasma". This quark-gluon plasma was present in the early Universe up to times $\sim 10^{-5}$s, but it is not easy to find observable consequences of this fact, while it may be possible (although not so easy!) to detect signals of a quark-gluon plasma created in the laboratory by heavy-ion collisions[1].

The main basis for our belief in this new phase of matter comes from lattice simulations. In the case of a pure $SU(3)$ gauge theory, experts now agree that they observe a first-order deconfinement phase transition at a temperature $T_c \simeq$230 MeV[2]. The character (and even the existence) of the transition depends in a complicated way on the number of flavours and the masses of the quarks, and the situation does not seem to be fully settled in lattice simulations. In the realistic case of two light quarks and one heavy quark (u, d, and s flavours), results are consistent with a first-order transition, corresponding to chiral symmetry restoration and deconfinement, at a critical temperature $T_c \simeq 150$ MeV.

In order to express T_c in MeV, one has to make a comparison with another prediction. In general the comparison is made with the calculation of the ρ-mass, m_ρ, and the quoted values for T_c have been given with that choice. Unfortunately the m_π/m_ρ ratio has not yet reached its physical value in lattice calculations, so that it is reasonable to allow for a rather large error bar on T_c; we shall take as a conservative estimate T_c in the range 150-200 MeV.

Another estimate of the critical temperature comes from low energy effective Lagrangians. The quark condensate can be computed in perturbation theory and its low temperature expansion is given by[3]

$$< q\bar{q} >_T = < q\bar{q} >_{T=0} \left(1 - \frac{T^2}{8F^2} - \frac{T^4}{384F^4} - \frac{T^6}{288F^6}\ln\frac{\Lambda_q}{T} + O(T^8)\right) \tag{1.1}$$

For simplicity the formula has been written for massless quarks. The parameter F is closely related to the pion decay constant f_π and Λ_q to the $\pi - \pi$ scattering lengths; their numerical values are

$$F = (1.057 \pm 0.012) f_\pi = 88.3 \pm 11 \text{ MeV} \quad ; \quad \Lambda_q = 470 \pm 110 \text{ MeV}$$

The scale of the low temperature expansion is set by $\sqrt{8}F \simeq 250$ MeV and the expansion seems to be reliable up to $T \simeq 150$ MeV. A tempting (but bold) extrapolation to higher values of T leads to a vanishing condensate at $T_c \simeq 190$ MeV. Various corrections (massive quarks and massive hadrons) do not modify this estimate in an essential way.

At temperatures much higher than T_c, one expects the quark-gluon plasma to behave almost as an ideal gas of free quarks and gluons: indeed, because of asymptotic freedom, the QCD coupling constant $g(T)$ tends to zero when T goes to infinity. As T is the only scale at our disposal we must have

$$\alpha_S(T) = \frac{g^2(T)}{4\pi} = \frac{6\pi}{(33 - 2N_f)\ln(cT/\Lambda_{\overline{MS}})} = \frac{1}{8\pi\beta_0 \ln(cT/\Lambda_{\overline{MS}})} \tag{1.2}$$

the constant c being unknown at present ; as usual β_0 denotes the first coefficient of the β-function and $\Lambda_{\overline{MS}}$ the QCD scale in the $\overline{MS}$ renormalization scheme. In the limit of infinite temperature, we thus get the Stefan-Boltzmann (SB) law for the energy density

$$\varepsilon_{SB} = \frac{\pi^2}{15}(N_c^2 - 1 + \frac{7}{4}N_c N_f)T^4 \tag{1.3}$$

where N_c is the number of colours and N_f the number of flavours. In order to get an order of magnitude estimate of the energy density around T_c, we simply plug in (1.3) the value of the critical temperature and find

$$\varepsilon \simeq 1.0 \text{ GeV}/(\text{fm})^3 \quad T_c = 150 \text{ MeV}$$

$$\varepsilon \simeq 3.3 \text{ GeV}/(\text{fm})^3 \quad T_c = 200 \text{ MeV}$$

The other extreme way of reaching the chirally symmetric phase is to increase the baryonic density, and thus the chemical potential μ at small values of T; for simplicity we take $T = 0$. Recall that the nuclear density $d_0 \simeq 0.15$ nucleon/(fm)^3, and that the chemical potential μ is given from d_0 by

$$d_0 = \frac{2}{3\pi^2}(\mu^2 - m^2) \tag{1.4}$$

where m is the nucleon mass. Unfortunately the region $T = 0$, μ increasing, is not accessible to present lattice calculations. In order to estimate the critical chemical potential μ_c, one has to rely on the bag model or on effective chiral Lagrangians[4]. It is likely that a transition occurs at a density $\sim 4 - 5$ times the nuclear density; taking into account (1.4) and the fact that the quark chemical potential $\mu_q = \mu/3$ at the transition, this corresponds to $\mu_q \simeq 300$ MeV. Thus we can guess that the phase diagram is roughly that drawn in fig.1. However we must stress that the exact shape of the curve in fig.1, as well as the character of the transition for $\mu \neq 0$, are still very uncertain. Let us only mention that the situation with $T \simeq 0$ and $d \simeq 5d_0$ could be reached in the core of some neutron stars, which could thus contain a quark-gluon plasma.

Let us come back to the case $\mu = 0$ and to the results of lattice calculations; one of the best signals of a first-order phase transition is the jump of the energy density ε at T_c (fig.2). Above T_c, it is interesting to compare lattice results with those of perturbative calculations, since for $T \gg T_c$ we expect the SB-law to hold. We can even do a little better since the perturbative corrections to the SB-law have been evaluated; one finds for $N_c = 3$

$$\varepsilon/T^4 = \frac{8\pi^2}{15} + \frac{7\pi^2}{20}N_f - (\frac{1}{2} + \frac{5}{24}N_f)g^2 + \frac{2g^3}{\pi}(1 + \frac{1}{6}N_f)^{3/2} + ... \tag{1.5}$$

Notice the term of order g^3, and not g^4, as could be expected from a perturbative expansion. This feature arises from the infrared behaviour of any (renormalizable) relativistic thermal field theory and is not typical of QCD; it happens for instance in the case of the φ^4 scalar field theory. Infrared divergent terms are resummed in the so-called "ring-diagrams" (see subsection 4.1). Terms of order $g^4 \ln g^2$ have also been computed, but they are not meaningful in the absence of a complete $O(g^4)$ calculation.

It is interesting to see how far down in T one can extrapolate the perturbatively corrected SB-law (1.5); however it is necessary to correct first (1.5) , which is a continuum formula, for lattice effects. In other words one has to make the comparison with a perturbative calculation on the lattice. Such a calculation has been performed[5] and one can see on fig.2 that there is a very good agreement between the lattice simulations and the perturbative calculation down to $T/T_c \simeq 1.2$. However the situation does not look so bright when one examines the pressure P. In an ideal ultrarelativistic gas one should have $\varepsilon = 3P$; the ratio $(\varepsilon - 3P)/T^4$ is shown in fig.3[6], and one can see that there are certainly strong non-perturbative effects up to $T/T_c \simeq 2$. In (continuum) perturbation theory the leading contribution to $(\varepsilon - 3P)/T^4$ is of order g^4

$$(\varepsilon - 3P)/T^4 = -\frac{2}{3}\beta_0 g^4 (\frac{1}{2} + \frac{5}{24}N_f) + O(g^5) \tag{1.6}$$

and this has the wrong sign if compared to lattice results.

Another physically meaningful quantity is the Debye screening length r_D, or its inverse the electric mass m_{el} ($m_{el} = r_D^{-1}$), which governs the screening of the heavy quark potential above T_c. One expects this potential to be given by

$$\begin{aligned} V(r,T) &= \frac{\alpha_S(T)}{r} + r\sigma(T) \quad T < T_c \\ V(r,T) &= \frac{T\alpha_S(T)}{(rT)^n} e^{-\mu(T)r} + c(T) \quad T > T_c \end{aligned} \tag{1.7}$$

with $n \simeq 2$ and $\mu \simeq 2m_{el}$ for $T \gg T_c$. One finds on the lattice in pure $SU(3)$

$$\mu(T)/T \sim 2 - 3 \;\text{ for }\; T_c \leq T \leq 2T_c \tag{1.8}$$

while perturbation theory (subsection 3.4) gives

$$m_{el}^2(T) \simeq (\frac{1}{3}N_c + \frac{1}{6}N_f)\, g^2(T)\, T^2 \tag{1.9}$$

With $g(T) \sim 1$, there is rough agreement between the lattice and perturbative calculations.

Thus our motivation for studying the perturbative aspects of FTQFT is that perturbation theory seems to be a good approximation for $T/T_c \geq 2$, and might even be a good qualitative guide for lower values of this ratio. In any case, comparison with lattice results, and matching of these results with the perturbative ones around $T \simeq 2T_c$ may give useful indications on the behaviour of the quark-gluon plasma, and in particular may give useful hints on the non-perturbative effects which occur at lower values of T/T_c.

2. Feynman rules at finite temperature

We shall limit ourselves to a brief account of Feynman rules at finite temperature and refer to [7], [8] or [9] for a detailed derivation of these rules.

2.1 Feynman rules in imaginary time.

The formalism of FTQFT relies on a formal analogy between inverse temperature and imaginary time. The imaginary time $\tau = it$ varies in the range $[0, \beta]$, where $\beta = 1/T$ (Fig.4). Because average values in statistical mechanics are calculated from a trace, the fields must be periodic (antiperiodic for fermions) in imaginary time with period β. Thus the most natural formalism in FTQFT is the so-called "imaginary-time formalism", or equivalently "Matsubara formalism". Because of periodicity in imaginary time, the energies (or frequencies) can take only discrete values, called "Matsubara frequencies"

$$\omega_n = \frac{2\pi n}{\beta} \tag{2.1}$$

Then the expression for the (free=F) propagator is

$$\Delta_F(p) = \Delta_F(i\omega_n, \vec{p}) = \frac{1}{\omega_n^2 + \vec{p}^2 + m^2} = \frac{1}{\omega_n^2 + \omega_p^2} \tag{2.2}$$

This expression is quite analogous to that of the propagator of an Euclidean $T = 0$ field theory, except that $i\omega_n = p_4$ is restricted to discrete values. Because of this, in a loop integral $\int dp_4$ will be replaced by a sum over ω_n :

$$\int \frac{d^4p}{(2\pi)^4} \to T\sum_n \int \frac{d^3p}{(2\pi)^3} \tag{2.3}$$

The rules for the vertices are the same as those of the Euclidean $T = 0$ field theory[8]; they can be easily deduced from a path-integral formalism for the correlation functions at finite temperature. For example in a $g^2\varphi^4$ scalar theory (the use of g^2 for the coupling constant, rather than g, allows a convenient comparison with other field theories), to each vertex will be associated a factor $-g^2$. For particles of spin 1/2 and for gauge particles, the spin structure of the propagator will be taken from the Euclidean $T = 0$ field theory.

In order to illustrate these rules, let us work out a simple example : the tadpole in $g^2\varphi^4$ (fig.5)

$$\Pi = \frac{g^2T}{2}\sum_n \int \frac{d^3p}{(2\pi)^3}\ \frac{1}{\omega_n^2 + \vec{p}^2 + m^2} \tag{2.4}$$

The standard trick, in order to perform the summation over n in (2.4) is to use a contour integral (fig.6)

$$T\sum_{n=-\infty}^{\infty} f(p_0 = i\omega_n) = \frac{T}{2i\pi}\int_C dp_0 f(p_0)\frac{\beta}{2}\coth\frac{\beta p_0}{2} \tag{2.5}$$

which is valid provided $f(p_0)$ has no singularities on the imaginary axis. After some simple algebra one finds

$$\frac{1}{2i\pi}\int_{-i\infty}^{i\infty} dp_0\frac{1}{2}(f(p_0) + f(-p_0)) + \frac{1}{2i\pi}\int_{C_2} dp_0(f(p_0) + f(-p_0))\frac{1}{e^{\beta p_0} - 1} \tag{2.6}$$

This allows to separate Π into a vacuum part, Π_{vac}, and a thermal part Π_β ; using the explicit value of $f(p_0)$:

$$f(p_0) = \frac{1}{\omega_p^2 - p_0^2} \qquad \omega_p^2 = \vec{p}^2 + m^2$$

we get :

$$\Pi_{vac} = \frac{g^2}{2}\int \frac{d^4p}{(2\pi)^4}\ \frac{1}{p_4^2 + \vec{p}^2 + m^2} \tag{2.7}$$

$$\Pi_\beta = \frac{g^2}{2}\int \frac{d^3p}{(2\pi)^3}\frac{1}{\omega_p(e^{\beta\omega_p} - 1)} \tag{2.8}$$

Π_{vac} is of course divergent, but this divergence will be taken care of by the usual $T = 0$ renormalization, while Π_β is easily evaluated for $T \gg m$:

$$\Pi_\beta = \frac{g^2T^2}{24} = \delta m_\beta^2 \tag{2.9}$$

Equation (2.9) shows that in a thermal bath, massless particles acquire a mass of order gT.

Instead of using the contour integration technique, it may be more elegant to start from a mixed representation of the propagator (2.2). We define the Fourier transform in imaginary time through

$$\Delta_F(i\omega_n, \vec{p}) = \frac{1}{\omega_n^2 + \vec{p}^2 + m^2} = \int_0^\beta d\tau e^{i\omega_n \tau} \Delta_F(\tau, \vec{p}) \tag{2.10}$$

and it is straightforward to check that $\Delta_F(\tau, \vec{p})$ is given by

$$\Delta_F(\tau, \vec{p}) = \frac{1}{2\omega_p}[(1 + n_B(\omega_p))e^{-\omega_p \tau} + n_B(\omega_p)e^{\omega_p \tau}] \tag{2.11}$$

where the Bose-Einstein distribution $n_B(x)$ is defined by

$$n_B(x) = (e^{\beta|x|} - 1)^{-1} \tag{2.12}$$

Using this representation of $\Delta_F(i\omega_n, \vec{p})$ allows to perform easily the sum over n. For example for the graph of fig.7 we have :

$$\sum_n \int d\tau d\tau' e^{i\omega_n \tau} e^{i(\omega - \omega_n)\tau'} = \beta \int d\tau e^{i\omega\tau} \tag{2.13}$$

This formula gives automatically the correct continuation to real energies. However there is a tricky point in more complicated loop calculations : if an integration variable does not belong to the interval $[0, \beta]$, one has to extend (2.11) by enforcing the periodicity condition on the τ-dependence of the propagator.

2.2 Real time formalism

Instead of imaginary time, it is also possible to work directly with a real time formalism, as at $T = 0$. The form of the propagator in real time was first worked out by Dolan and Jackiw[10]

$$D_F(p) = \frac{i}{p^2 - m^2 + i\epsilon} + 2\pi n_B(p_0)\delta(p^2 - m^2) \tag{2.14}$$

One might hope that it would be enough to take the $T = 0$ Feynman rules and use the real-time propagator (2.14) in order to compute in perturbation theory at $T \neq 0$. Unfortunately this is not the whole story: when one wants to write Feynman rules for real-time Green's functions, namely for Green's functions whose time-arguments are real, one runs into problems[9]. Let us assume that we want to compute

$$D(x, y) = \frac{1}{Z}\mathrm{Tr}(e^{-\beta H} T(\varphi(x)\varphi(y))) \tag{2.15}$$

where the times x^0 and y^0 are real. The problem is that x^0 and y^0 belong to the real axis, and if we want to write a path-integral, we know that if the initial time is t_i, the final time must be $t_i - i\beta$. We have thus to choose a time-path starting from some time t_i, which goes through x^0 and y^0, and lands into $t_i - i\beta$. Taking into account the domain of definition of D_+ and D_-, we see that the imaginary part of t must be non-increasing along the contour (fig.8). The standard choice is that of fig.9 , with $t_i \to -\infty$ and $t_f \to +\infty$. This leads to a doubling of the number of degrees of freedom, because one has to introduce a ghost field φ_2 which lives on the contour C_2. The propagator takes a matrix form, the indices 1 and 2 corresponding to the normal (φ_1) and ghost (φ_2) field respectively :

$$D^F_{ab}(p) = U_{ac}(p) \begin{pmatrix} D^F_0(p) & 0 \\ 0 & D^{F*}_0(p) \end{pmatrix}_{cd} U_{db}(p) \tag{2.16}$$

where $D_0^F(p)$ is the $T = 0$ Feynman propagator

$$D_0^F(p) = \frac{i}{p^2 - m^2 + i\epsilon} \tag{2.17}$$

and the matrix U is given by

$$U(p) = \begin{pmatrix} \sqrt{1 + n_B(p)} & \sqrt{n_B(p)} \\ \sqrt{n_B(p)} & \sqrt{1 + n_B(p)} \end{pmatrix} \tag{2.18}$$

We have suppressed the bar over D^F for notational simplicity. The Dolan-Jackiw propagator (2.14) is nothing but the (11)-matrix element of D^F :

$$D_{11}^F(p) = \frac{i}{p^2 - m^2 + i\epsilon} + 2\pi n_B(p_0)\delta(p^2 - m^2) \tag{2.19}$$

There are two types of vertices : in a scalar theory, to normal field vertices are associated factors $(-ig)$, while to ghost vertices are associated factors ig. Note that the fields are mixed through propagators, and not through vertices (fig.10)

We shall not attempt to give a derivation of these Feynman rules, since this has been worked out in detail in the literature[9], but we shall illustrate the necessity of the doubling of degrees of freedom on a simple example. Assume that we have a free Lagrangian describing particles of mass $m^2 + \mu^2$, but that we treat μ^2 as a perturbation :

$$\delta\mathcal{L} = -\frac{1}{2}\mu^2\varphi_1^2 - \frac{1}{2}\mu^2\varphi_2^2 \tag{2.20}$$

where φ_1 is the normal field and φ_2 the ghost field. To first order in perturbation theory we have (fig. 11)

$$D_{11} = D_{11}^F(m^2) - i\mu^2[(D_{11}^F(m^2))^2 - (D_{12}^F(m^2))^2] + ... \tag{2.21}$$

where D_{12}^F is given from (2.16) :

$$D_{12}^F = 2\pi n_B(p_0)e^{\beta|p_0|/2}\delta(p^2 - m^2) \tag{2.22}$$

In order to interpret (2.21), we need a regularization of δ-functions ; the correct regularization is[9]

$$\delta(x) = \lim_{\epsilon \to 0} \frac{1}{\pi} \frac{\epsilon}{x^2 + \epsilon^2} \tag{2.23}$$

which gives :

$$\frac{1}{x + i\epsilon}\delta(x) = -\frac{1}{2}\delta'(x) \; - \; i\pi(\delta(x))^2 \tag{2.24}$$

This regularization is also useful at $T = 0$, and is not typical of finite temperature. One discovers that the unwanted δ^2 disappear in (2.21), which may be rewritten as

$$D_{11} = D_{11}^F(m^2) + \mu^2\frac{\partial D_{11}^F}{\partial m^2} + ... = D_{11}^F(m^2 + \mu^2)$$

as it should. This formula is known as the mass-derivative formula, and has been used as a heuristic trick before the doubling of degrees of freedom was discovered.

2.3 The self-energy in the real-time formalism.

The above example was given in order to illustrate the doubling of the number of degrees of freedom. There is of course a much faster (and better) way of handling the calculation, which makes use of the diagonal form (2.16) of the propagator. We start from Dyson's equation in matrix form and write the full propagator $D_{ab}(p)$ by introducing the self-energy $\Sigma_{ab}(p)$

$$D_{ab}(p) = D^F_{ab}(p) + D^F_{ac}(p)(-i\Sigma_{cd}(p))D_{db}(p) \tag{2.25}$$

One can show that the same matrix $U(p)$ which diagonalizes the free propagator $D^F_{ab}(p)$ also diagonalizes the full propagator $D_{ab}(p)$[11]; then $U^{-1}(p)$ diagonalizes Σ_{ab}

$$-i\Sigma_{ab}(p) = U^{-1}_{ac}(p)\begin{pmatrix} -i\Sigma(p) & 0 \\ 0 & (-i\Sigma(p))^* \end{pmatrix}_{cd} U^{-1}_{db}(p) \tag{2.26}$$

The calculation in the end of the previous subsection could have been performed at once by noting that $\Sigma = \mu^2$ and by making in the diagonal form of the propagator the substitution

$$\frac{i}{p^2 - m^2 + i\epsilon} \to \frac{i}{p^2 - (m^2 + \mu^2) + i\epsilon}$$

Furthermore the matrix equation (2.26) leads to relations between various matrix elements

$$\begin{aligned} \mathrm{Re}\Sigma_{11}(p) &= \mathrm{Re}\Sigma(p) \\ \mathrm{Im}\Sigma_{11}(p) &= \coth(\frac{1}{2}\beta|p_0|)\mathrm{Im}\Sigma(p) \\ \Sigma_{12}(p) &= \frac{-i}{\sinh(\beta|p_0|/2)}\mathrm{Im}\Sigma(p) \end{aligned} \tag{2.27}$$

One can derive the following relation between the real- and imaginary-time self-energies[7],[9]

$$\Sigma(p) = \overline{\Sigma}(p_0 + i\epsilon p_0, \vec{p}) \tag{2.28}$$

where the imaginary-time self-energy $\overline{\Sigma}(z,\vec{p})$ is related to the imaginary-time propagator through

$$\Delta^{-1}(z,\vec{p}) = \Delta_F^{-1}(z,\vec{p}) + \overline{\Sigma}(z,\vec{p}) \tag{2.29}$$

It is very useful to express $\Delta(z,\vec{p})$ in terms of a spectral function $\rho(p_0,\vec{p})$

$$\Delta(z,\vec{p}) = \int_{-\infty}^{+\infty} \frac{dp_0}{2\pi}\frac{\rho(p_0,\vec{p})}{p_0 - z} \tag{2.30}$$

The function $\rho(p_0,\vec{p})$ obeys the positivity condition $\varepsilon(p_0)\rho(p_0,\vec{p}) \geq 0$, where *varepsilon* is the sign function, and the sum rule[7],[9]

$$\int_{-\infty}^{+\infty} \frac{dp_0}{2\pi} p_0\rho(p_0,\vec{p}) = 1 \tag{2.31}$$

which follows from the equal-time commutation relations of the field $\hat{\varphi}(\vec{x},t)$ and its time derivative. In a relativistic context sum rules like (2.31) are in general useless because of renormalization problems: $\Sigma(p)$ has to be subtracted in order to be made finite. In relativistic thermal field theories, such sum rules may hold in the limit of high temperatures, because in many important cases (see the next two sections) the leading term of Σ is proportional to

T^2 and the $T = 0$ terms are negligible. This leading term is finite in perturbation theory, at least to one-loop order, and the sum rule can then be applied.

Finally we may note that the relation between real and imaginary time Green's functions has been investigated recently by Kobes and relations generalizing (2.28) have been established for 3- and 4-point Green's functions[12].

3. Collective excitations in a quark-gluon plasma

In this section and in the following one, we shall always assume that the temperature is much larger than the masses of the particles which we consider : $T \gg m$. This is of course true for gluons, and also for u- and d-quarks. In what follows, we shall simply take massless quarks and gluons, in order to avoid unnecessary notational complications. It turns out that one must distinguish three scales for the momenta of the external particles

- momenta $\sim g^2T$: magnetic mass scale
- momenta $\sim gT$: soft momenta
- momenta $\sim T$: hard momenta.

In what follows, we shall adopt the following convention: four-momenta will be denoted by upper case letters, while lower case letters will denote the corresponding energies and three-momenta: $P_\mu = (p_0, \vec{p})$. When we write $P \sim gT$, we mean that p_0 and $p \sim gT$. The quasi-particle spectrum will be especially interesting for soft external lines, since, as we already saw in the $g^2\varphi^4$-theory, thermal masses are of order gT.

3.1 Fermionic excitations (QED or QCD).

The quasi-particle spectrum is found from the fermion self-energy $\Sigma(P)$[13]-[14] ; we shall be interested by the leading term in T only, which has the remarkable property of being gauge-invariant. This is a particular case of a general statement which will be discussed in the next section, in the framework of the effective expansion. We perform in QCD the one-loop calculation of Σ in the real-time formalism ; we only need the (11)-component of the fermion and gluon propagators

$$S_F(P) = \not{P} \left(\frac{i}{P^2 + i\epsilon} - 2\pi n_F(p_0)\, \delta(P^2) \right) \tag{3.1}$$

$$D_F^{\mu\nu}(K) = -g^{\mu\nu} \left(\frac{i}{K^2 + i\epsilon} + 2\pi n_B(k_0)\, \delta(K^2) \right) \tag{3.2}$$

The Fermi-Dirac and Bose-Einstein factors are given by

$$n_F(p_0) = \frac{1}{e^{\beta|p_0|} + 1} \;\; ; \;\; n_B(k_0) = \frac{1}{e^{\beta|k_0|} - 1}$$

The gluon propagator has been taken in the Feynman gauge ; one has to be careful when using other gauges[15]. From (3.1) and (3.2) we immediately get for the real part of the thermal self-energy Σ_β (fig.12)

$$\mathrm{Re}\Sigma_\beta = 2g^2 C_F \mathbf{P} \int \frac{d^4K}{(2\pi)^3} \frac{1}{(P+K)^2} \left((\not{P} + \not{K}) n_B(k_0) + \not{K} n_F(k_0) \right) \delta(K^2) = C\not{P} + \not{D} \tag{3.3}$$

where $\mathbf{P}$ denotes a principal value. Since Lorentz invariance has been lost, we can build Lorentz scalars other than $\not{P}$; indeed we have at our disposal $u_\mu = (1, \vec{0})$, which defines the rest-frame of the plasma. It is very important to remark that chiral invariance is maintained since obviously

$$\{\gamma_5, \Sigma_\beta\} \;=\; 0 \tag{3.4}$$

With massive fermions, we would get in addition a scalar term (B1) and a term of the form $EP\!\!\!/ \!\!\not{u}$, which of course do not anticommute with γ_5 and break chiral invariance. Equation (3.3) leads to a chiral invariant mass ; indeed the full propagator $S(P)$ is :

$$-iS(P) \;=\; \frac{1}{P\!\!\!/(1-C) - D\!\!\!\!/} \;=\; \frac{(1-C)P\!\!\!/ - D\!\!\!\!/}{P^2(1-C)^2 - 2P.D(1-C) + D^2} \tag{3.5}$$

and to first order in perturbation theory we discover that the thermal mass is $2m_f^2 = 2P.D$. Let us evaluate m_f^2 from (3.3) ; it is easy to find that

$$m_f^2 \;=\; \frac{g^2 C_F}{2\pi^2}\int_0^\infty \omega d\omega\,(n_B(\omega) + n_F(\omega)) \;+\; O(g^2P^2)$$

namely

$$m_f^2 \;\simeq\; \frac{g^2 C_F T^2}{8} \tag{3.6}$$

In order to complete the evaluation of Σ_β, we need the integral

$$\int_{-1}^{1} \frac{d(\cos\theta)}{P^2 + 2\eta p_0 k + 2pk\cos\theta} \;=\; \frac{1}{2pk}\ln\frac{E_\eta(E_{-\eta}+k)}{E_{-\eta}(E_\eta+k)} \tag{3.7}$$

where

$$\eta = \pm 1 \;\;,\;\; E_\eta = \frac{1}{2}(p_0 + \eta p) \tag{3.8}$$

One then notices that a sum of terms of the form $[\eta = +1] + [\eta = -1]$ leads to a logarithmic divergence in the k-integration, while a difference $[\eta = +1] - [\eta = -1]$ leads to a quadratic divergence, resulting in a $T^2 \ln E_+/E_-$ factor in the final result. Such a quadratic divergence will appear in the calculation of D_0, since k_0 can be either positive or negative. We find from (3.3)

$$D_0 \;=\; \frac{g^2 C_F}{4\pi^2 p}\ln\frac{E_+}{E_-}\int_0^\infty d\omega\;\omega(n_B(\omega) + n_F(\omega)) \;=\; \frac{m_f^2}{p}\,Q_0(\frac{p_0}{p}) \tag{3.9}$$

where Q_0 is a Legendre function of the second kind :

$$Q_0(x) = \frac{1}{2}\ln\frac{x+1}{x-1} \tag{3.10}$$

We shall also need $Q_1(x)$

$$Q_1(x) \;=\; xQ_0(x) - 1 \tag{3.11}$$

Knowing D_0 and $P.D$, it is straightforward to compute the space components D_i, using rotational invariance :

$$|\vec{D}| \;=\; \frac{1}{p}\,(p_0 D_0 - P.D) \;=\; \frac{m_f^2}{p} Q_1(\frac{p_0}{p}) \tag{3.12}$$

On the other hand it is obvious that the terms proportional to $P\!\!\!/$ are non-leading, and we may write the final result, to leading order in T ($\hat{p} = \vec{p}/p$) :

$$iS^{-1}(P) = \not{P} - \Sigma = A_0\gamma_0 \; - \; A_S\vec{\gamma}.\hat{p} \tag{3.13}$$

$$A_0 = p_0 - \frac{m_f^2}{p}\, Q_0(p_0/p) \tag{3.14}$$

$$A_S = p - \frac{m_f^2}{p}\, Q_1(p_0/p) \tag{3.15}$$

In order to find the quasi-particle spectrum, we must locate the poles of $S(P)$. Now we have to be a bit careful. Indeed if we write

$$-iS(P) \simeq \frac{1}{\not{P} - \not{D}} = \frac{\not{P} - \not{D}}{P^2 - 2P.D + D^2} \tag{3.16}$$

naive perturbation theory allows to neglect D^2, which is of order g^4, and gives a pole at $p^2 = 2m_f^2$, in agreement with the previous discussion. However this is no longer the case if the external momentum is of order gT ; because of the m_f^2/p factor in (3.9) and (3.12), all terms in the denominator of (3.16) are now of order g^2T^2, and all of them are to be taken into account. A convenient way of rewriting (3.13) is

$$-iS(P) = \frac{1}{2(A_0 + A_S)}\,(\gamma_0 + \vec{\gamma}.\hat{p}) \; + \; \frac{1}{2(A_0 - A_S)}\,(\gamma_0 - \vec{\gamma}.\hat{p}) \tag{3.17}$$

There are two dispersion laws[14],[16]: the first one corresponds to the solutions of $A_0 + A_S = 0$. The corresponding quasi-particles have a negative helicity/chirality ratio ($\chi = -1$), while those corresponding to $A_0 = A_S$ have $\chi = +1$. We denote the dispersion laws by $\omega^{\pm}(p)$; the two functions are displayed in fig.13. It is interesting to give their behaviour for $p \to 0$ and for $p \to \infty$, as well as that of the residue $Z_2(p)$ at the pole ; in the case $\chi = +1$ we have

$$p \to 0 : \omega(p) \simeq m_f + \frac{1}{3}p \;\; ; \;\; Z_2 \simeq \frac{1}{2} + \frac{p}{6m_f} \tag{3.18a}$$

$$p \to \infty : \omega(p) \simeq p + m_f^2/p \;\; ; \;\; Z_2 \simeq 1 + \frac{m_f^2}{2p^2} \tag{3.18b}$$

while for $\chi = -1$ one finds :

$$p \to 0 : \omega(p) \simeq m_f - \frac{1}{3}p \;\; ; \;\; Z_2 \simeq \frac{1}{2} - \frac{p}{6m_f} \tag{3.19a}$$

$$p \to \infty \quad Z_2 \simeq \exp(-2p^2/m_f^2) \tag{3.19b}$$

One sees that the mode with $\chi = -1$ decouples at large p. The physical interpretation is clear : $\chi = +1$ corresponds to ordinary $T = 0$ Dirac particles ; at large values of p the propagation in the plasma induces only weak effects, in the form of a thermal mass $\sqrt{2}m_f$, while the $\chi = -1$ quasi-particles decouple. On the contrary, when $p \to 0$, helicity is no more defined and one cannot distinguish between $\chi = +1$ and $\chi = -1$ excitations: they both have the same mass m_f. Both $\chi = +1$ and $\chi = -1$ quasiparticles are collective excitations, but the latter is more collective than the former!

Let us conclude this subsection with three remarks

(i) We have not computed the imaginary part of Σ: it can be shown that this imaginary part is non-zero only below the light-cone, i.e. for $p_0^2 < \vec{p}^2$, and it is just given by the imaginary part of the Q_l functions in (3.10) and (3.11)

$$\ln\frac{x+1}{x-1} = \ln|\frac{x+1}{x-1}| - i\pi\theta(1-|x|) \tag{3.20}$$

We shall see in section 4 that this is a general feature of the leading term in T which is related to the so-called "Landau damping" mechanism.

(ii) It has been argued by Lebedev and Smilga[17] that one cannot observe in fact the behaviour (3.19b): for $p \geq gT$ the one-loop calculation is unstable and the $\chi = -1$ excitation disappears below the light-cone.

(iii) In general one does not expect $Z_2(\chi = +1) + Z_2(\chi = -1) = 1$, as could be concluded from (3.18)-(3.19). Indeed one can derive a sum rule analogous to (2.31), which tells us that $Z_2(\chi = +1) + Z_2(\chi = -1) \leq 1$. The continuum contribution is negligible for $p \to 0$ or $p \to \infty$, but in the general case one finds numerically[16]

$$0.8 \leq Z_2(\chi = +1) + Z_2(\chi = -1) \leq 1 \tag{3.21}$$

3.2 The photon propagator in a QED plasma

Before we study the gluon propagator, we first examine the simpler case of the photon propagator in a QED plasma. We write the photon propagator $D^{\mu\nu}$ in a covariant gauge, exhibiting the transverse, longitudinal and gauge-fixing parts[18]

$$iD^{\mu\nu} = \frac{1}{G-K^2}P_T^{\mu\nu} + \frac{1}{F-K^2}P_L^{\mu\nu} + \frac{\rho}{K^2}\frac{K^\mu K^\nu}{K^2} \tag{3.22}$$

with

$$P_T^{00} = P_T^{0i} = 0 \quad P_T^{ij} = \delta^{ij} - \frac{k^i k^j}{\vec{k}^2}$$

$$P_L^{\mu\nu} = \frac{K^\mu K^\nu}{K^2} - g^{\mu\nu} - P_T^{\mu\nu} \tag{3.23}$$

and ρ is the gauge parameter; in fact one can build a fourth tensor structure which cannot play any role in QED and seems also unimportant in QCD; thus we shall neglect this structure for simplicity. The polarization operator $\Pi^{\mu\nu}$ is defined as the difference between the full and free inverse propagators

$$\Pi_{\mu\nu} = -i(D^{-1}_{\mu\nu} - D^{-1}_{F,\mu\nu}) = FP^L_{\mu\nu} + GP^T_{\mu\nu} \tag{3.24}$$

Since Lorentz invariance is lost at finite T, the advantage of working in a covariant gauge are much less evident than at $T = 0$: popular choices of non-covariant gauges are the Coulomb and time-axial (TAG) gauges.

The results of a one-loop calculation have a rather simple expression in the high-temperature limit ($T \gg m$); keeping only terms proportional to T^2 one finds ($k = |\vec{k}|$)[19],[20]

$$F(k_0,k) = \frac{e^2T^2}{3}\left[1 - \frac{k_0^2}{k^2}\right]\left[1 - \frac{k_0}{2k}(\ln\left(\frac{k_0+k}{k_0-k}\right) - i\pi\theta(k^2-k_0^2))\right] \tag{3.25a}$$

$$G(k_0,k) = \frac{e^2T^2}{6}\left[\frac{k_0^2}{k^2} + (1 - \frac{k_0^2}{k^2})\frac{k_0}{2k}(\ln\left(\frac{k_0+k}{k_0-k}\right) - i\pi\theta(k^2-k_0^2))\right] \tag{3.25b}$$

Let us apply to the QED plasma a (weak) external classical electric field $\vec{E}_{cl}$; the perturbation is then

$$V = \int d^3x\ \vec{E}.\vec{E}_{cl} \tag{3.26}$$

The response of the plasma is a modification of the average external field $\vec{E}$, which of course vanishes in the absence of $\vec{E}_{cl}$. From linear response theory[21],[22], this average is given by the retarded commutator of electric fields

$$\delta < E_i(\vec{x},t) >= -i \int d^4x'\ E_{j,cl}(\vec{x}',t')\theta(t-t') < [E_i(x), E_j(x')] > \tag{3.27}$$

The retarded commutator is to be computed from the corresponding retarded commutator of the A_μ, namely

$$D^R_{\mu\nu} = \theta(t-t')\ < [A_\mu(x), A_\nu(x')] > \tag{3.28}$$

A straightforward calculation gives

$$E_i^{tot} = E_{i,cl} + \delta < E_i >= -i \int d^4x'\ E_{j,cl}(x')\left(-\partial_i\partial'_j D^R_{00} + \partial_i\partial'_0 D^R_{0j} + \partial_0\partial'_j D^R_{i0} - \partial_0\partial'_0 D^R_{ij}\right) \tag{3.29}$$

For a <u>static</u> electric field, this formula is easily translated into

$$E_i^{tot}(\vec{k}) = \frac{k_i k_j E_{j,cl}(\vec{k})}{\vec{k}^2 + F(k_0 = 0, \vec{k})} \tag{3.30}$$

In the $k_0 = 0, k \to 0$ limit, we get from (3.25a) the electric mass, which controls the screening of the Coulomb potential of a static charge Q

$$m_{el}^2 = F(k_0 = 0, k \to 0) = -\Pi_{00}(k_0 = 0, k \to 0) \simeq \frac{1}{3}e^2T^2 \tag{3.31}$$

Let us now turn to the plasma excitations, which are controlled by the transverse (G) and longitudinal (F) components of the polarization operator. We have for the transverse modes

$$\omega_T^2 = \vec{k}^2 + \mathrm{Re}G(\omega_T, k)$$

$$\gamma_T = -\frac{1}{2\omega}\mathrm{Im}G(\omega_T, k) \tag{3.32}$$

From the explicit expression (3.41b) of G one derives the following results

$$\omega, k \gg eT: \quad \omega_T^2 \simeq \vec{k}^2 + m_P^2 \qquad m_P^2 = \frac{1}{6}e^2T^2 \tag{3.33}$$

$$\omega, k \sim eT: \quad \omega_T^2 \simeq \omega_P^2 + \frac{6}{5}\vec{k}^2 \qquad \omega_P^2 = \frac{1}{9}e^2T^2 \tag{3.34}$$

where ω_P is the <u>plasma frequency</u>, and for $k \to 0$[23]

$$\gamma_T = \frac{e^2}{24\pi}\omega_P n_F\left(\frac{\omega_P}{2}\right) \simeq \frac{e^2}{24\pi}\omega_P \tag{3.35}$$

This last result is easy to understand on dimensional grounds, since γ_T is proportional to the available phase space, which is nothing but ω_P. Furthermore since $\omega_P \ll T$, the Fermi-Dirac distribution can be approximated by one; the linear dependence of (3.35) on n_F will be explained in the next section.

The dispersion law for longitudinal modes is given by

$$\omega_L^2 = \vec{k}^2 + \mathrm{Re}F(\omega_L, k)$$

We find that they do not propagate when $\omega, k \gg eT$, as the residue of the pole tends to zero exponentially; this is quite analogous to what happens for $\chi = -1$ fermions (see (3.21)). For $\omega, k \sim eT$ one finds

$$\omega_L^2 \simeq \omega_P^2 + \frac{3}{5}\vec{k}^2 \qquad \gamma_L = \gamma_T \tag{3.36}$$

The physical interpretation of the above results is quite analogous to that found for fermions: at $T = 0$, the physical modes are the transverse modes, and these are the only modes which propagate at $T \neq 0$ for large values of k; however the photons acquire a (gauge invariant) mass$\sim eT$: this behaviour is quite reminiscent of that found for $\chi = 1$ fermions. On the other hand the longitudinal modes, which are unphysical at $T = 0$, do not propagate at $T \neq 0$ for large values of the momenta. On the contrary, for $k = 0$, we have $\omega_T = \omega_L$ since one cannot distinguish between transverse and longitudinal modes, and for the same reason we must have $\gamma_T = \gamma_L$. We can see that we have three different masses, all of order eT, for dimensional reasons. However it must be understood that they have a quite different physical interpretation as they are obtained by taking different limits:

$$(i) \text{ the electric mass}: \ m_{el}^2 = \frac{1}{3}e^2T^2 \quad (k_0 = 0, k \to 0)$$

$$(ii) \text{ the T-dependent mass}: \ m_P^2 = \frac{1}{6}e^2T^2 \quad (k_0, k \gg eT)$$

$$(iii) \text{the plasma frequency}: \ \omega_P^2 = \frac{1}{9}e^2T^2 \quad (k_0 \sim eT, k = 0)$$

3.3 The gluon propagator in the QCD plasma

At first sight the situation in QCD looks very similar to that in QED, except for colour factors and the substitution $e \to g$, where g is the strong coupling constant. At one-loop order one finds in the high T-limit that the gluon polarization tensor is given by (3.41) with the substitution

$$e^2 \to g^2(N_c + \frac{1}{2}N_f)$$

where N_c is the number of colours and N_f the number of flavours. Thus one finds

(*i*) $m_{el}^2 = \frac{1}{3}g^2(N_c + N_f/2)T^2$. This electric mass gives the screening of the heavy quark potential, which may play a crucial role in the J/ψ suppression, if this suppression comes from the formation of a plasma. As already discussed, it can also be calculated on the lattice, with results in rough agreement with those of the perturbative calculation.

(*ii*) $m_P^2 = \frac{1}{6}g^2(N_c + N_f/2)T^2$, which gives the temperature-dependent mass of high energy, transversely polarized gluons in the plasma.

(*iii*) $\omega_P^2 = \frac{1}{9}g^2(N_c + N_f/2)T^2$, which gives the frequency of long wave-length transverse or longitudinal collective excitations, or plasma oscillations.

The crucial difference between QED and QCD is that the functions F and G in (3.25) are gauge dependent; fortunately the results for m_{el}, m_P and ω_P turn out to be gauge independent, at least to one-loop order and to leading order in T. However the plasmon damping rate γ does not seem to be gauge independent: this problem has been known as the "plasmon puzzle". Furthermore the linear response theory is complicated by the fact that the chromoelectric field is not linear in A

$$E_i^a = \partial_i A_0^a - \partial_0 A_i^a - g f_{abc} A_i^b A_0^c \tag{3.37}$$

where f_{abc} represents the Lie algebra structure constants. Thus the retarded commutator $[E, E]$ will involve not only terms bilinear in A, but also A^3 and A^4 terms. This problem is overcome in the TAG-gauge where

$$E_i^a = -\partial_0 A_i^a \tag{3.38}$$

but the TAG-gauge has problems of its own. The plasmon decay rate was first computed in this gauge and found to be[23]

$$\gamma = \frac{g^2 N_c T}{24\pi} \tag{3.39}$$

With respect to (3.35), the absence of a factor ω_P can be understood as follows : the main decay mechanism is the decay into two gluons, which implies a factor $n_B(\omega_P) \sim T/\omega_P$, because $\omega_P \sim gT \ll T$. On the other hand the contribution of $q\overline{q}$ pairs is negligible, and the result (3.55) is independent of N_f.

The calculation has been repeated in the Coulomb gauge [24], with the same result as in (3.28), and in covariant or background gauges, where it is found that (3.39) should be multiplied by a factor ~ -5 to -10. This would lead to plasma instability (as γ would be negative!), and some authors have argued that perturbation theory is unstable, because the expansion is performed around the wrong vacuum. However we shall see in the next section that the one-loop calculation is incomplete, as the final state gluons are bare transverse gluons, while they should be gluons dressed by their interactions in the plasma. Actually the decay of the plasmon into two plasmons (dressed gluons) is kinematically forbidden !

At this point it may be useful to add some comments on the gauge independence of perturbative calculations at finite temperature. At $T = 0$ it is well-known that physical quantities like S-matrix elements can be proven to be gauge independent order by order in the loop expansion, at least if one is dealing with a broken gauge theory in order to avoid infrared divergences. However in the calculation of a dispersion law the loop expansion breaks down, as will be explained in detail in the next section. Nevertheless Kobes et al.[25] have been able to prove that dispersion laws are gauge independent: the locations of the physical poles in the gluon propagator are gauge independent in spite of the gauge dependence of the propagator itself. It follows that in any consistent approximation scheme, the dispersion laws should be found to be gauge independent. The loop expansion is certainly not such a scheme, since contributions from higher order loops are of the same order in g as the lower order terms. As will be shown in the next section, the effective expansion seems to provide such a consistent scheme: at least the results for the plasmon damping rate have been found to be identical in a general covariant gauge and in the Coulomb gauge. Athough this result is certainly very encouraging, it cannot be considered as a proof that the approximation scheme does give all the leading order terms!

4. Resummation methods in perturbation theory

In this section we shall examine resummation methods which are necessary in order to supplement perturbation theory, because of the singular infrared behaviour.

4.1 The φ^4-model revisited

Let us come back to the massless $g^2\varphi^4$- model which was already studied in subsection 2.1 ; it was shown there that the first order correction to the propagator is of order g^2T^2. This result has the following consequence: assume that we look at the propagator of a particle with a "soft momentum", which means that its energy p_0 and its momentum p are $\sim gT$ (we recall our convention that 4-momenta are denoted by upper case letters, energies and 3-momenta by the corresponding lower case letters: $P^\mu = (p_0, \vec{p})$). For such a "soft particle", the inverse propagator is of order g^2T^2, but the first order correction Π_1 to Π_β (see (2.9)) is also of order g^2T^2. This means that one cannot limit oneself to a naïve perturbative expansion, but that one must use some kind of resummation. The solution in this simple case is obvious : one has to use the summation drawn in fig.14[26], or even better one can use a self-consistent equation[27]:

$$\Pi = \frac{g^2}{2} T \sum_n \int \frac{d^3p}{(2\pi)^3} \frac{1}{\omega_n^2 + \vec{p}^2 + \Pi} \tag{4.1}$$

Having performed the summation over n and the $T = 0$ mass renormalization, we are led for the T-dependent part to the equation

$$1 = \frac{g^2}{4\pi^2} \int_1^\infty \frac{\sqrt{x^2-1}\, dx}{\exp(\beta \Pi_\beta^{1/2} x) - 1} \tag{4.2}$$

The expansion in powers of g is obtained from that of the function $F(u)$:

$$F(u) = \int_1^\infty dx \frac{\sqrt{x^2-1}}{e^{ux}-1} = \frac{2\pi^2}{u^2}\left(\frac{1}{12} - \frac{u}{4\pi} + O(u^2 \ln u)\right) \tag{4.3}$$

The results are plotted in fig.15[27], where one can see that the deviation with respect to the first order calculation is rather small. However the expansion in powers of g reveals an interesting feature :

$$\Pi_\beta = \frac{g^2T^2}{24}[1 - 3(\frac{g^2}{24\pi^2})^{1/2} + ...] \tag{4.4}$$

We see that the next term in Π_β is not in g^4, as one would expect from naïve perturbation theory, but in g^3. This is what could be called a "mild violation" of perturbation theory, to be contrasted with the strong violations which will be seen in the next section.

The same feature also appears in the perturbative expansion of the partition function. The same resummation method leads to the so-called "ring diagrams" (fig.16), which give the following contribution

$$\frac{1}{2}\beta V T \sum_n \int \frac{d^3p}{(2\pi)^3} \sum_{N=2}^\infty \frac{1}{N} (-\Pi_1(\omega_n, \vec{p}) \Delta_F(\omega_n, \vec{p}))^N$$

$$= -\frac{1}{2}\beta V T \sum_n \int \frac{d^3p}{(2\pi)^3} (\ln(1 + \Pi_1(\omega_n,\vec{p})\Delta_F(\omega_n,\vec{p})) - \Pi_1(\omega_n,\vec{p})\Delta_F(\omega_n,\vec{p})) \tag{4.5}$$

The summation over N begins at $N = 2$, because the graph of fig.17 has already been taken into account at first order ; $\Pi_1 = g^2T^2/24$ is the first order temperature-dependent self-energy, and the explicit expression of (4.5) is, with $\lambda = g^2/24$

$$-\frac{1}{2}\beta V T \sum_n \int \frac{d^3p}{(2\pi)^3} \left(\ln(1 + \frac{\lambda T^2}{\omega_n^2 + \vec{p}^2}) - \frac{\lambda T^2}{\omega_n^2 + \vec{p}^2} \right) \tag{4.6}$$

The mode $\omega_n = 0$ is easily seen to give a contribution proportional to $\lambda^{3/2}$, corresponding again to a mild violation of perturbation theory. From the explicit evaluation of (4.6) and of the graph in fig.17, one obtains for the pressure[26]

$$P = \frac{\pi^2 T^4}{90}[1 - \frac{15}{8}(\frac{\lambda}{\pi^2}) + \frac{15}{2}(\frac{\lambda}{\pi^2})^{3/2} + ...] \tag{4.7}$$

The first term corresponds to the free boson gas, the second one to the graph of fig.17, and the last one to the ring diagrams.

In φ^4-theory, this use of a dressed propagator is all we have to do in order to deal with the IR behaviour. Indeed the dominant correction at one-loop order is the tadpole graph of fig.5 which behaves as T^2. The one-loop correction to the coupling constant is only logarithmically divergent, and, as we shall see below, its thermal part is in fact proportional to $\ln T$.

4.2 Remarks on the one-loop correction to the propagator

Before going to QCD, it is useful to examine in some detail a one-loop graph more complicated than a tadpole. Still in the case of a scalar theory, we consider a $\varphi\varphi_1\varphi_2$ coupling and the one-loop correction to the φ-propagator : see fig.7, which also defines the kinematics. From the Feynman rules and the techniques described in section 2, we get[28]

$$\Pi = g^2 \int \frac{d^3k}{(2\pi)^3}\frac{1}{4E_1E_2}\left(\frac{1+n_1+n_2}{p_0 - E_1 - E_2} + \frac{n_2 - n_1}{p_0 - E_1 + E_2} + 1 \leftrightarrow 2 - \frac{1+n_1+n_2}{p_0 + E_1 + E_2}\right) \tag{4.8}$$

where :

$$E_1 = E_{\vec{k}} = k \;\; ; \;\; E_2 = E_{|\vec{p}-\vec{k}|} = |\vec{p} - \vec{k}| \tag{4.9}$$

while n_1 and n_2 are the corresponding Bose-Einstein factors. The physical interpretation of the various terms in Π is made clearer if one looks at the discontinuity of Π; since Π is the continuation to real-time of the imaginary-time self-energy, it is in fact a retarded Green's function, and p_0 in (4.9) should be interpreted as $p_0 + i\epsilon$. Thus

$$\begin{aligned}\text{Disc}\,\Pi = -2i\pi g^2 \int \frac{d^3k}{(2\pi)^3}\frac{1}{4E_1E_2}[(1+n_1+n_2)\delta(p_0 - E_1 - E_2) \\ + (n_1 - n_2)\delta(p_0 - E_1 + E_2) + 1 \leftrightarrow 2 - (1+n_1+n_2)\delta(p_0 + E_1 + E_2)]\end{aligned} \tag{4.10}$$

Let us look at the Bose-Einstein factor $1+n_1+n_2$ in the first term of (4.10). The discontinuity of Π is proportional to the <u>difference</u> (the sum for fermions)

$$\Gamma = \Gamma_d - \Gamma_c \tag{4.11}$$

between the decay and creation rates of the particle φ; the decay process, for $p_0 = E_1 + E_2$, is $\varphi \to \varphi_1 + \varphi_2$, and the creation process is the inverse process $\varphi_1 + \varphi_2 \to \varphi$. The Bose-Einstein factor is thus :

$$(1+n_1)(1+n_2) - n_1 n_2 = 1 + n_1 + n_2 \tag{4.12}$$

At zero temperature, one has of course $\Gamma_c = 0$ and Γ_d is proportional to the factor of one in (4.12). Similarly the second term in (4.10), with $p_0 = E_1 - E_2$, corresponds to the reactions $\varphi_1 \to \varphi + \overline{\varphi}_2$ (creation) and $\overline{\varphi}_2\varphi \to \varphi_1$ (decay), the Bose-Einstein factor being

$$n_2(1+n_1) - n_1(1+n_2) = n_2 - n_1 \tag{4.13}$$

In all cases we have of course the detailed balance condition

$$\Gamma_d = e^{\beta p_0} \Gamma_c \tag{4.14}$$

In the case of a small deviation from thermal equilibrium, Γ should be interpreted as the inverse of the relaxation time which governs the approach to equilibrium[28].

The terms where $p_0 = E_1 - E_2$ or $p_0 = E_2 - E_1$ correspond to the so-called "Landau damping" mechanism: particles disappear or are created through scattering in the bath, and not via the processes which are avaible at zero temperature.

We are going to be interested in a situation where external particles are soft: p_0, $p \sim gT \ll T$. Momenta of order T are called hard momenta : since the T-dependent part of loop integrations is limited by the thermal Bose-Einstein or Fermi-Dirac factors, these are typical momenta running around a loop. We need kinematical approximations for loop momenta $k \sim T$ and external momental $p \sim gT \ll T$. Take as an example the expression (4.8) ; we have

$$\begin{gathered} E_1 = k \quad ; \quad E_2 = |\vec{p} - \vec{k}| \simeq k - p\cos\theta \\ p_0 - E_1 - E_2 \simeq -2k \quad ; \quad p_0 - E_1 + E_2 \simeq p_0 - p\cos\theta \\ n_1 \simeq n_2 \simeq \frac{1}{e^{\beta k} - 1} \sim 1 \\ n_1 - n_2 \simeq p\cos\theta \, \frac{dn}{dk} = -\beta p \cos\theta \; n(k)(1 + n(k) \sim -\beta p\cos\theta \end{gathered} \tag{4.15}$$

The last approximations for n_1, n_2 and $n_1 - n_2$ are to be used for power counting only. Using these approximations, it is clear that the first and last term of (4.8) diverge as $\ln T$, while the second and third term behave as a constant.

4.3 Hard thermal loops in QCD and the effective expansion

In what follows we shall call hard thermal loops (HTL) one-loop integrals which diverge as T^2[29],[30]. The simplest example is of course the tadpole in $g^2\varphi^4$. From naïve power counting one could imagine that the only HTL correspond to loops which diverge quadratically when $T = 0$. Thus the gluon propagator for instance should have HTL. However this reasoning misses most of the HTL : even diagrams which are UV-convergent, such as the N-gluons Green's function with $N \geq 5$ do have HTL. Except in the case $N = 2$, HTL come uniquely from terms analogous to the 2nd and 3rd term in (4.8). In (4.8), one can see that one power of k is gained for soft external momenta since

$$p_0 - E_1 + E_2 \simeq p_0 - p\cos\theta$$

Thus the integral diverges linearly, and not logarithmically.

As we just mentioned, the case of the gluon propagator is special because one finds quadratic divergences at $T = 0$. The 4-gluon vertex will contribute via the tadpole graph, and there will be a further term coming from the combination $(n_1 + n_2)$ in (4.8). This term is tadpole like, in the sense that its imaginary part is zero at order T^2. On the other hand there will be terms corresponding to the combination $n_1 - n_2$: since

$$\int_{-1}^{1} \frac{\mathrm{d}(\cos\theta)}{p_0 - p\cos\theta} = \frac{1}{p} \ln \frac{p_0 - p}{p_0 + p} \tag{4.16}$$

these terms will have an imaginary part of order T^2 in the region $-p \leq p_0 \leq p$, i.e. below the light cone. This property is typical of Landau damping.

The quark propagator is linearly divergent at $T=0$; thus the terms of order T^2 which we found in subsection 3.1 were typical of Landau damping. Let us now examine the one-loop correction Γ_1 to the 3-gluon vertex (fig.18). Since the external momenta are supposed to be soft, the 3-gluon coupling can be simplified, by retaining only terms proportional to K^μ. After summation over n, the leading contribution to Γ_1 is given by a sum of terms of the form :

$$\Gamma_1 \sim g^3 \int \frac{\mathrm{d}^3k}{(2\pi)^3} \frac{K^{\mu_1}K^{\mu_2}K^{\mu_3}}{E_k E_{p_1+k} E_{p_2-k}} (n(E_k) - n(E_{p_1+k}))$$
$$\frac{1}{(p_{10} - E_k + E_{p_1+k})(p_{20} - E_k + E_{p_2-k})} \qquad (4.17)$$

Power counting is very simple (see (4.15))

$$\Gamma_1 \sim g^3 T^3 \frac{T^3}{T^3} \frac{P}{T} \frac{1}{P^2} \sim g^2 T$$

Since $\Gamma_0 \sim g(gT) = g^2 T$, we see that the HTL is of the same order of magnitude as the zero-loop approximation. This means for example that the 2-loop correction to the gluon propagator in fig.19 is of the same order as the one-loop correction, and the effective expansion will use an effective vertex in order to sum Γ_0 and Γ_1.

Power counting for HTL has been established in[29]

- $\int \mathrm{d}^3k \to T^3$

- 1st propagator $\to (T\sum_n) : T^{-1}$

- each additional propagator : $(PT)^{-1}$

- powers of K^μ : T

- Statistical factor : P/T

Hard thermal loops enjoy remarkable properties[29]-[32]

(i) HTL exist only for Green's functions with N external gluons lines or $(N-2)$ external gluon and 2 external quark lines. There are no HTL with external ghost lines.

(ii) HTL are totally symmetric with respect to external gluon indices; the colour factors are $(N_c + N_f/2)$ for HTL with external gluons only and C_F for HTL with two external quark lines.

(iii) HTL do not involve the 4-gluon vertex in the Feynman and Coulomb gauges, except in the case of the gluon propagator (contribution to the tadpole)

(iv) HTL are gauge independent: at least it has been shown in [29] and [30] that they are the same in general covariant, Coulomb and axial gauges; a general proof has been given in[25].

(v) HTL obey Ward identies analogous to those of QED; calling $\delta\Pi$ etc. the HTL correction and ignoring the colour factors for notational simplicity we have :

$$\text{propagator} \ : P^\mu \delta\Pi_{\mu\nu}(P) = 0 \qquad (4.18)$$
$$3 - \text{gluon vertex} \ : K^\mu \delta\Gamma_{\mu\nu\rho}(K,P,Q) = \delta\Pi_{\nu\rho}(P) - \delta\Pi_{\nu\rho}(Q) \qquad (4.19)$$
$$\text{quark gluon vertex} \ : K^\mu \delta\tilde{\Gamma}_\mu(K,P,Q) = i\delta\Sigma(P) - i\delta\Sigma(Q) \qquad (4.20)$$

where Σ is the quark self-energy, Γ the 3-gluon vertex and $\tilde{\Gamma}$ the quark-gluon vertex. In general, K^μ dotted into a N-point HTL gives a combination of $(N-1)$-point HTL.

From our study of HTL we can establish the rules for an effective expansion which gives consistently the lowest order term in the coupling constant for soft external momenta. The effective expansion is similar to the ordinary perturbation theory, except that for soft external momenta bare propagators and vertices are replaced by propagators and vertices which include the contribution of HTL. Topologically many diagrams are the same as in the bare expansion, but one may also find diagrams built out of effective vertices which do not have bare counterparts. The effective propagators resum all insertions of HTL: $\delta\Pi$ for gluons and $\delta\Sigma$ for quarks. The effective vertices are formed by adding HTL to the bare vertices: $\Gamma \to \Gamma + \delta\Gamma$.

4.4 The plasmon damping rate

The gluon inverse propagator is written

$$iD^{-1}_{\mu\nu} = P^2 g_{\mu\nu} - P_\mu P_\nu + G.F.T - \delta\Pi_{\mu\nu} - \Pi^*_{\mu\nu} \tag{4.21}$$

where the first three terms correspond to the bare inverse propagator, $\delta\Pi_{\mu\nu}$ is the HTL correction and $\Pi^*_{\mu\nu}$ is given by the effective expansion (fig.20). One notes that $\delta\Pi_{\mu\nu}$ does not contribute to the damping rate, because $\mathrm{Im}\delta\Pi = 0$ "on mass-shell", from energy-momentum conservation : a plasmon of mass ω_P cannot decay into two plasmons of mass ω_P. Thus the damping rate will be down by one power of g with respect to the HTL, namely

$$\Pi^* \sim g(gT)^2 \quad ; \quad \mathrm{Im}\Pi^* \neq 0 \tag{4.22}$$

To leading order the plasmon polarisation $e_\mu(P)$ obeys :

$$(D^{-1}_{F,\mu\nu} - \delta\Pi_{\mu\nu})e^\nu(P) = 0 \tag{4.23}$$

and

$$P^\mu e_\mu(P) = 0 \ \text{(covariant gauge)}$$
$$p_i e_i(P) = 0 \ \text{(Coulomb gauge)}$$

The gauge invariant decay rate is given by

$$\mathrm{Im} e^\mu(P)\Pi^*_{\mu\nu}e^\nu(P)$$

in a covariant gauge, and by a similar expression in the Coulomb gauge. The calculation is done by using the spectral representation (2.30) of propagators

$$\rho(k_0, \vec{k}) = Z(\vec{k})\left(\delta(k_0 - \omega(\vec{k})) - \delta(k_0 + \omega(\vec{k}))\right) + \beta(k_0, \vec{k})\theta(k^2 - k_0^2) \tag{4.24}$$

and the result for the damping rate γ (cf. (3.39)) is[33]

$$\gamma = a\,\frac{g^2 N_c T}{24\pi} \quad ; \quad a \simeq 6.63 \tag{4.25}$$

the result being the same in a general covariant and in the Coulomb gauge. The factor of a in (4.25) is given by a rather long formula

$$a = 9\int_0^\infty dk \int_{-\infty}^{+\infty} \frac{dk_0}{k_0} \int_{-\infty}^{+\infty} \frac{dk_0'}{k_0'} \delta(\omega_P - k_0 - k_0')$$
$$(2(k^2 + k_0 k_0')^2 \rho_T(k_0, k)\rho_L(k_0', k) + k^4 \rho_L(k_0, k)\rho_L(k_0', k) -$$

$$-\frac{k^2}{\omega_P^2}(k^2-k_0^2)^2\rho_T(k_0,k)\rho_L(k_0',k)+\frac{k_0}{6k^3}(k^2-k_0^2)^2\theta(k^2-k_0^2)\rho_T(k_0',k)) \qquad (4.26)$$

where ρ_T and ρ_L represent the spectral functions of the transverse and longitudinal gluons respectively.

4.5 Soft photon production in a plasma

The same techniques have been used in order to compute the production rate of lepton pairs, via virtual photons in a plasma, when the virtual photons have a mass$\sim gT$[34]. The result (fig.21) has a number of interesting features, due to the opening of various thresholds for $\chi = +1$ and $\chi = -1$ fermionic quasi-particles (see subsection 3.1). In the low-momentum region, the result is larger by several order of magnitudes than the naïve Drell-Yan expectation.

5. Conclusions and outlook

We have seen that the quark-gluon plasma displays a very rich structure: one finds Debye screening, plasma oscillations, collective excitations with quark quantum numbers etc. Unfortunately there does not seem to exist a direct (or even indirect) way of detecting these structures in present experiments. Nevertheless it is clear that a good knowledge of these structures is essential if we want to make accurate predictions on possible signals of the plasma as well as on its evolution between formation and hadronization.

From a theoretical point of view, although important progress has been made recently in our understanding of the infrared structure of thermal QCD[29]-[34], much remains to be done before we can really claim to understand fully this IR behaviour. Also it is clear that the perturbative predictions can be at best qualitative since the region in T where it could be trusted ($T/T_c \sim 2$), the coupling constant g is of order one, and there is really no hierarchy of scales.

Finally, for lack of time and/or competence, a number of interesting topics had to be left out. Among these topics one could quote

- the problem of infrared divergences of the partition function in perturbative QCD[35]

- the role of screening in transport phenomena[36],[37]

- the generalization of the Kinoshita-Lee-Nauenberg theorem at finite temperature[38]-[44]

- the problem of chiral anomalies at finite temperature[45]

- the perturbative calculation of interface tension[46]

- scalar particle emission (axions) from supernovae[47]

- the general theory of transport phenomena in the quark-gluon plasma[37]

- the role of instantons at finite temperature[48],[49].

In any case we hope to have convinced the reader that hot QCD is an exciting topic with a lot of open problems.

References

[1] For a recent review see e.g. H. Satz, Plenary Talk at the ECFA Workshop on Large Hadron Colliders, Aachen, CERN preprint TH.5917/90 (1990).

[2] For a recent review see e.g. A. Ukawa, Plenary talk at the 25th International Conference on High Energy Physics, Singapore, preprint UTHE-P213 (1990)

[3] P. Gerber and H. Leutwyler, Nucl. Phys. **B321** (1989) 387.

[4] A. Barducci, R. Casalbuoni, S. De Curtis, R. Gatto and G. Pettini, Phys. Rev. **D41** (1990) 1610.

[5] U. Heller and F. Karsch, Nucl. Phys. **B251**[FS13] (1985) 254; **B258** (1985) 29; F. Karsch, CERN preprint TH.4851/87 (1987); F. Karsch, CERN preprint TH.5498/89 (1989) to appear in "Quark Gluon Plasma" R. C. Hwa ed. World Scientific.

[6] J. Engels et al., Phys. Lett. **252** (1990) 625; B. Petersson, invited talk at Quark Matter'90, Menton, France, Nucl. Phys. **A525** (1991) 237.

[7] M. Le Bellac, Lectures at the XXXth Schladming Winter School, INLN preprint 1991/8, to be published in Lectures Notes in Physics.

[8] For a review of the imaginary time formalism see e.g.: J. Kapusta, "Finite temperature field theory", Cambridge University Press (1989).

[9] For a review of the real time formalism, see e.g.: N. Landsman and Ch. van Weert, Phys. Rep. **145** (1987) 142; this review contains a complete list of references prior to 1987.

[10] L. Dolan and R. Jackiw, Phys Rev **D9** (1974) 3320.

[11] R. Kobes and G. Semenoff, Nucl. Phys. **260** (1985) 714; Nucl. Phys. **B272** (1986) 329.

[12] R. Kobes, Phys. Rev. **D42** (1990) 562; University of Winnipeg preprint, to be published in Phys. Rev. (1990).

[13] H. A. Weldon, Phys. Rev. **D26** (1982) 2789.

[14] V. Klimov, Sov. Journ. Nucl. Phys. **33** (1981) 934; O. Kalashnikov, Forsch. Phys. **32** (1984) 525.

[15] K. Kowalski, Zeit. Phys. **C36** (1987) 665.

[16] H. A. Weldon, Phys. Rev. **D40** (1990) 2410.

[17] V.Lebedev and A. Smilga Ann. Phys. (N-Y) **202** (1990) 229; Bern preprint BUTP-90/38 (1990).

[18] Kapusta, ref[15] chap.5.

[19] V. Silin, Sov. Phys. JETP **11** (1960) 1136.

[20] H. A. Weldon, Phys. Rev. **D26** (1982) 1394.

[21] A. Fetter and J. D. Walecka, "Quantum Theory of Many Particle Systems", Mc Graw Hill (1971), chap.5.

[22] Kapusta, ref[15] chap.6.

[23] K. Kajantie and J. Kapusta, Ann. Phys. (N-Y) **160** (1985) 477.

[24] U. Heinz, K.Kajantie and T. Toimela, Ann. Phys. (N-Y) **176** (1987) 218.

[25] R. Kobes, G. Kunstatter and A. Rebhan, Phys. Rev. Lett. **64** (1990) 2992; CERN preprint TH.5937/90 (1990).

[26] Kapusta, ref[15] chap.4.

[27] T. Altherr, Phys. Lett. **B238** (1990) 360.

[28] H. A. Weldon, Phys. Rev. **D28** (1983) 2007.

[29] R. Pisarski, Nucl. Phys. **B309** (1988) 476; R. Pisarski, Phys. Rev. Lett. **63** (1989) 1129; E. Braaten and R. Pisarski, Nucl. Phys.**B337** (1990) 569.

[30] J.Frenkel and J.C. Taylor, Nucl. Phys. **B334** (1990) 199.

[31] E. Braaten and R. Pisarski, Nucl. Phys. **B339** (1990) 310.

[32] J. C. Taylor and S. M. Wong, Nucl. Phys. **B346** (1990) 115.

[33] E. Braaten and R. Pisarski, Phys. Rev. **D42** (1990) 2156.

[34] E. Braaten, R. Pisarski and T. Yuan, preprint NUHEP-90-1(1990), to be published

[35] A. D. Linde, Phys. Lett. **B96** (1980) 289.

[36] G. Baym, H. Monien, C. Pethick and D. Ravenhall Phys. Rev. Lett. **64** (1990) 1867.

[37] For a review of transport phenomena in the quark-gluon plasma, see e.g. H.-T. Elze and U. Heinz, Phys. Rep. **183** (1989) 81.

[38] T. Grandou, M. Le Bellac and J.L. Meunier, Z. Phys. **C43** (1989) 575.

[39] R. Baier, B. Pire and D. Schiff, Phys. Rev. **D38** (1988) 2814.

[40] T. Altherr, P.Aurenche and T. Becherrawy, Nucl. Phys. **B315** (1989) 436; T. Altherr and T. Becherrawy, Nucl. Phys. **B330** (1990) 174; T. Altherr and P.Aurenche, Z. Phys. **C45** (1990) 99.

[41] J. Cleymans and I. Dadic, Zeit. Phys. **C42** (1989) 133.

[42] Y. Gabellini, T. Grandou and D. Poizat, Ann. Phys. (N.Y.) **202** (1990) 436.

[43] T. Grandou, M. Le Bellac and D. Poizat, Phys. Lett. **B249** (1990) 478; INLN preprint 1991/1 to be published in Nucl. Phys. **B**.

[44] T. Altherr, Annecy preprint LAPP-Th-325/91(1991).

[45] A. Das and A. Karev, Phys. Rev. **D36** (1987) 623; Y. Liu and G. Ni, Phys. Rev. **D38** (1988) 3840; C. Contreras and M. Loewe, Zeit. Phys. **C40** (1988) 253.

[46] T. Bhattacharaya, A. Gocksch, C. Korthals-Altes and R. Pisarski, Phys. Rev. Lett. **66** (1991) 998.

[47] J. Ellis and P. Salati, Nucl. Phys. **B342** (1990) 317; T. Altherr, Zeit. Phys. **C47** (1990) 559 and LAPP preprint (1990), to be published in Ann. Phys. (N-Y).

[48] D. Gross, R. Pisarski and L. Yaffe, Rev. Mod. Phys. **53** (1981) 43.

[49] E. Shuryak, "The QCD Vacuum, Hadrons and the Superdense Matter" World Scientific (1988) chap. 10; invited talk at Quark Matter'90, Menton, France, Nucl. Phys. **A525** (1991) 3.

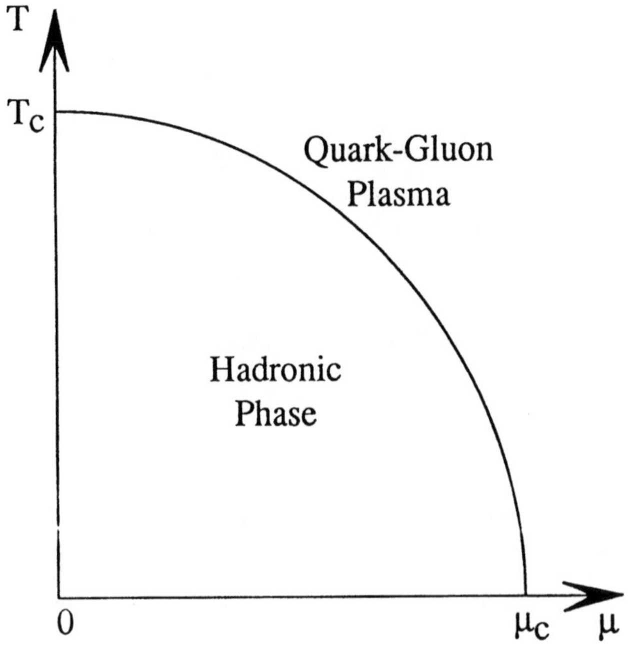

Fig.1 The phase diagram in the $\mu - T$ plane.

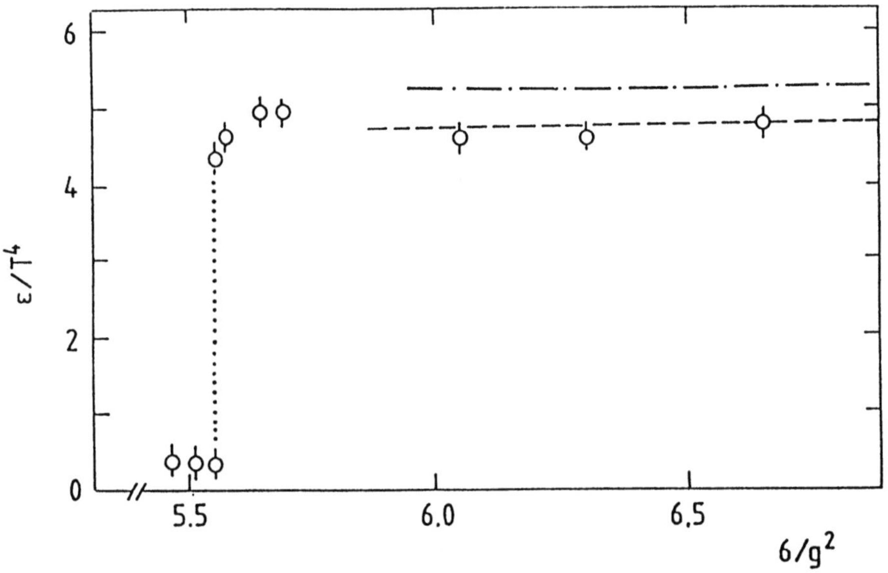

Fig.2 The ratio ε/T^4 as a function of $6/g^2$ in pure $SU(3)$[5].

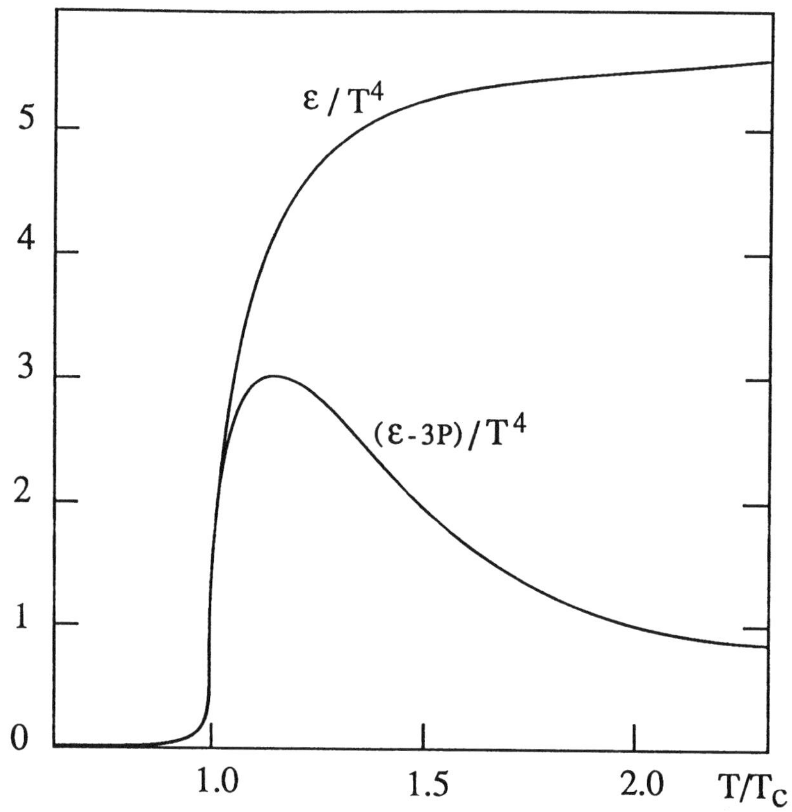

Fig.3 The ratios ε/T^4 and $(\varepsilon - 3P)/T^4$ as a function of T/T_c[6].

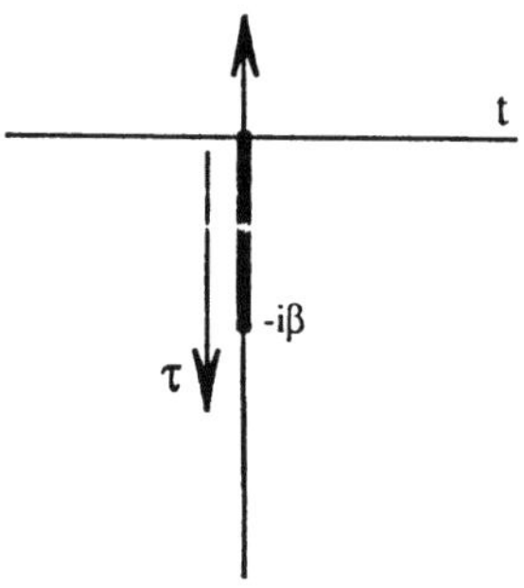

Fig.4 The imaginary time axis.

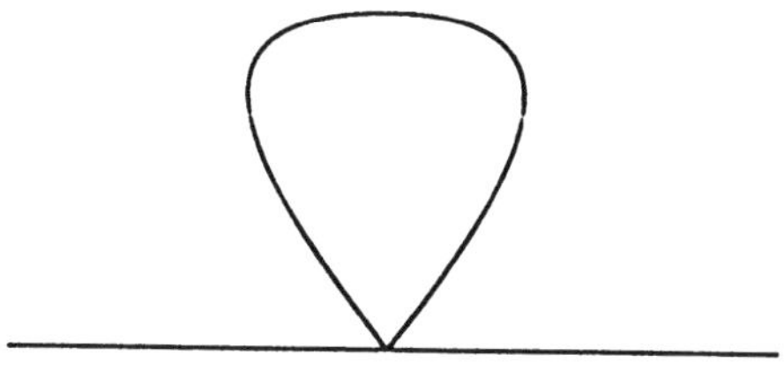

Fig.5 The tadpole in $g^2\varphi^4$.

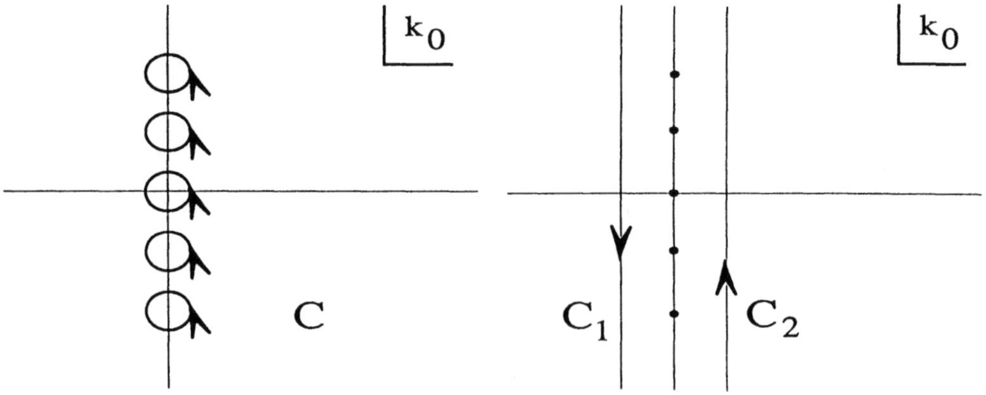

Fig.6 The integration contour in the p_0 complex plane.

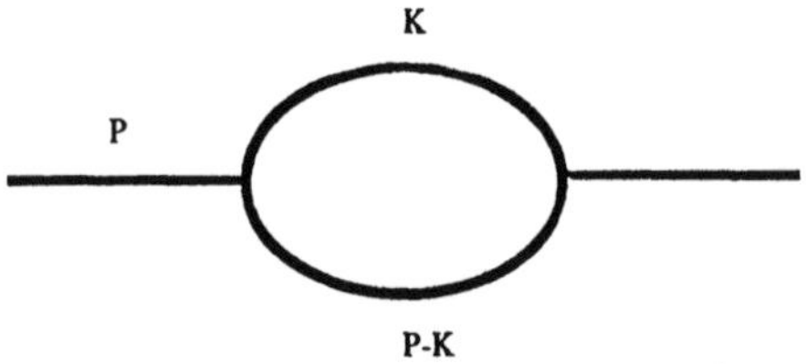

Fig.7 A one-loop contribution to the self-energy.

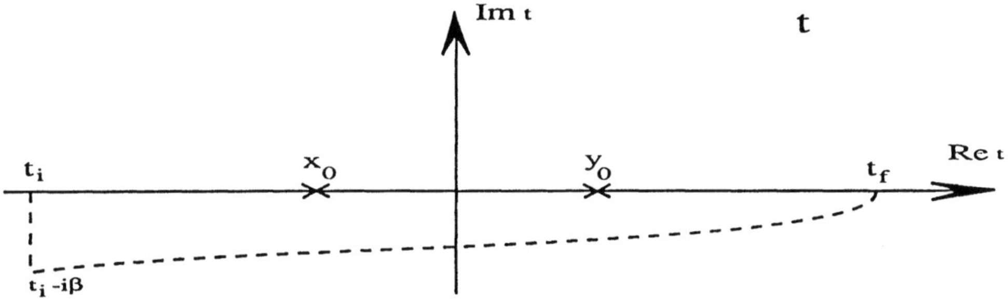

Fig.8 A possible time-path in the real-time formalism.

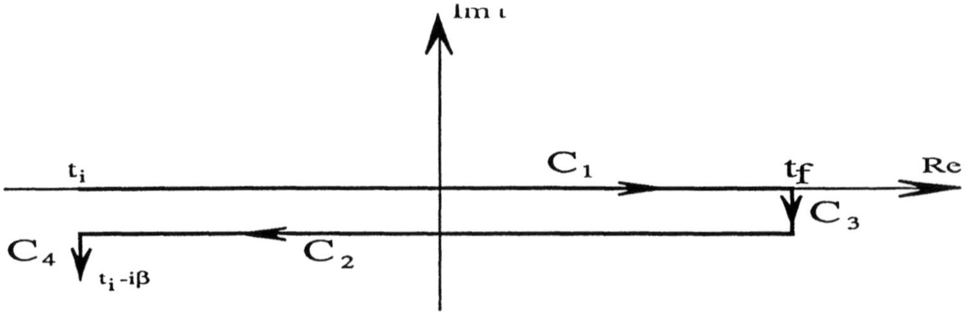

Fig.9 The usual time-path in the real-time formalism.

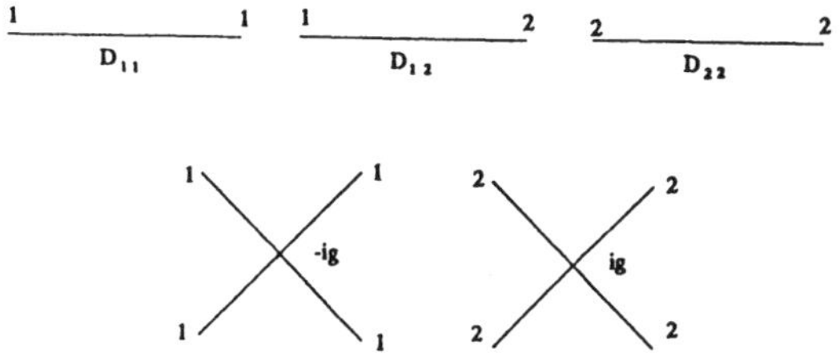

Fig.10 Feynman rules in real-time.

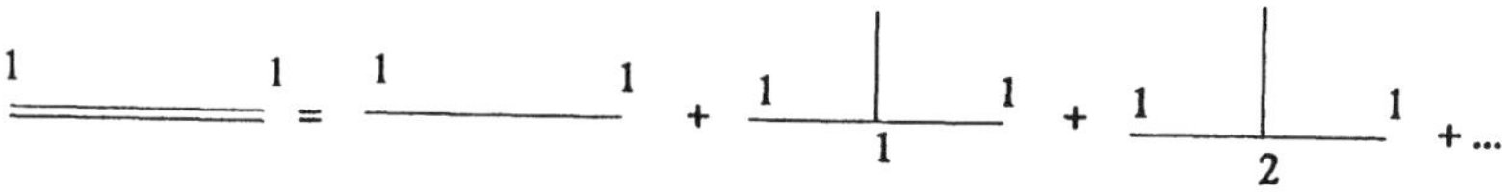

Fig.11 The perturbative expansion of the propagator.

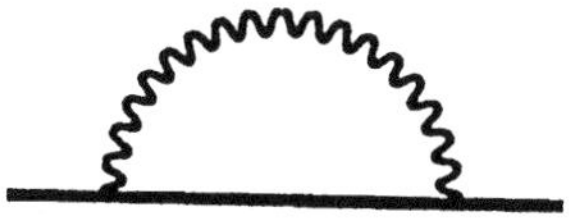

Fig.12 The fermion self-energy to one-loop order (full lines: quarks; wavy lines: gluons).

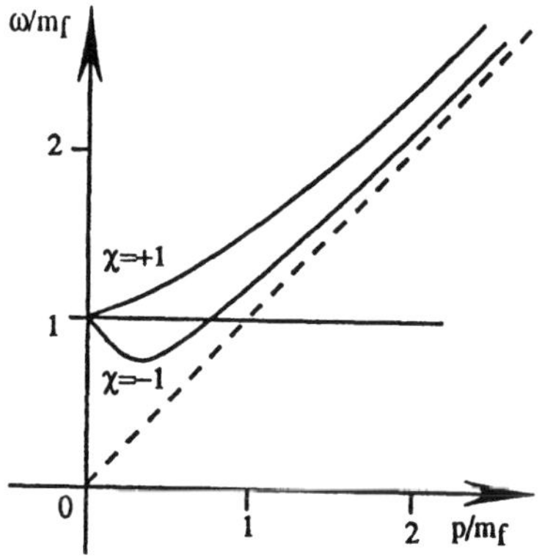

Fig.13 The dispersion laws for the fermionic excitations.

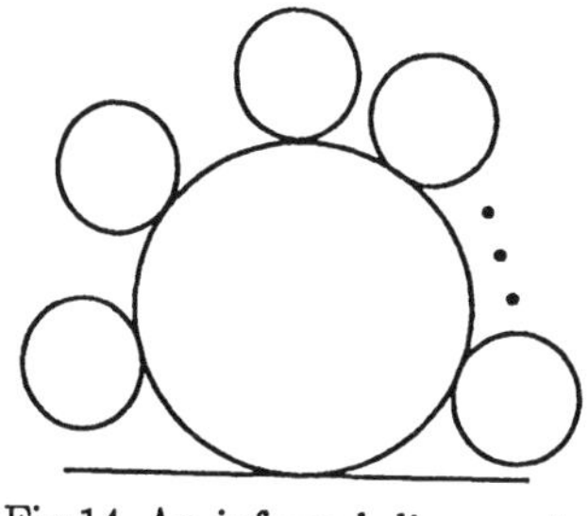

Fig.14 An infrared divergent contribution to the propagator.

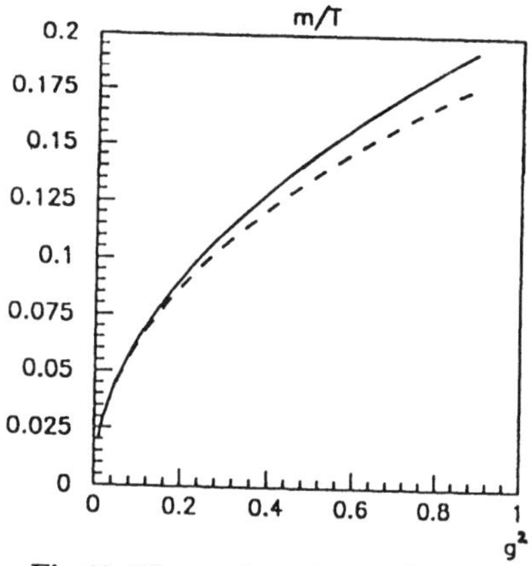

Fig.15 The ratio m/T in $g^2\varphi^4$[27].

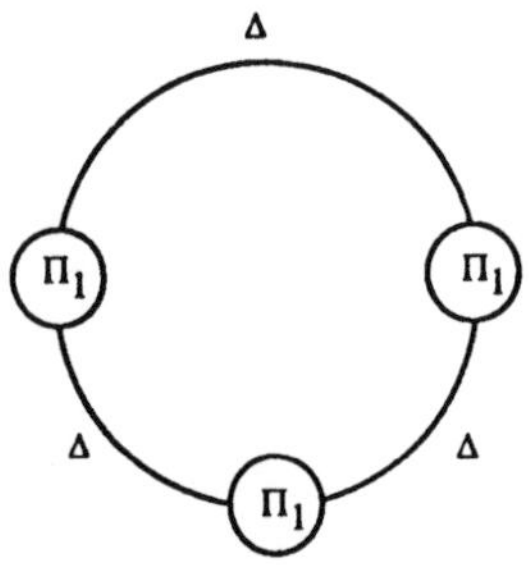

Fig.16 An example of ring diagram contributing to order g^3 to the partition function.

Fig.17 The graph contributing to order g^2 to the partition function.

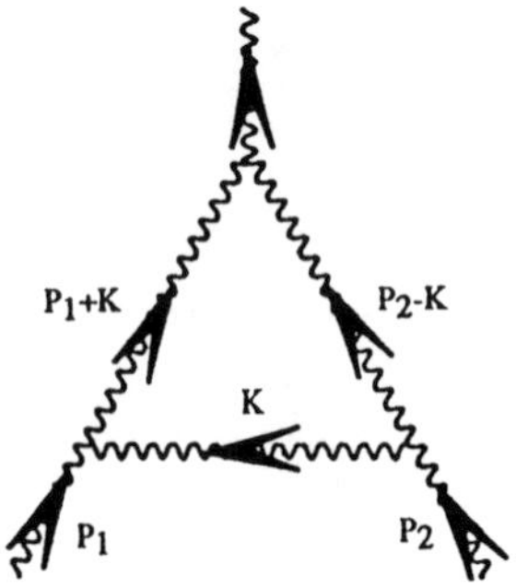

Fig.18 The one-loop correction to the three-gluon vertex (wavy lines: gluons).

Fig.19 An effective vertex.

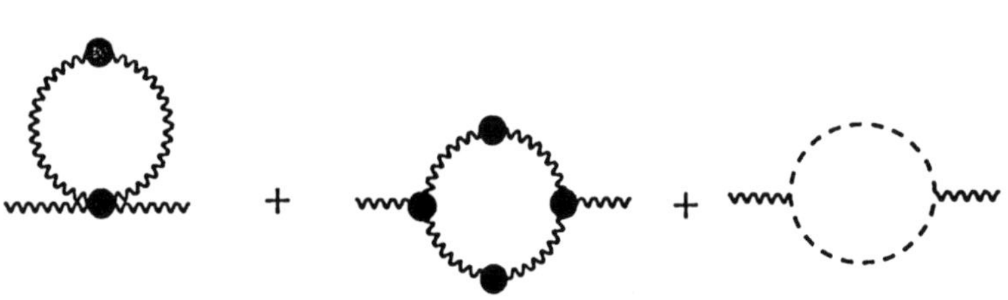

Fig.20 Effective expansion graphs for the plasmon damping rate (dotted lines: ghosts; black blobs: effective vertices and propagators).

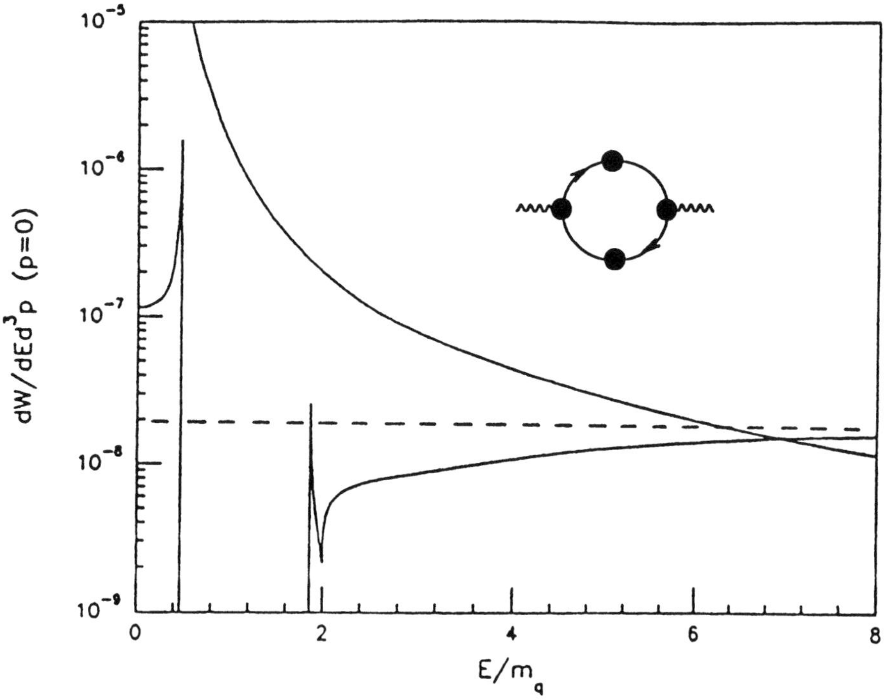

Fig.21 The cross-section for lepton-pair production (wavy lines: external photons)

NEW APPROACH TO THE SOLUTION OF NONLINEAR PROBLEMS

Carl M. Bender
Department of Physics
Washington University
St. Louis, MO 63130
USA

ABSTRACT

In this talk we introduce a new technique, called the delta expansion, which can be used to solve nonlinear problems in both classical and quantum physics. The idea of the delta expansion is to expand in the power of a nonlinear term. For example, to treat a y^4 term, we introduce a small parameter delta and consider a $(y^2)^{(1+\delta)}$ term. When we expand in powers of delta, the resulting perturbation series appears to have a finite radius of convergence and numerical results are superb. We illustrate the delta expansion by applying it to various difficult nonlinear equations taken from classical physics, the Lane-Emden, Thomas-Fermi, Blasius, Duffing, Burgers, and Kordeweg-de Vries equations. In the study of quantum field theory, the delta expansion is a powerful calculational tool that provides nonperturbative information. The basic idea is to expand in the power of the interaction term. For example, to solve a $\lambda\phi^4$ theory in d-dimensional space-time, we introduce a small parameter δ and consider a $\lambda(\phi^2)^{1+\delta}$ field theory. We show how to expand such a theory as a series in powers of δ using graphical methods. The resulting perturbation series appears to have a finite radius of convergence and numerical results for low-dimensional models are good. We compute the Green's functions for a scalar quantum field theory to first order in delta, renormalize the theory, and see that when the space-time dimension is four or more, the theory is free. This conclusion remains valid to second order in delta, and we believe that it remains valid to all orders in delta. The delta expansion is consistent with global supersymmetry invariance. We examine a supersymmetric quantum field theory for which we do not know of any other means for doing analytical calculations and compute the ground-state energy density and the fermion-boson mass ratio to second order in delta. Last, we show how to use the delta expansion to solve theories having a local gauge invariance. We compute the anomaly in two-dimensional electrodynamics and discuss the calculation of g-2 in four-dimensional electrodynamics.

INTRODUCTION

In recent papers[1,2] we introduced a new analytical technique for solving self-interacting scalar quantum field theories such as $g\phi^4$ theory. This technique is perturbative in character; it relies upon the introduction of an artificial perturbation parameter δ and expresses field-theoretic quantities such as the n-point Green's function $G^{(n)}(x_1,x_2,\ldots,x_n,\delta)$ as a series in powers of δ:

$$G^{(n)}(x_1,x_2,\ldots,x_n,\delta)=\sum_{k=0}^{\infty}\delta^k g^{(n)}{}_k(x_1,x_2,\ldots,x_n)\ . \qquad (1)$$

This technique has many advantages. There is strong evidence that the δ series has a finite radius of convergence. This is a dramatic advantage over conventional weak-coupling perturbation series which are rigorously known to have zero radius of convergence. Second, the δ expansion is nonperturbative in the sense that the functional dependence on the physical parameters of the theory (such as coupling constants and masses) is nontrivial. In the δ expansion it is only the parameter δ which is considered small. Thus, like the $1/N$ expansion, the results are nonperturbative in the physical parameters such as the mass and the coupling constant. Third, there is a well-defined orderly diagrammatic procedure for obtaining the coefficients in the δ expansion. Moreover, while the coefficients in the δ expansion can be infinite and, in general, must be renormalized, they are typically much less divergent than the coefficients in a conventional weak-coupling series. Fourth and finally, in a theory where there is no natural perturbation parameter, the δ expansion provides a straightforward and practical route to an analytic solution. All of these features are evident in the models that we discuss in this talk.

To review the key ideas of the δ expansion let us consider a self-interacting scalar field theory whose interaction Lagrangian has the form

$$L=g\phi^4\ .$$

Rather than following the conventional approach in which we expand in powers of g (or in powers of $1/g$) we replace L by a new interaction Lagrangian containing a dimensionless artificial perturbation parameter δ:

$$L(\delta)=g(\phi^2)^{1+\delta}\ .$$

It is important to point out that $L(\delta)$ is a *positive* operator. For example, when $\delta=1/2$, $L(\delta)$ means $|\phi|^3$ and not ϕ^3.

Clearly, expanding $L(\delta)$ in powers of δ produces a nonpolynomial Lagrangian involving logarithms of ϕ^2. However, as is shown in Refs. 1 and 2, the Green's functions for $L(\delta)$ can be expressed as convergent power series in δ. The novelty of our work is that we have discovered how to calculate the coefficients in these series using conventional graphical techniques.

An additional attractive feature of the δ expansion is that it preserves global supersymmetry invariance. We will see that it is possible to introduce the parameter δ into a supersymmetric Lagrangian in such a way that the resulting Lagrangian is exactly supersymmetric for all δ.

ILLUSTRATIVE EXAMPLE

Here is a simple problem which we solve by introducing an artificial perturbation parameter. Consider the problem of finding the (unique) real positive root $x=0.75487767\cdots$ of the fifth-degree polynomial

$$x^5+x=1 \ .$$

We can introduce a small artificial perturbation parameter δ in three possible ways:

(i) Weak coupling:

$$\delta x^5+x=1 \ ;$$

(ii) Strong coupling:

$$x^5+\delta x=1 \ ;$$

(iii) Delta expansion:

$$x^{1+\delta}+x=1 \ .$$

We then compute the root $x(\delta)$ in the form of a power series:

$$x(\delta)=\sum_{n=0}^{\infty} a_n \delta^n \ .$$

Such series are easy to determine. The weak-coupling series begins

$$x(\delta)=1-\delta+5\delta^2-35\delta^3+285\delta^4-2530\delta^5+23751\delta^6\cdots \ .$$

This series has the extremely small radius of convergence 0.08196. (In quantum field theory most weak-coupling series have a vanishing radius of convergence.) Thus, when we try to recover the root $x(1)$ by evaluating the above series at $\delta=1$ we get 21476. However, the (3,3)-Padé gives the good result 0.76369.

The strong-coupling series begins

$$x(\delta)=1-\frac{\delta}{5}-\frac{\delta^2}{25}-\frac{\delta^3}{125}+\frac{21\delta^5}{15625}+\frac{78\delta^6}{78125}\cdots \ .$$

If we evaluate this series at $\delta=1$ we get $x(1)=0.75434$, an extremely good approximation to the exact root. The radius of convergence of the strong-coupling series is 1.64938.

The coefficients of the delta series are slightly more complicated because they involve the constant log2. This series begins

$$x(\delta)=\frac{1}{2}+\frac{\delta}{4}\log 2-\frac{\delta^2}{8}\log 2+\cdots \ .$$

This series diverges when $\delta=4$ but a (3,3)-Padé gives the good result 0.754479 and a (6,6)-Padé gives the excellent result 0.75487654.

SOME NONLINEAR CLASSICAL DIFFERENTIAL EQUATIONS

Next, we show how to use the delta expansion to solve some difficult problems taken from classical physics.[3] First, we consider the Lane-Emden equation

$$y''(x) + \frac{2}{x}y'(x) + [y(x)]^n = 0, \quad y(0)=1, \quad y'(0)=0,$$

which describes a self-gravitating ball of fluid (a star). The objective is to determine the radius of the star by finding the first zero ξ of y: $y(\xi)=0$. Following the delta expansion approach we insert a small parameter δ in the exponent of the Lane-Emden equation by replacing n by $1+\delta$:

$$y''(x) + \frac{2}{x}y'(x) + [y(x)]^{1+\delta} = 0 \quad ,$$

and expand the solution $y(x)$ as a series in powers of δ:

$$y(x) = y_0(x) + \delta y_1(x) + \delta^2 y_2(x) + \delta^3 y_3(x) + \cdots \quad .$$

Notice that when $\delta=0$ the nonlinear differential equation becomes linear and therefore extremely easy to solve. The first two functions in the delta series are

$$y_0(x) = \frac{\sin x}{x}$$

and

$$y_1(x) = \frac{\cos x}{2x}\int_0^x ds \, \ln(\sin s) - \frac{\sin x}{2x}\ln\left(\frac{\sin x}{x}\right) + \frac{3}{4}\cos x$$
$$+ \frac{\sin x}{4x} - \frac{1}{2}\cos x \ln x - \frac{\cos x}{4x}Si(2x) - \frac{\sin x}{4x}Cin(2x) \quad ,$$

where

$$Si(x) = \int_0^x dt \, \frac{\sin t}{t} \quad \text{and} \quad Cin(x) = \int_0^x dt \, \frac{1-\cos t}{t} \quad .$$

Using just three terms in the delta series for $y(x)$ we obtain spectacular predictions for ξ as we see in the following table:

δ	(1,1) Padé Prediction for ξ	Exact value of ξ
0	π	π
−0.5	2.4465	2.4494
0.5	4.3603	4.3529
1.0	7.0521	6.8969
1.5	17.967	14.972

We can use the (1,1)-Padé to predict the value of δ for which $\xi=\infty$. The zero of the denominator gives $\delta=3.65$ while the exact value of δ is 4. Thus, with only three terms in the delta series, we have a 9% error associated with this extremely large value of δ.

Next, we consider the Thomas-Fermi equation

$$y''(x) = [y(x)]^{3/2} x^{-1/20}, \quad y(0)=1, \quad y(\infty)=0.$$

This nonlinear boundary-value problem describes the charge distribution in a nucleus. Our objective is to determine the value of $y'(0)$. The exact solution to the Thomas-Fermi equation is shown in the figure below:

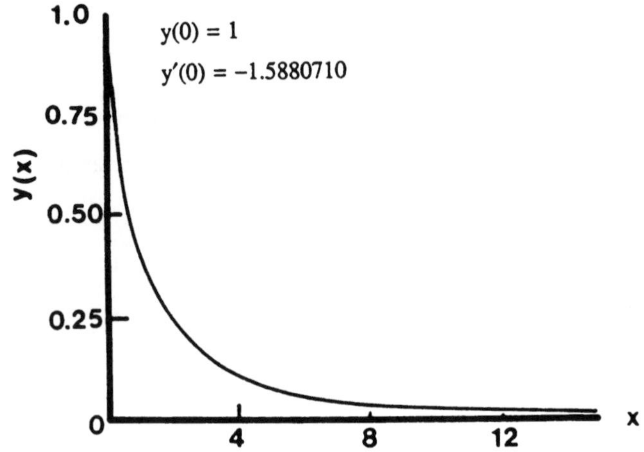

To use the delta expansion method to solve the Thomas-Fermi equation we insert a small parameter δ in the exponent:

$$y''(x) = [y(x)]^{1+\delta} x^{-\delta}/20, \quad y(0)=1, \quad y(\infty)=0.$$

Observe that at $\delta=0$ this equation is linear and trivial to solve. We have carried out the delta expansion to order δ^3 (four terms). Our numerical predictions for the value of $y'(0)$ are as follows (the exact numerical value is -1.5880710): In zeroth order we get $y'(0)=-1$, in first order we get $y'(0)=-1.1926$, in second order we get $y'(0)=-1.3843$, and in third order we get $y'(0)=-1.4789$. We can improve our results even further by converting the delta series to a Padé; a (2,1)-Padé of the series gives $y'(0)=-1.5712$ (1.1% error).

Next, we consider the Blasius equation

$$y'''(x) + y''(x) y(x) = 0, \quad y(0)=y'(0)=1, \quad y'(\infty)=0.$$

This third-order nonlinear boundary-value problem describes the fluid flow in the boundary layer that develops when fluid is flowing along a flat plate. Our objective here is to use the delta expansion to determine the value of $y''(0)$. The numerical value of $y''(0)$ is 0.46960.

To apply the delta expansion to the Blasius equation we insert a small parameter δ in the exponent as follows:

$$y'''(x) + y''(x)[y(x)]^{\delta} = 0, \quad y(0)=y'(0)=1, \quad y'(\infty)=0.$$

It is crucial that at δ=0 this equation is linear and trivial to solve. We have computed three terms in the delta series and find that to zeroth order in delta $y''(0)=1$, to first order in delta $y''(0)=0.319$ (32% error), to second order in delta $y''(0)=0.429$ (8.7% error).

Finally, we consider the Duffing equation

$$y''(t) + y'(t) + \varepsilon[y(x)]^3 = 0, \quad y(0)=1, \quad y'(0)=0.$$

This second-order nonlinear initial-value problem describes the classical anharmonic oscillator. Here, our objective is to use the delta expansion to determine the period of the oscillator. We obtain extremely accurate results from just the first two terms in the delta expansion (order δ) as shown in the table below:

	exact	first-order prediction:
ε=1	4.768	4.918 (3%)
ε=3	3.521	3.674 (4%)
ε=8	2.413	2.529 (5%)

The above four nonlinear classical ordinary differential equations are discussed in detail in Ref. 3. Observe that in all four equations, after we have inserted a parameter δ the differential equations become linear and easily solvable as δ tends to 0. As δ increases from 0, the nonlinearity of the differential equations smoothly turns on.

NONLINEAR PARTIAL DIFFERENTIAL EQUATIONS

We can also use the delta expansion methods to treat nonlinear partial differential equations.[4] Consider, for example, the Burgers' equation

$$u^t + u\, u_x = \nu\, u_{xx} \ .$$

We insert the parameter δ in the usual way:

$$u^t + u^{\delta}\, u_x = \nu\, u_{xx}$$

and expand in powers of δ. If we take a Gaussian initial condition

$$u(x, 0 = \exp(-ax\ 2) \ ,$$

we obtain a perturbation series which that is trivial to calculate to any given order in delta. We have determined the delta series to sixth order in powers of delta and formed the main sequence of Padé approximants. In the Figure below we compare the exact solution at $u(x,t=1)$ with the first six Padé approximants at δ=1.

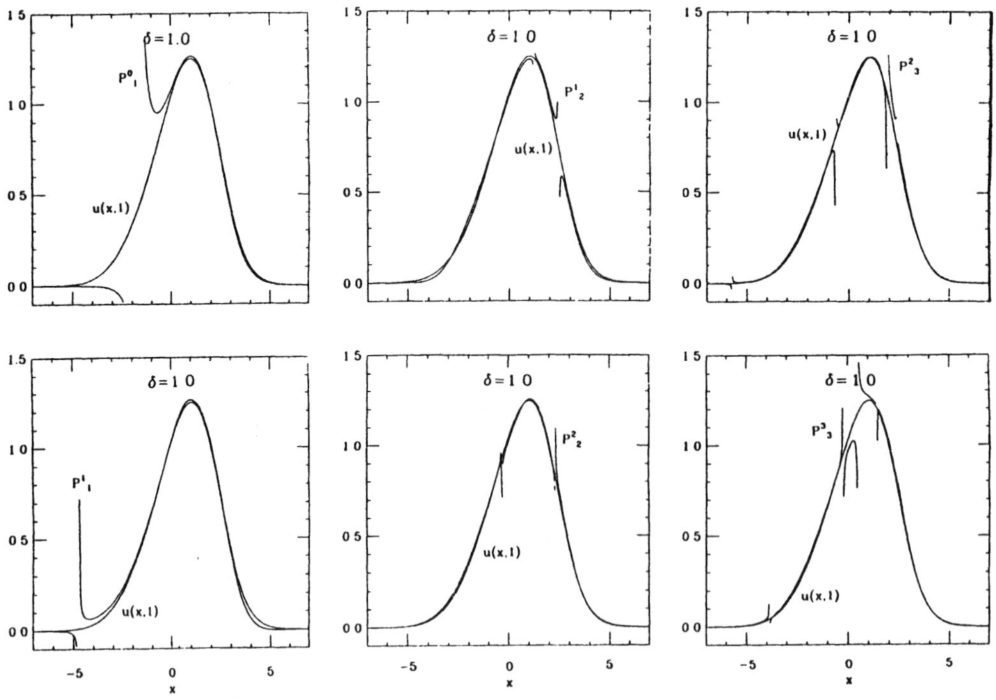

QUANTUM FIELD THEORY

In quantum field theory the problem of course is to find a method for expanding the Green's functions as perturbation series in powers of δ. Consider the Lagrangian

$$L=\frac{1}{2}(\partial\phi)^2+\frac{1}{2}\mu^2\phi^2+\lambda M^2\phi^2(\phi^2M^{2-d})^\delta \tag{2}$$

in d-dimensional Euclidean space. In (2), μ is the bare mass, λ is the *dimensionless* bare coupling constant, and M is a fixed mass parameter that allows the interaction to have the correct dimensions. The problem is that if we expand the Lagrangian in (2) as a series in powers of δ using the identity

$$x^\delta=e^{\delta\ln x}=1+\delta\ln x+\frac{\delta^2}{2!}(\ln x)^2+\frac{\delta^3}{3!}(\ln x)^3+\cdots\ ,$$

we obtain a horrible-looking nonpolynomial Lagrangian:

$$L=\frac{1}{2}(\partial\phi)^2+\frac{1}{2}(\mu^2+2\lambda M^2)\phi^2+\delta\lambda\phi^2M^2\ln\left[\phi^2M^{2-d}\right]$$
$$+\frac{\delta^2}{2}\lambda\phi^2M^2\left[\ln\left[\phi^2M^{2-d}\right]\right]^2+\frac{\delta^3}{6}\lambda\phi^2M^2\left[\ln\left[\phi^2M^{2-d}\right]\right]^3+\cdots\ . \tag{3}$$

We have devised a very simple and orderly procedure for calculating the n-point Green's function of the Lagrangian in (3) as series in powers of δ. It consists of three steps:

(i) Replace the Lagrangian L in (3) with a new Lagrangian L_{whacko} having *polynomial* interaction terms.

(ii) Using L_{whacko}, compute the Green's function $G^{(n)}{}_{whacko}$ using ordinary Feynman diagrams.

(iii) Apply a derivative operator D to $G^{(n)}{}_{whacko}$ to get the delta expansion for $G^{(n)}$.

The new Lagrangian L_{whacko} and the derivative operator D depend on the number of terms in the δ series that we intend to compute. For example, if we need one term in the δ series we take

$$L_{whacko}=\frac{1}{2}(\partial\phi)^2+\frac{1}{2}(\mu^2+2\lambda M^2)\phi^2+\delta\lambda M^d\left[\phi^2M^{2-d}\right]^{\alpha+1} .$$

Then we compute the n-point Green's function $G^{(n)}{}_{whacko}$ to order δ, apply

$$D=\frac{\partial}{\partial\alpha} ,$$

and set $\alpha=0$.

Now suppose we need two terms in the δ series expansion of $G^{(n)}$. We take

$$L_{whacko}=\frac{1}{2}(\partial\phi)^2+\frac{1}{2}(\mu^2+2\lambda M^2)\phi^2$$
$$+(\delta+\delta^2)\lambda M^d\left[\phi^2M^{2-d}\right]^{\alpha+1}+(-\delta+\delta^2)\lambda M^d\left[\phi^2M^{2-d}\right]^{\beta+1} .$$

Then we compute $G^{(n)}{}_{whacko}$ to order δ^2, apply

$$D=\frac{1}{2}(\frac{\partial}{\partial\alpha}-\frac{\partial}{\partial\beta})+\frac{1}{4}(\frac{\partial^2}{\partial\alpha^2}+\frac{\partial^2}{\partial\beta^2}) ,$$

and set $\alpha=\beta=0$.

For three terms in the δ series expansion of $G^{(n)}$, we take

$$L_{whacko}=\frac{1}{2}(\partial\phi)^2+\frac{1}{2}(\mu^2+2\lambda M^2)\phi^2$$
$$+[\delta+\frac{\delta^2}{2}(1+\alpha)+\delta^3]\lambda M^d\left[\phi^2M^{2-d}\right]^{\alpha+1}$$
$$+[\delta\omega+\frac{\delta^2}{2}(\omega^2+\beta)+\delta^3]\lambda M^d\left[\phi^2M^{2-d}\right]^{\beta+1}$$
$$+[\delta\omega^2+\frac{\delta^2}{2}(\omega+\gamma)+\delta^3]\lambda M^d\left[\phi^2M^{2-d}\right]^{\gamma+1} .$$

As above, we compute $G^{(n)}{}_{whacko}$ to order δ^3, apply

$$D=\frac{1}{3}(\frac{\partial}{\partial\alpha}+\omega^2\frac{\partial}{\partial\beta}+\omega\frac{\partial}{\partial\gamma})+\frac{1}{6}(\frac{\partial^2}{\partial\alpha^2}+\omega\frac{\partial^2}{\partial\beta^2}+\omega^2\frac{\partial^2}{\partial\gamma^2})+\frac{1}{18}(\frac{\partial^3}{\partial\alpha^3}+\frac{\partial^3}{\partial\beta^3}+\frac{\partial^3}{\partial\gamma^3}) ,$$

and set $\alpha=\beta=\gamma=0$.

If we need four terms in the delta expansion, we take

$$L_{whacko}=\frac{1}{2}(\partial\phi)^2+\frac{1}{2}(\mu^2+2\lambda M^2)\phi^2$$

$$+[\delta+\delta^2\frac{2+\alpha+\beta+\gamma+\nu+3(\alpha^2-i\beta^2-\gamma^2+i\nu^2)}{6}+\delta^3\frac{4+5\alpha}{9}+\delta^4]\,\lambda M^d\left[\phi^2 M^{2-d}\right]^{\alpha+1}$$

$$+[i\delta+\delta^2\frac{-2-i(\alpha+\beta+\gamma+\nu)+3(i\alpha^2+\beta^2-i\gamma^2-\nu^2)}{6}+\delta^3\frac{-4i+5\beta}{9}+\delta^4]\,\lambda M^d\left[\phi^2 M^{2-d}\right]^{\beta+1}$$

$$+[-\delta+\delta^2\frac{2-\alpha-\beta-\gamma-\nu+3(-\alpha^2+i\beta^2+\gamma^2-i\nu^2)}{6}+\delta^3\frac{-4+5\gamma}{9}+\delta^4]\,\lambda M^d\left[\phi^2 M^{2-d}\right]^{\gamma+1}$$

$$+[-i\delta+\delta^2\frac{-2+i(\alpha+\beta+\gamma+\nu)+3(-i\alpha^2-\beta^2+i\gamma^2+\nu^2)}{6}+\delta^3\frac{4i+5\nu}{9}+\delta^4]\,\lambda M^d\left[\phi^2 M^{2-d}\right]^{\nu+1} \; .$$

We compute the Green's function $G^{(n)}{}_{whacko}$ to order δ^4, apply

$$D=\frac{1}{4}\left(\frac{\partial}{\partial\alpha}-i\frac{\partial}{\partial\beta}-\frac{\partial}{\partial\gamma}+i\frac{\partial}{\partial\nu}\right)+\frac{1}{8}\left(\frac{\partial^2}{\partial\alpha^2}-\frac{\partial^2}{\partial\beta^2}+\frac{\partial^2}{\partial\gamma^2}-\frac{\partial^2}{\partial\nu^2}\right)$$

$$+\frac{1}{24}\left(\frac{\partial^3}{\partial\alpha^3}+i\frac{\partial^3}{\partial\beta^3}-\frac{\partial^3}{\partial\gamma^3}-i\frac{\partial^3}{\partial\nu^3}\right)+\frac{1}{96}\left(\frac{\partial^4}{\partial\alpha^4}+\frac{\partial^4}{\partial\beta^4}+\frac{\partial^4}{\partial\gamma^4}+\frac{\partial^4}{\partial\nu^4}\right) ,$$

and set $\alpha=\beta=\gamma=\nu=0$.

We do not have the general form of the Lagrangian L_{whacko} needed to obtain N terms in the delta series. However, we *do* have the form of the derivative operator D:

$$D=\frac{1}{N}\sum_{j=1}^{N}\sum_{k=1}^{N}\frac{e^{2\pi i(1-k)j/N}}{j!}\left[\frac{\partial}{\partial\alpha_k}\right]^j \; .$$

LOW-DIMENSIONAL MODELS

To examine the form of the delta expansion and to verify its numerical accuracy, we consider a zero-dimensional and a one dimensional field theory. The functional integral for the vacuum-vacuum amplitude Z of a ϕ^4 field theory in zero-dimensional space-time is an ordinary Riemann integral:

$$Z=\int_{-\infty}^{\infty}\frac{dx}{\pi^{1/2}}e^{-x^4} \; .$$

Now we insert the expansion parameter δ:

$$Z=\int_{-\infty}^{\infty}\frac{dx}{\pi^{1/2}}e^{(-x^2)^{1+\delta}} \; . \tag{4}$$

Recall that the ground-state energy E is given in terms of Z: $E(\delta)=-\ln Z$. For this simple theory we can, of course, evaluate directly the integral in (4):

$$E(\delta)=-\ln\left[\frac{2}{\pi^{1/2}}\Gamma\left[\frac{2\delta+3}{2\delta+2}\right]\right] . \tag{5}$$

To find the delta series we merely expand the right side of (5) in a Taylor series in powers of δ:

$$E(\delta)=\frac{\delta}{2}\psi(\frac{3}{2})-\frac{\delta^2}{8}[4\psi(\frac{3}{2})+\psi'(\frac{3}{2})]+\frac{\delta^3}{48}[24\psi(\frac{3}{2})+12\psi'(\frac{3}{2})+\psi''(\frac{3}{2})]$$
$$-\frac{\delta^4}{384}[192\psi(\frac{3}{2})+144\psi'(\frac{3}{2})+24\psi''(\frac{3}{2})+\psi'''(\frac{3}{2})]+\cdots . \tag{6}$$

Notice that the structure of the delta series in (6) is rather strange in that the coefficients all depend on polygamma functions evaluated at $3/2$. The polygamma function $\psi(x)$ is defined as the logarithmic derivative of a gamma function:

$$\psi(x)\equiv\frac{\Gamma'(x)}{\Gamma(x)} .$$

There is a general formula in terms of zeta functions for the nth derivative of a polygamma function evaluated at $3/2$:

$$\psi^{(n)}(3/2)=(-1)^n n![(1-2^{n+1})\zeta(n+1)+2^{n+1}] .$$

The first two such polygamma functions are $\psi(3/2)=2-\gamma-2\ln 2$ and $\psi'(3/2)=\frac{\pi^2}{2}-4$. We list below the numerical values of the first few polygamma functions:

$\psi(3/2) = 0.0364899740$;
$\psi'(3/2) = 0.9348022005$;
$\psi''(3/2) = -0.8287966442$;
$\psi'''(3/2) = 1.4090910340$.

It is crucial to determine for which δ the series in (6) converges. Note that $E(\delta)$ in (5) is singular whenever the argument of the gamma function vanishes. There are an infinite number of such singular points δ_k in the complex-δ plane given by the formula

$$\delta_k=-\frac{2k+3}{2k+2} , \quad k=0,1,2,3,\ldots .$$

Each of these singular points is a logarithmic branch point. Note that these singular points form a monotone sequence on the negative-δ axis beginning at the point $\delta=-3/2$ and converging to the point $\delta=-1$. We conclude that the delta series in (6) has a radius of convergence of 1.

A ϕ^4 theory corresponds to $\delta=1$, which is situated on the circle of convergence. Thus, to compute the delta series with high numerical accuracy we use Padé summation. Here are the results: The exact value of the energy is $E(1)=-0.0225104$. Because we are on the circle of convergence we do not expect that a direct summation of the delta series will give a good result, and indeed it does not: ten terms in the power series give -0.367106 and twenty terms in the power series give -0.517356. However, a (3,2) Padé gives -0.02252 and a (5,4) Padé gives -0.0225103.

Now let us see how well the delta expansion works in one-dimensional field theory (quantum mechanics). The Hamiltonian for the anharmonic oscillator is:

$$H=-\frac{d^2}{2dx^2}+\frac{1}{2}x^4 \ .$$

Our strategy is to insert the parameter δ in the x^4 term:

$$H=-\frac{d^2}{2dx^2}+\frac{1}{2}(x^2)^{1+\delta} \ .$$

The ground-state energy E for this Hamiltonian has the delta expansion:

$$E(\delta)=\frac{1}{2}+\frac{\delta}{4}\psi(\frac{3}{2})$$
$$-\frac{\delta^2}{128}\left[\psi''(\frac{3}{2})+8\psi'(\frac{3}{2})\ln 2-8[\psi(\frac{3}{2})]^2+16\psi(\frac{3}{2})-32+32\ln 2\right]+\cdots \ .$$

This series is extremely accurate numerically. The exact value of $E(1)$ is 0.530176, while the sum of the above series to order δ^2 is 0.534385. Notice that the form of the series is similar to that in (6); the coefficients are all constructed out of polygamma functions evaluated at 3/2.

RENORMALIZATION AND TRIVIALITY OF ϕ^4 QUANTUM FIELD THEORY

We now consider the problem of how to renormalize the δ-expansion. It was pointed out in Refs. 1 and 2 that when $d\geq 2$ the coefficients of δ^k in the expansions of the Green's functions are less divergent (as functions of the ultraviolet cutoff Λ in momentum space) than the terms in the conventional weak-coupling expansion in powers of λ. However, the coefficients $g^{(n)}{}_k(x_1,x_2,\cdots,x_n)$ in the δ expansion are still divergent and it is necessary to use a renormalization procedure.

We will show how to regulate the theory by introducing a short distance cutoff a (which is equivalent to an ultraviolet cutoff $\Lambda=1/a$) and we compute the renormalized coupling constant G_R in terms of the bare mass μ and the bare coupling constant λ. We then show that if we hold the renormalized mass M_R fixed at a finite value, then as the cutoff a is allowed to tend to 0 ($\Lambda\to\infty$), G_R can remain finite and nonzero only when $d<4$. When $d\geq 4$, $G_R\to 0$ as $a\to 0$. This result is the continuum analog of the numerical nonperturbative results already obtained in lattice Monte Carlo calculations.[5]

We have computed the d-dimensional two-point Euclidean Green's function $G^{(2)}(p^2)$ to second order in powers of δ. From $G^{(2)}(p^2)$ we can obtain the wave-function renormalization constant Z and the renormalized mass M_R. The conventional definitions are

$$Z^{-1} \equiv 1+\frac{\partial}{\partial(p^2)}[G^{(2)}(p^2)]^{-1}\,|_{p^2=0} \ , \tag{7}$$

and

$$M_R{}^2 \equiv Z[G^{(2)}(p^2)]^{-1}\,|_{p^2=0} \ . \tag{8}$$

We have also computed $G^{(4)}(p_1,p_2,p_3,p_4)$, the connected d-dimensional Euclidean Green's function with legs amputated, to second order in powers of δ. From $G^{(4)}$ we can obtain the dimensionless renormalized coupling constant G_R in the usual way:

$$G_R \equiv -Z^2 G^{(4)}(0,0,0,0) M_R{}^{d-4} \ . \tag{9}$$

We do not discuss the calculations of $G^{(2)}$, $G^{(4)}$, and the higher Green's functions such as $G^{(6)}$ here; the calculation is long and detailed and it is presented elsewhere.[6] It is sufficient to state that the calculation follows exactly the rules set down in Refs. 1 and 2. Here are the results for Z, $M_R{}^2$, and G_R to first order in δ:

$$Z=1+O(\delta^2) \ , \tag{10}$$

$$M_R{}^2=\mu^2+2\lambda M^2+2\lambda\delta M^2\left[1+\psi(\frac{3}{2})+\ln[2\Delta(0)M^{2-d}]\right]+O(\delta^2) \ , \tag{11}$$

$$G_R=4\lambda\delta\frac{M^{d-2}}{\Delta(0)}+O(\delta^2) \ . \tag{12}$$

In (10)-(12), $\Delta(x)$ represents the free propagator in d-dimensional coordinate space; $\Delta(x)$ can be expressed as an associated Bessel function:

$$\begin{aligned}\Delta(x) &= (2\pi)^{-d}\int d^d p \frac{e^{i\mathbf{x}\cdot\mathbf{p}}}{p^2+m^2} \\ &= (2\pi)^{-d/2}(x/m)^{1-d/2}K_{1-d/2}(mx),\end{aligned} \tag{13}$$

where $m^2=\mu^2+2\lambda M^2$.

The function $\Delta(x)$ is finite at $x=0$ when $d<2$:

$$\Delta(0)=2^{-d}\pi^{-d/2}m^{d-2}\Gamma(1-d/2) \ . \tag{14}$$

However, we are concerned with quantum field theory, in which $d \geq 2$. For these values of d, $\Delta(0)=\infty$, and it is clearly necessary to regulate the expressions for the renormalized quantities in (10)-(12) because of this divergence.

To regulate the theory we introduce a short-distance (ultraviolet) cutoff a; to wit, we replace $\Delta(0)$ in (11) and (12) with $\Delta(a)$, where

$$\Delta(a)=(2\pi)^{-d/2}(a/m)^{1-d/2}K_{1-d/2}(ma) \ . \tag{15}$$

Apparently, there are three distinct cases which we must consider:

Case 1: $am \ll 1$ $(a \to 0)$. Here we can approximate the Bessel function in (15) for small argument:

$$\Delta(a) \approx \frac{1}{4\pi} \Gamma(\frac{d}{2} - 1)(\pi a^2)^{1-d/2} \ . \tag{16}$$

Case 2: $am = O(1)$ $(a \to 0)$. Here,

$$\Delta(a) \approx (constant)\ m^{d-2} \ . \tag{17}$$

Case 3: $am \gg 1$ $(a \to 0)$. Here we can approximate the Bessel function in (15) for large argument:

$$\Delta(a) \approx \frac{1}{2m} \cdot (\frac{2\pi a}{m})^{(1-d)/2} e^{-ma} \ . \tag{18}$$

Now we consider each of these three cases in turn. In *case* 1 we substitute (16) into (12) to obtain

$$\lambda\delta = (constant)\ G_R\ (aM)^{2-d} \ . \tag{19}$$

Then we use (19) to eliminate $\lambda\delta$ from (11). The result is

$$M_R{}^2 = m^2 + (constant)\ [logarithm\ term]\ G_R\ M^2\ (aM)^{2-d} \ . \tag{20}$$

The renormalized mass must be finite. But as $a \to 0$ the second term on the right side of (20) becomes infinite when $d > 2$. Thus, *both* terms on the right side of (20) must be infinite and must combine to produce a finite result. Hence they must be of the same order of magnitude as $a \to 0$:

$$(constant)\ [logarithm\ term]\ G_R\ M^2\ (aM)^{2-d} \simeq m^2 \ . \tag{21}$$

If we multiply (21) by a^2 we obtain

$$(constant)\ [logarithm\ term]\ G_R\ (aM)^{4-d} \simeq (am)^2 \ll 1 \tag{22}$$

by the assumption of *case* 1. Thus, when $d < 4$, G_R can remain finite and nonzero as $a \to 0$, but when $d \geq 4$, $G_R \to 0$ as $a \to 0$ and the theory is free.

Next, we consider *case* 2. We substitute (17) into (12) to obtain

$$\lambda\delta = (constant)\ G_R\ (m/M)^{d-2} \ . \tag{23}$$

We use (23) to eliminate $\lambda\delta$ from (11) and obtain

$$M_R{}^2 = m^2 + (constant)\ \ln(m/M)\ G_R\ M^2\ (m/M)^{d-2} \ . \tag{24}$$

As above, we argue that the left side of (24) must be finite so the two (infinite) terms on the right side of (24) must be of equal magnitude:

$$(constant)\ \ln(m/M)\ G_R\ M^2\ (m/M)^{d-2} \simeq m^2\ . \tag{25}$$

We divide (25) by m^2 and solve for G_R:

$$G_R=(constant)\ (m/M)^{4-d}/\ln(m/M)\ . \tag{26}$$

Again we observe that when $d \geq 4$, $G_R \to 0$ as $m \to \infty$.

Finally, we consider *case* 3. We substitute (18) into (12) to obtain

$$\lambda\delta=(constant) G_R (aM)^{(1-d)/2} (m/M)^{(d-3)/2} e^{-ma}\ . \tag{27}$$

We use (27) to eliminate $\lambda\delta$ from (11) and obtain

$$M_R{}^2=m^2+\frac{[logarithm\ term]}{(constant)} G_R M^2 (aM)^{(1-d)/2} (m/M)^{(d-3)/2} e^{-ma}\ . \tag{28}$$

Once again, we observe that the two terms on the right side of (28) are divergent and must be of the same magnitude:

$$\frac{[logarithm\ term]}{(constant)} G_R M^2 (aM)^{(1-d)/2} (m/M)^{(d-3)/2} e^{-ma} \simeq m^2\ . \tag{29}$$

From (29) we then have

$$G_R \simeq \frac{(constant)}{[logarithm\ term]} e^{ma} (ma)^{(d-1)/2} (m/M)^{4-d}\ . \tag{30}$$

Thus, when $d \leq 4$, $G_R \to \infty$ as $am \to \infty$. Hence, *case* 3 may be excluded when $d \leq 4$. It is interesting that when $d>4$, G_R *can* remain finite as $am \to \infty$ so long as m/M grows exponentially with am:

$$m/M \simeq (constant) e^{ma/(d-4)} (am)^{(d-3)/(2d-8)}\ . \tag{31}$$

However, this possibility can be ruled out by computing the $2n$-point Green's functions $G^{(2n)}$. To order δ we have

$$G^{(2n)}(0,0,\cdots,0)=\delta\lambda(n-2)!M^2 2^n[-\Delta(0)]^{1-n}+O(\delta^2)\ . \tag{32}$$

If (31) holds, then (32) implies that for all $n>2$, $G^{(2n)} \to 0$ as $am \to \infty$ and the theory becomes trivial.

We have been able to generalize these arguments to second order in powers of δ. However, we do not give the argument here. We merely present the result for the renormalized mass to second order in delta:

$$\begin{aligned} M_R{}^2 = {} & \mu^2+2\lambda M^2+2\lambda\delta M^2 S+\delta^2 \{\lambda M^2[S^2+1+\psi'(3/2)]-4\lambda^2\Delta(0)M^4 S\int d^dx\ z \\ & -4\lambda^2\Delta(0)M^4\int d^dx\int_0^1 dt\frac{\sqrt{1-t}}{t^2}[zt+\ln(1-zt)] \\ & +4\lambda^2\Delta(0)M^4\int d^dx\int_0^1 dt\frac{\sqrt{z-zt}}{t}\ln(1-zt)\}+O(\delta^3)\ , \end{aligned} \tag{33}$$

where $S=\psi(3/2)+\ln[2\Delta(0)M^{2-d}]+1$ and $z=[\Delta(x)/\Delta(0)]^2$. We cannot evaluate the integrals in

(33) in closed form except in particular space-time dimensions; namely, when $d=1$ and when d is even and negative semidefinite ($d=0,-2,-4,-6,\cdots$). For these special values of d we give the explicit evaluation of these integrals in Ref. 6.

SUPERSYMMETRIC THEORIES

Now consider a generic two-dimensional Wess-Zumino Lagrangian[7]

$$L=\frac{1}{2}(\partial\phi)^2+\frac{i}{2}\bar{\psi}\partial\psi+\frac{1}{2}\lambda S'(\phi)\bar{\psi}\psi+\frac{1}{2}\lambda^2[S(\phi)]^2 \ , \tag{34}$$

where ψ is a Majorana spinor. In the following we take $S(\phi)=(\phi^2)^{(1+\delta)/2}$. This gives

$$L=\frac{1}{2}(\partial\phi)^2+\frac{i}{2}\bar{\psi}\partial\psi+\frac{1}{2}\lambda(1+\delta)\,(\phi^2)^{\delta/2}\bar{\psi}\psi+\frac{1}{2}\lambda^2(\phi^2)^{1+\delta} \ . \tag{35}$$

Some models of this type exhibit spontaneous breaking of supersymmetry as manifested by the fact that the ground-state energy is nonzero. For example, a $\phi\bar{\psi}\psi+\phi^4$ theory exhibits spontaneous symmetry breaking, while $\phi^2\bar{\psi}\psi+\phi^7$ remains unbroken. The apparent reason that $\phi\bar{\psi}\psi+\phi^4$ has a broken supersymmetry is that ϕ is not a positive operator. However, if we set $\delta=1$ in Eq. (35) we obtain a $|\phi|\bar{\psi}\psi+\phi^4$ interaction term which *is* positive as we emphasized above. Thus, we expect that L in Eq. (35) has an *unbroken* supersymmetry for all values of δ.

We demonstrate the power of the δ-expansion technique by computing the ground-state energy density E of the theory in Eq. (35) to second order in δ. To this order we obtain the supersymmetric result $E=0$. We know of no other perturbative way to compute the physical quantities of this system.[8]

SUPERSYMMETRIC QUANTUM MECHANICS

To prepare for the field-theory calculation we first examine a toy model in quantum mechanics in which the complexity of fermions is absent. Like a supersymmetric quantum field theory, this simple model is designed to have a ground-state energy which is identically zero. It is quite easy to construct such a theory. One first chooses a normalizable function $\psi(x)=\exp(-\lambda|x|^{\delta+2})$, which has no nodes, to be the ground-state wave function and then derives the coordinate-space Schrödinger equation for which this wave function has zero energy:

$$-\frac{1}{2}\psi''(x)+[\frac{1}{2}(\delta+2)^2\lambda^2|x|^{2\delta+2}-\frac{1}{2}(\delta+1)(\delta+2)\lambda|x|^{\delta}]\psi(x)=0 \ .$$

The corresponding Euclidean-space Lagrangian is

$$L=\frac{1}{2}\dot{x}^2-\frac{1}{2}g^2(\delta+2)^2(x^2)^{1+\delta}+\frac{1}{2}g(\delta+1)(\delta+2)(x^2)^{\delta/2} \ . \tag{36}$$

Following the procedure outlined in Refs. 1 and 2, we can obtain the Green's functions of Eq. (36) correct to any order in the δ expansion from the appropriate *provisional* (whacko) Lagrangian $\bar{L}$ For example, if we wish to compute to first order in δ we take the provisional Lagrangian $\bar{L} = \bar{L}_0 + \bar{L}_\alpha$, where

$$\bar{L}_0 = \frac{1}{2}(\dot{x}^2) + \frac{1}{2}\lambda^2(4+4\delta)x^2 - \frac{1}{2}\lambda(2+3\delta) \tag{37a}$$

and

$$\bar{L}_\alpha = 2\delta\lambda^2(x^2)^{1+\alpha} - \frac{1}{2}\delta\lambda(x^2)^\alpha \ . \tag{37b}$$

Note that the Lagrangian $\bar{L}_0$ in Eq. (37a) is quadratic and that the interaction terms in Eq. (37b) are proportional to the perturbation parameter δ. Therefore, we can use conventional graphical perturbation theory to calculate any Green's function $\bar{G}(\alpha)$ for $\bar{L}$ to order δ (in this calculation it is necessary to assume that α is a positive integer). As we explained in Refs. 1 and 2 we obtain the Green's function G for L to order δ by first applying the derivative operator $D = d/d\alpha$ to $\bar{G}$ and then evaluating the result at $\alpha=0$. To this result we then must add the corresponding order-δ Green's functions $\bar{G}_0$ computed from $\bar{L}_0$ in Eq. (37a):

$$G_{\text{order }\delta} = \bar{G}_0 + \frac{d}{d\alpha}\bar{G}(\alpha)\,|_{\alpha=0} \ . \tag{38}$$

Note that by differentiating with respect to α and setting $\alpha=0$ we have analytically continued in α to the point at which $\bar{L}$ in Eq. (37) is free.

The Feynman rules for this calculation are easy to obtain and we find the ground-state energy from the connected graphs having no external legs. To order δ, these graphs possess only a single vertex and the result is

$$E_{\text{ground state}} = -\frac{1}{2}\lambda\delta + \frac{3}{2}\lambda\delta + \frac{d}{d\alpha}\left[-\frac{\delta\lambda(2\alpha+2)!}{4(8\lambda)^\alpha(\alpha+1)!} + \frac{\delta\lambda(2\alpha)!}{2(8\lambda)^\alpha\alpha!}\right]_{\alpha=0} = 0 \ .$$

Now we show how to compute the ground-state energy to order δ^2. The provisional Lagrangian for this computation now has two integer parameters α and β: $\bar{L} = \bar{L}_0 + \bar{L}_{\alpha,\beta}$, where

$$\bar{L}_0 = \frac{1}{2}(\dot{x})^2 + \frac{1}{2}\lambda^2 x^2(2+\delta)^2 - \frac{1}{2}\lambda(2+3\delta+\delta^2) \tag{39a}$$

and

$$\bar{L}_{\alpha,\beta} = (2\delta+4\delta^2)\lambda^2(x^2)^{1+\alpha} - 2\delta\lambda^2(x^2)^{1+\beta} - (\frac{\delta}{2}+\delta^2)\lambda(x^2)^\alpha + \frac{1}{2}(\delta+\delta^2)\lambda(x^2)^\beta \ . \tag{39b}$$

The Green's functions for L are obtained by applying the differential operator

$$D = \frac{1}{2}(\frac{\partial}{\partial\alpha} - \frac{\partial}{\partial\beta}) + \frac{1}{4}(\frac{\partial^2}{\partial\alpha^2} + \frac{\partial^2}{\partial\beta^2}) \tag{40}$$

at the point $\alpha = \beta = 0$. The Feynman rules assume that α and β are integers. Note that there are five vertices. Graphs contributing to the ground-state energy to order δ^2 have either one or two vertices. We must apply the derivative operator in Eq. (40) to each graph amplitude. Some of the two-vertex graphs give rise to infinite series which we can sum:[9]

$$\frac{\sqrt{\pi}}{2}\sum_{l=2}^{\infty}\frac{(l-2)!}{(l-1)\Gamma(l+1/2)}=\psi'(3/2)\ , \tag{41}$$

where $\psi(x)=\frac{d}{dx}\log\Gamma(x)$ is the digamma function. To obtain the ground-state energy we must combine the contributions from 20 different graphs. The final result is $E_{\text{ground state}}=0$ through order δ^2.

FIELD-THEORY CALCULATION

Now we generalize these computational procedures to compute the ground-state energy density E of the supersymmetric Lagrangian in Eq. (35). We begin by writing the provisional Lagrangian $\bar{L}$appropriate for doing calculations to second order in δ: $\bar{L}=\bar{L}_0+\bar{L}_{\alpha,\beta}$,

$$\bar{L}_0=\frac{1}{2}(\partial\phi)^2+\frac{i}{2}\bar{\psi}\partial\psi+\frac{1}{2}\lambda^2\phi^2-\frac{1}{2}\bar{\psi}\psi\lambda(1+\delta)\ , \tag{42a}$$

$$\bar{L}_{\alpha,\beta}=\frac{1}{2}\lambda^2(\delta+\delta^2)(\phi^2)^{\alpha+1}+\frac{1}{2}\lambda^2(\delta^2-\delta)(\phi^2)^{\beta+1}$$
$$-\frac{1}{8}\lambda(2\delta+3\delta^2)(\phi^2)^{\alpha}\bar{\psi}\psi+\frac{1}{8}\lambda(2\delta+\delta^2)(\phi^2)^{\beta}\bar{\psi}\psi\ . \tag{42b}$$

The Feynman rules appropriate for $\bar{L}$in Eq. (42) are a slight generalization of those used in the quantum-mechanical model because there are now fermion lines as well as boson lines and we are now working in two- rather than in one-dimensional space-time. In coordinate space the propagator for the boson is

$$\Delta(x)=\frac{1}{(2\pi)^2}\int\frac{d^2p}{p^2+\lambda^2}e^{i\mathbf{p}\cdot\mathbf{x}}$$

and the propagator for the fermion is

$$\Delta_F(x)=\frac{1}{(2\pi)^2}\int\frac{d^2p}{\not{p}-\lambda}e^{i\mathbf{p}\cdot\mathbf{x}}=(i\not{\partial}-\lambda)\Delta(x)\ .$$

To second order in δ there are 19 graphs depending on α and β. Some of these graphs give rise to infinite series after the operator D in Eq. (40) is applied. All other graphs give rise to single expressions involving polygamma functions and two integrals:

$$I=\frac{1}{(2\pi)^2}\int\frac{d^2p}{(p^2+\lambda^2)}=\Delta(0)$$

and

$$J=\int d^2x\Delta^2(x)\ .$$

Note that the integral I is logarithmically divergent.

If we combine the infinite series mentioned above into a single sum we obtain

$$\frac{\sqrt{\pi}}{16}\sum_{l=2}^{\infty}\frac{\lambda^2\delta^2(l-2)!}{(l-1)\Gamma(l-1/2)}\left[\frac{4l\lambda^2}{2l-1}\int d^2x\frac{\Delta^{2l}(x)}{I^{2l-2}}-\int d^2x\frac{\Delta^{2l-2}(x)}{I^{2l}-2}\operatorname{tr}\Delta_F(x)\bar{\Delta}_F(x)\right]. \tag{43}$$

To simplify this sum, we have found a lovely identity which is easy to derive using integration by parts:

$$\operatorname{tr}\int d^2x\Delta^{2l-2}(x)\Delta_F(x)\bar{\Delta}_F(x)=-\frac{2}{2l-1}I^{2l-1}+\frac{4l\lambda^2}{2l-1}\int d^2x\Delta^{2l}(x)\ . \tag{44}$$

Substituting Eq. (44) into Eq. (43) gives a sum proportional to I containing no coordinate space integrals. This sum is precisely that in Eq. (41). When this result is combined with all remaining terms we obtain $E=0$.[10]

We have also computed R, the ratio of the physical boson mass to the physical fermion mass, to second order in δ and obtain the supersymmetric result $R = 1$. The calculation of R is described in Ref. 11.

RECENT RESULTS IN FIELD THEORY

I conclude this talk with a brief mention of some recent results on the delta expansion applied to quantum field theory. At the beginning of this talk I showed how to apply the delta expansion to nonlinear classical differential equations. In the context of quantum field theory it is natural to try to solve the Langevin equation, a classical differential equation used to perform stochastic quantization. The delta expansion techniques work extremely well.[12,13] The delta expansion can also be used in high temperature field theory.[14] Also, the delta expansion has been used to examine the strong-coupling limit of quantum field theories (a very nonperturbative regime of quantum field theory).[15] The large-δ limit has also been studied.[16]

Finally, I mention that work is in progress on understanding how to use the techniques of the delta expansion to solve theories having a local Abelian gauge invariance. These are electrodynamic theories whose interaction has the form

$$\bar{\psi}(i\partial\!\!\!/ - eA\!\!\!/)\psi\ .$$

Such theories are invariant under local gauge transformations in which we replace ψ by $\psi\, exp\,(-ieLAMBDA)$, $\bar{\psi}$ by $\bar{\psi}\, exp\,(ieLAMBDA)$, and A_μ with $A_\mu\ \partial_\mu\Lambda$. To apply the techniques of the delta expansion to such theories we simply modify the interaction term so that it reads

$$\bar{\psi}(i\partial\!\!\!/ - eA\!\!\!/)^{\delta}\psi\ .$$

Introducing the parameter δ in this way preserves the local gauge invariance.

A first paper on the Schwinger model (two-dimensional massless quantum electrodynamics) has been published[17] in which we show that the delta expansion truncates after one term. The anomaly is given by $\delta e^2/\pi$. Note that when δ=0 there is no anomaly and that as δ increases from 0 to 1 the anomaly smoothly attains it known value.

In such theories we have shown that the parameter δ has a very simple interpretation: Imagine a theory of quantum electrodynamics in which there are δ species of electrons, all coupled minimally to a single photon field A^μ. In the delta expansion we are expanding in the number of these electron species in the limit as the number of such species is small. Two additional papers on electrodynamics[18,19] have been submitted. We are currently working on the problem of calculating $g-2$ in four-dimensional quantum electrodynamics.

We do not yet know how useful our methods will ultimately be in non-Abelian gauge theories. Much more research is required. However, it is already clear at this early stage that the delta expansion has very wide applicability.

REFERENCES

1. C. M. Bender, K. A. Milton, M. Moshe, S. S. Pinsky, and L. M. Simmons Jr., Phys. Rev. Lett. **58,** 2615 (1987).

2. C. M. Bender, K. A. Milton, M. Moshe, S. S. Pinsky, and L. M. Simmons Jr., Phys. Rev. D **37,** 1472 (1988).

3. C. M. Bender, K. A. Milton, S. S. Pinsky, and L. M. Simmons Jr., J. Math. Phys. **30,** 1447 (1989).

4. C. M. Bender, S. Boettcher, and K. A. Milton, to be published in the Journal of Mathematical Physics.

5. B. Freedman, P. Smolensky, and D. Weingarten, Phys. Lett. **113B,** 481 (1982), and references therein.

6. C. M. Bender and H. F. Jones, Phys. Rev. **38,** 2526 (1988) and J. Math. Phys. **29,** 2659 (1988).

7. L. Alvarez-Gaumé, D. Z. Freedman, and M. T. Grisaru, Harvard University Report No. HUTHP 81/B111, 1981 (unpublished).

8. A perturbative expansion in powers of λ is not fruitful here because it is plagued by infrared divergences. Moreover, for dimensional reasons, the energy density E is not a series in powers of λ, but rather is a numerical multiple of λ^2. One can, of course, perform lattice calculations [see, for example, C. M. Bender, P. H. Burchard, A. Das, H.-A. Lim, and J. A. Shapiro, Phys. Rev. Lett. **54**, 2481 (1985)]. However, introducing a lattice violates global supersymmetry invariance. The results are numerical and approximate rather than analytical, as in the present talk.

9. The two-vertex graphs give rise to finite sums over l, where the upper limit depends on the integers α and β. This finite sum is a hypergeometric function ${}_2F_1$ whose indices depend on α and β. We define this sum for noninteger values of α and β by analytically continuing in the indices. Applying the derivative operator D in Eq. (40) and setting α and β to zero, we obtain the infinite series in Eq. (41). A shortcut to obtain this infinite series consists of replacing the upper limit of the sum on l by infinity. This is permissible because, so long as α and β

are integers, the summand vanishes because of factorials of negative integers in the denominator. Applying the derivative operator D in Eq. (40) under the summation sign and setting α and β equal to zero again yields the infinite series in Eq. (41).

10. C. M. Bender, K. A. Milton, S. S. Pinsky, and L. M. Simmons Jr., Phys. Lett. B **205,** 493 (1988).

11. C. M. Bender and K. A. Milton, Physical Review D **38,** 1310 (1988).

12. C. M. Bender, F. Cooper, and K. A. Milton, Physical Review D **39,** 3684 (1989).

13. C. M. Bender, F. Cooper, G. Kilkup, and P. Roy, L. M. Simmons, Jr., to be published in the Journal of Statistical Physics.

14. C. M. Bender and T. Rebhan, Physical Review D **41,** 3269 (1990).

15. C. M. Bender, K. A. Milton, S. S. Pinsky, and L. M. Simmons, Jr., Journal of Mathematical Physics **31,** 2722 (1990)

16. C. M. Bender and S. Boettcher, Journal of Mathematical Physics **31,** 2579 (1990)

17. C. M. Bender, F. Cooper, and K. A. Milton, Physical Review D **40,** 1354 (1989).

18. C. M. Bender, F. Cooper, K. Milton, M. Moshe, S. S. Pinsky, and L. M. Simmons, Jr., submitted to Physical Review D.

19. C. M. Bender, K. A. Milton, and M. Moshe, submitted to Physical Review D.

NUMERICAL STUDY OF THE TRANSITION TO CHAOTIC CONVECTION INSIDE SPHERICAL SHELLS.

Lorenzo Valdettaro[1] and **Michel Rieutord**[1,2]

[1] CERFACS, 42 Av. G. Coriolis, 31057 Toulouse CEDEX, France.
[2] Observatoire Midi-Pyrénées, 14 Av. E. Belin, 31400 Toulouse, France.

ABSTRACT The properties of convection inside a spherical shell heated from within are studied by direct numerical simulations. A pseudo-spectral method is used. Both the compressible and the incompressible (Boussinesq) case are treated.
We consider first a non rotating configuration. It is well known that the solutions of the linear problem are degenerate, due to the spherical symmetry, and that their angular behaviour is in the form of spherical harmonics Y_{lm} with a given l and any m. The degeneracy is removed by taking into account the nonlinear terms. This selects a particular value of the wavenumber m [Bus75]. We observe the expected pattern very near the critical Rayleigh number. However when we increase the Rayleigh number the solution undergoes transitions to other steady configurations.
We then study the transition to chaotic convection by increasing the Rayleigh number, both in a non rotating and in a moderately rotating (Taylor number of 100) configuration. In both cases we observe at first the onset of a periodic behaviour, then the appearance of a second frequency, followed by a chaotic regime. The behaviour of the convection cells in the 1-frequency and 2-frequency regimes is presented.

1 Introduction.

The study of convection in spherical shells is motivated by astrophysical interest (convective envelope of stars) and geophysical interest (convection in the Earth's deep interior).

The case of a compressible fluid has already been considered by Bercovici et al. [BSGZ89]. However these authors considered the case of an infinite Prandtl number suitable to the Earth's mantle. Here, we consider the case of a finite Prandtl number (namely unity) for which, except the general work of Busse [Bus75] or the one of Gilman and Miller [GM86], nothing was known.

Independently, we also worked on the case of a Boussinesq fluid in the same configuration but including a mild rotation.

For these two cases, we studied the transition from marginal convection to the chaotic one. In both cases, the Ruelle-Takens scenario appeared as the good one. In the Boussinesq case, we also demonstrated the origin of the two frequencies which are the key of the scenario.

2 Stationary convection

2.1 Parameters of the runs

The numerical code we use in the stationary convection regime is the compressible one [VM89]. We fix the following equilibrium parameters:

- The stratification rate $\chi \equiv \rho_{Top}/\rho_{Bottom}$. We take a constant density, i.e. $\chi = 1$. Consequently the polytropic index m is equal to zero.

- The entropy gradient

$$\varepsilon \equiv - \left. \frac{dS/C_p}{d \log r} \right|_{Top}$$

This quantity measures the relative importance of compressiblity, since it is proportional to the square of the Mach number ([GG81]). We choose a value of order 1 ($\varepsilon = 3$) in order to have non negligible compressible effects. It turns out from the computations that the average Mach number is slightly less than 1.

- The Taylor number: We have considered the case without rotation, i.e. $T_a = 0$.

- The Prandtl number: we have assumed the kinematic viscosity and the thermal diffusivity constant and equal. Therefore $P_r \equiv \nu/\kappa$ is equal to 1 throughout the shell.

- The aspect ratio β, i.e. the ratio between the outer and the inner radius of the shell; we have chosen the value $\beta = 2$.

- Finally we have assumed a perfect monoatomic gas, i.e. $c_p = 5/2$, $c_v = 3/2$.

The boundary conditions we impose are the following: we specify the temperature at the bottom and at the top of the shell. The ratio between the inner and outer temperature is fixed by the parameters we have defined above: $T_{Bottom}/T_{Top} = 1 + 2\varepsilon\beta c_p/(c_v - m) = 21$. We have thus slightly less than 3 temperature scale heights. For the velocity we consider rigid boundary conditions, $\vec{v} = 0$.

The above parameters are all fixed. The control parameter we use to change regime is the Rayleigh number, defined as

$$R_a \equiv - \left. \frac{dS/C_p}{dr} \frac{gh^3}{\kappa\nu} \right|_{Top}$$

(g being the gravity, h the depth of the shell and r the radial coordinate).

2.2 Numerical results.

At the critical Rayleigh number $R_{ac}$the linearly unstable modes are those with $l = 4$, and there is degeneracy in the wavenumber m. According to [Bus75] the only stationary patterns which are possible are the axisymmetric solution (Figure 1a) and the "cubic solution", which displays 6 convection cells with a cubic symmetry (Figure 1b); moreover from the analysis of [Bus75] the expected pattern near the threshold is the 6-cells pattern.

We observed that the stationary convection regime extends to Rayleigh numbers up to 16 times the critical number. In this range We have searched systematically for the stationary solutions by perturbing the conductive profile with a random noise. For each solution we studied the stability to a variation of the Rayleigh number.

At very slightly supercritical Rayleigh numbers, between $R_{ac}$and $1.6R_{ac}$, the wavenumber $l = 4$ is the only one which is linearly unstable. Two types of solutions were observed in this regime:

- The 6-cells solution predicted by [Bus75] (Figure 1b). We have observed both the solution with ascending currents inside the cells and that with descending currents.
- The axisymmetric solution which displays 2 rising rolls, 2 sinking cells and 1 sinking roll (Figure 1a), together with its reciprocal (2 sinking rolls, 2 rising cells and 1 rising roll).

We point out that none of these solutions is a pure $l = 4$ mode, since by non-linear interaction other l-modes are excited. However the energy contained in the $l = 4$ mode represents more than 80% of the total energy.

These solutions are stable in the range $R_{ac} \rightarrow 1.6R_{ac}$, but their basin of attraction is very sensitive to the Rayleigh number. For very slightly supercritical Rayleigh numbers we have obtained most of the time the 6-cells solution, and no preference is found between the ascending-cells and the descending-cells configuration; when we increase the Rayleigh number howewver we observe a preference for the axisymmetric solutions, and also for the configurations which have downwelling cells and upwelling rolls. This means that the basin of attraction of the 6-cells solution shrinks when the Rayleigh number is increased, and that the symmetry between rising-cells and sinking-cells configurations is broken in favour of the sinking-cells configurations.

In the range $1.6R_{ac} \rightarrow 6R_{ac}$the 6-cells and $l = 4$ axisymmetric solutions were very difficult to maintain: a small change in the Rayleigh number was enough to produce a transition to other stationary solutions; we are not able however to say if they just become unstable or if their basin of attraction becomes so small that a small increase in the Rayleigh number is sufficient to leave it and enter another basin of attraction, leading to another solution. New solutions are obtained in this range: the $l = 3$ modes become linearly unstable and we find solutions which are dominated by this wavenumber. The new solutions we found are the following:

- A mixture of $l = 4$ and $l = 3$ modes, with a dominance of $l = 4$, which displays 4 convection cells (Figure 2); we will call it the "$l = 4$, 4-cells" solution.
- A mixture of $l = 4$ and $l = 3$ modes, with a dominance of $l = 3$, which also displays 4 convection cells (Figure 2); we will call it the "$l = 3$, 4-cells" solution.
- An almost pure $l = 3$ mode with 3 convection cells (Figure 3b)
- The $l = 3$ axisymmetric solution (Figure 3a) In this range we found almost only patterns with downwelling cells.

To illustrate the kind of transitions between the modes that we observe, we show in Figure 4 an example of computation, in which we started from the "$l = 3$ 3-cells" configuration at a Rayleigh number of $2.11R_{ac}$ and we progressively decreased the Rayleigh number to the critical value. We see that the "$l = 3$, 3-cells" solution undergoes at first a transition to the $l = 3$ axisymmetric pattern, then at $1.6R_{ac}$the $l = 3$ mode is stabilised and there is a transition to the $l = 4$ axisymmetric solution, which remains stable down to R_{ac}.

At Rayleigh numbers greater than $6R_{ac}$we see the apparition of the $l = 2$ axisymmetric modes, displaying two descending convection cells and one rising roll (Figure 5a). The

"$l = 4$, 4-cells" solution has totally disappeared in this range, the $l = 3$ mode is observed only in the form of the "$l = 3$, 3-cells" pattern.

At slightly supercritical Rayleigh numbers the shape of the convection cells and of the convection rolls is essentially symmetric with respect to the axis which pass through their center and is directed radially. When we increase the Rayleigh number, typically for values $\geq 10R_{ac}$, we observe that this symmetry breaks down. Every cell and roll is distorted with a characteristic wavenumber, and the amplitude of the distortion is increasing with the Rayleigh number. Figures 5b, 6 and 8 are representatives for the distorted patterns of the stationary solutions we observe at these Rayleigh numbers. For the $l = 2$ axisymmetric mode we find two types of distortions; the first type is an $m = 3$ symmetry breaking which produces cells with three legs (Figure 5b). The length of every leg is equal. The equatorial symmetry of the pattern is destroyed, due to the fact that the distortions of the two cells are shifted by $\pi/3$. In the second type of distortion we have a $m = 2$ symmetry breaking, which produces cells with 2 legs (Figure 6); the equatorial symmetry is maintained.

3 Time-dependent convection

The simulations of compressible convection were performed at higher Rayleigh number, together with simulations in the Boussinesq regime, to study the generic route from stationary to chaotic convection in spherical shells. In both cases the transition may be identified to a Ruelle-Takens scenario.

3.1 The compressible case without rotation.

We consider the same equilibrium parameters as in Section 2 and we increase the Rayleigh number until we start to have time dependency, for $R_a \geq 16R_{ac}$. At these Rayleigh numbers the patterns we observe are those of Figures 5b, 6 and 8. At the onset of time periodic convection the periodic motions have an amplitude consistent with a $(R_a - R_{ac1})^{1/2}$ law, R_{ac1} being the Rayleigh number for the onset of time dependent behaviour. This suggests that we are in presence of a Hopf bifurcation from the equilibrium state. Further increase in the Rayleigh number leads to an aperiodic motion which exhibits two uncorrelated frequencies. We describe here the periodic and aperiodic motions.

For the $l = 2$ solution with the $m = 3$ distortion of Figure 5b the periodic motion is in the form of a rigid rotation of the whole pattern about an axis which pass through the centers of the 2 convection cells. The pattern conserves the $\pi/3$ symmetry about that axis. At the onset of 2-periodic motion this symmetry is broken, due to the fact that the legs of the convection cells begin to pulse incoherently, with a frequency which is related to the second frequency (Figure 5c).

In the case of the $l = 2$ solution with the $m = 2$ distortion of Figure 6, the periodic motion is in the form of two counter rotating waves propagating around the axis which pass through the centers of the 2 convection cells. The equatorial symmetry is maintained. The second frequency breaks this symmetry: there appear new waves at the second frequency, which propagate around the same axis but are incoherently distributed (Figure 7).

Finally we consider the case of the "l = 3, 3-cells" solution. The periodic motion is more clearly understood by looking at the interstices between the convection cells, rather than at the cells themselves: the three branches of the interstices oscillate in phase (Figure 8). The 2-periodic motion superimposes waves which have the second frequency and destroy the phase-locking of the three branches (Figure 9).

3.2 THE BOUSSINESQ CASE WITH ROTATION.

We considered then the model of a Boussinesq fluid inside a spherical shell containing heat sources. Such a model has been used many times in the past and we refer the reader to Zhang and Busse (1987) [ZB87] for an exhaustive description. We studied the transition to chaotic convection at a moderate Taylor number $Ta = (2\Omega d^2/\nu)^2 = 100$, where Ω is the angular velocity of the frame, d is the thickness of the layer and ν is the kinematic viscosity. The Prandtl number was set to unity and the ratio of the radius of the inner and outer shells is 0.4 .

At the threshold of linear stability, the mode first destabilized is the 2^+ one[1]. In the frame of the shell, this mode is a wave moving westward, however the pattern of the wave is steady and so is the kinematic energy of it. This pattern contains all the modes excited by 2^+ through the nonlinear coupling and the spectrum is the set $\{(2p)^+\}_{p=0,1,...}$. The kinetic energy of the modes entering this chain is constant once the transitories are damped. This is valid in the interval of Rayleigh numbers $Ra_c < Ra < Ra_{c1}$; when this new critical Rayleigh number (Ra_{c1}) is passed the pattern starts to oscillate with a period of ~ 0.6 viscous time. Then, when further increased (beyond $\sim 2.0 Ra_c$) another frequency is seen in the time series. This second frequency has not been identified to a particular pattern motion. Then, if the Rayleigh number is still increased time series become irregular or chaotic; this is when $Ra > 2.4 Ra_c$.

Such a transition may be identified to a Ruelle-Takens scenario for transition to chaos. Analyzing more thouroughly the results we could make out that the two frequencies are present already in the range $Ra_c < Ra < Ra_{c1}$ of Rayleigh numbers.

Indeed, if one analyses the linear stability of the nonlinear solution developing when $Ra_c < Ra < Ra_{c1}$, that is the set $\{(2p)^+\}_{p=0,1,...}$ then one can establish the following results. Three sets of modes may be used to describe the linear perturbations; these are:

$$\{(2p+1)^+\}_{p=0,1,...}$$

$$\{(2p)^-\}_{p=0,1,...}$$

$$\{(2p+1)^-\}_{p=0,1,...}$$

In this range of Rayleigh numbers all these sets of modes are damped and the solution $\{(2p)^+\}_{p=0,1,...}$ is stable; when $Ra = Ra_{c1}$ then the set $\{(2p)^-\}_{p=0,1,...}$ becomes marginal. Then, only the damping rate of $\{(2p+1)^-\}_{p=0,1,...}$ is real, for the two other sets ($\{(2p)^-\}_{p=0,1,...}$ and $\{(2p+1)^+\}_{p=0,1,...}$) the damping rates are complex and exhibit a frequency of oscillation (say ω_- and ω_+).

[1]We classify the modes according to their symmetry which is referred to by a symbol m^+ or m^- which refers to the azimuthal dependence ($e^{im\phi}$) and its symmetry (+) or antisymmetry (−) with respect to equator.

It is these two frequencies which show up when the Rayleigh number is further increased (see Figures 10a and 10b).

4 Conclusion

Our extensive computations in the stationary compressible convective regime has shown some general trends, which have also been observed by other authors ([MR86] for the Boussinesq case and [BSGZ89] for the compressible infinite Prandtl number case); namely that increasing the Rayleigh number the axisymmetric configurations, and those which display a smaller number of cells, tend to be favoured. Indeed, while at slightly supercritical Rayleigh numbers the typical configuration is that with 6 convection cells and a cubic symmetry (Figure 1b), at the onset of time periodic convection we had almost always the axisymmetric configuration diplaying 2 descending convective cells of Figure 5a. This phenomenon still require a satisfactory explanation. As was pointed out in [MR85] and [BSGZ89], the assumption that convection acts to maximize the heat flow, with the consequence that the pattern chosen is the one which maximize the Nusselt number, is not supported. We confirm this in the compressible shell at finite Prandtl number. Another explanation was given in [BSGZ89], where it was inferred that the convective solutions tend to minimize the viscous shear of upwelling and/or downwelling regions. This would explain why there is a tendency to decrease the number of cells. However, they pointed out that the configuration which minimize the shear is that in which the upwelling regions are in form of cells and the downwelling regions in form of sheets, which is the contrary of what we obtain. We propose here that the transition could be due to an AKA instability ([SKR66, FSS87, SSSF89], which produces a cascade towards smaller wavenumbers.

The time dependent convection displays a Ruelle-Takens scenario, both in the compressible and the Boussinesq regime. This is consistent with previous sudies (for example [MY86] for Boussinesq infinite Prandtl number).

It appears that the time dependency of the convective patterns is a very simple one, namely the superposition of long wavelength waves on the original pattern; moreover in the periodic regime these waves maintain several symmetries. This simple result constrasts with the fact that the amplitude equations are very complicate, since they contain the interaction between all the modes which are linearly excited, which are from $l = 1$ to $l \simeq 15$ in our runs. This leaves a hope to undertake a semi analytic approach to the problem, once the leading terms in the amplitude equations have been detected. In the Boussinesq case the problem is even more simple, since, as we have shown, there are 4 sets of modes which are almost independently excited and display their own frequency.

5 References

[BSGZ89] D. Bercovici, G. Schubert, G. A. Glatzmaier, and A. Zebib. Three-dimensional thermal convection in a spherical shell. *J. Fluid Mech.*, 206:75, 1989.

[Bus75] F. H. Busse. Patterns of convection in spherical shells. *J. Fluid Mech.*, 72:67, 1975.

[FSS87] U. Frisch, Z. S. She, and P. L. Sulem. *Physica D*, 28:382, 1987.

[GG81] P. A. Gilman and G. A. Glatzmaier. Compressible convection in a rotating spherical shell. I. Anelastic equations. *The Astroph. J. Suppl. Series*, 45:335, 1981.

[GM86] P. A. Gilman and J. Miller. Nonlinear convection of a compressible fluid in a rotating spherical shell. *The Astrophysical Journal Supplement Series*, 61:585, 1986.

[MR85] P. Machetel and M. Rabinowicz. Transitions to a two mode axisymmetrical spherical convection: Application to the earth's mantle. *Geophys. Res. Lett.*, 12:227, 1985.

[MR86] P. Machetel and M. Rabinowicz. Three-dimensional convection in spherical shells. *Geophys. Astrophys. Fluid Dynamics*, 37:57, 1986.

[MY86] P. Machetel and D. Yuen. Infinite Prandtl number spherical-shell convection. Technical report, University of Minnesota Supercomputer Institute, 1986.

[SKR66] M. Steenbeck, F. Krause, and K. H. Rädler. *Z. Naturforsch.*, 21:369, 1966.

[SSSF89] P. L. Sulem, Z. S. She, H. Scholl, and U. Frisch. Generation of large-scale structures in three-dimensional flow lacking parity-invariance. *J. Fluid Mech.*, 205:341, 1989.

[VM89] L. Valdettaro and M. Meneguzzi. Compressible MHD in spherical geometry. In *Proceedings from the Workshop on Supercomputing Tools for Science and Engineering*, page 573, Pisa, December 1989.

[ZB87] Zhang K.-K. and Busse F. On the onset of convection in rotating spherical shells. *Geophys. Astrophys. Fluid Dyn.*, 39:119–147, 1987.

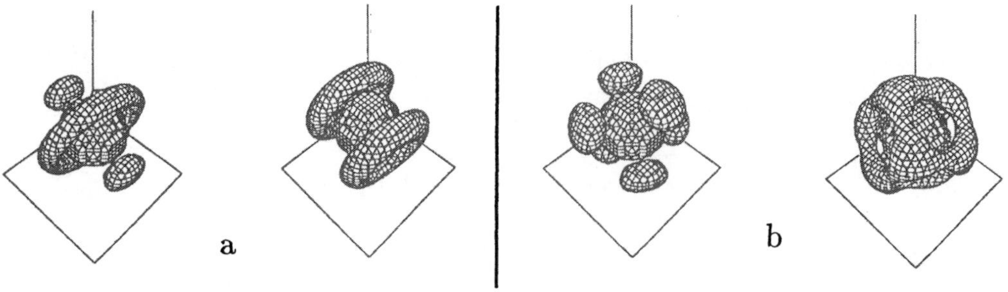

FIGURE 1. (a) Isosurfaces of constant radial velocity for the $l = 4$ axisymmetric pattern of convection. The first figure refers to downwelling motions, the second figure to upwelling motions. The inner bounding sphere is also shown.

(b) Isosurfaces of constant radial velocity for the $l = 4$ cubic pattern of convection. The first figure refers to downwelling motions, the second figure to upwelling motions. The inner bounding sphere is also shown.

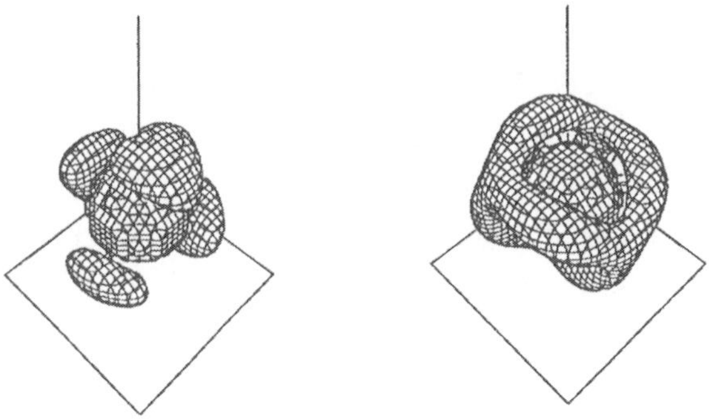

FIGURE 2. Isosurfaces of constant radial velocity for the "$l = 3$, 4-cells" pattern of convection, which has the tetrahedral symmetry. The first figure refers to downwelling motions, the second figure to upwelling motions. There exists another stationary solution, similar to this, but dominated by the $l = 4$ mode, which we refer to the "$l = 4$, 4-cells" solution.

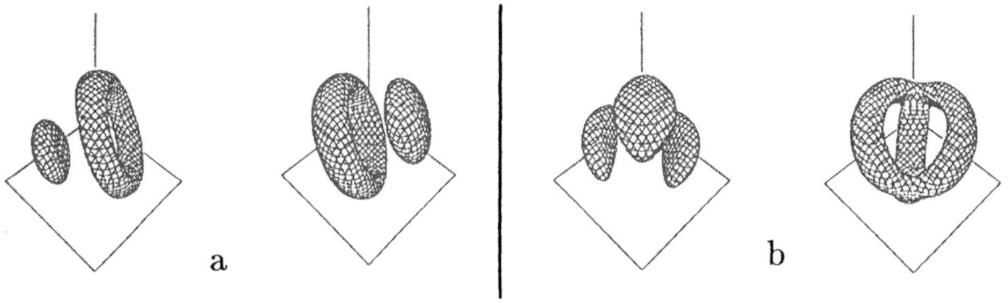

FIGURE 3. (a) Isosurfaces of constant radial velocity for the $l = 3$ axisymmetric pattern of convection. The first figure refers to downwelling motions, the second figure to upwelling motions.

(b) Isosurfaces of constant radial velocity for the "$l = 3$, 3-cells" pattern of convection. The first figure refers to downwelling motions, the second figure to upwelling motions.

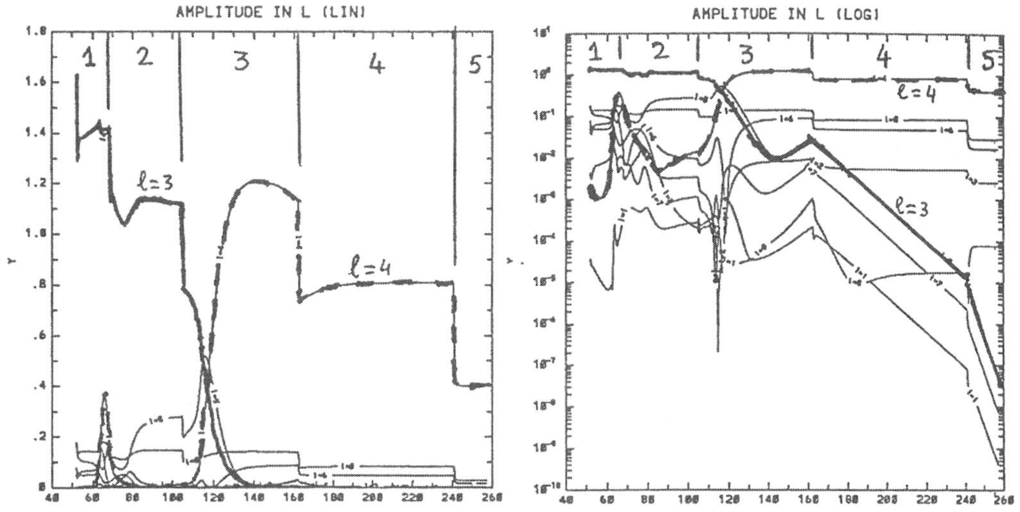

FIGURE 4. Energy in the different l—modes versus time for a typical run, starting from a stationary solution. The Rayleigh number is progressively decreased down to the critical value.
Region 1: $R_a = 2.11 R_{ac}$. The solution tends to the "l=3, 3-cells" pattern.
Region 2: $R_a = 1.9 R_{ac}$. The solution tends to the l=3 axisymmetric pattern.
Region 3: $R_a = 1.6 R_{ac}$. The solution tends to the l=4 axisymmetric pattern.
Region 4: $R_a = 1.4 R_{ac}$. The solution stays on the l=4 axisymmetric pattern.
Region 5: $R_a = 1.02 R_{ac}$. The solution stays on the l=4 axisymmetric pattern.
The first figure is in linear scale, the second in logarithmic scale. The time unit is the viscous time.

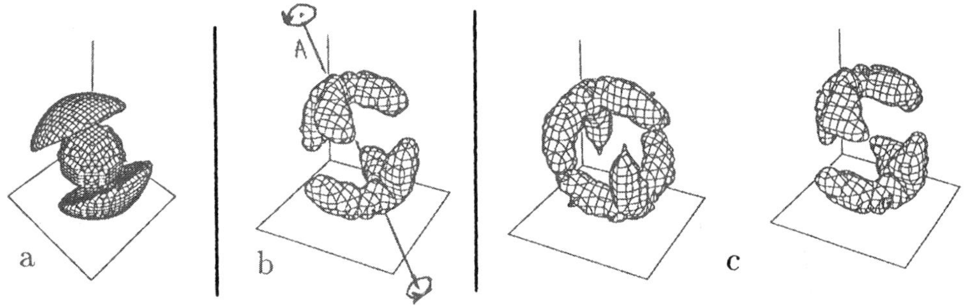

FIGURE 5. (a) Isosurfaces of constant radial velocity for the $l = 2$ axisymmetric pattern of convection, which has two downwelling cells (shown in the figure) and one upwelling roll. The inner bounding sphere is also plotted.
(b) l=2 axisymmetric solution with an $m = 3$ distortion in the time periodic regime: the motion corresponds to a rigid rotation of the pattern around the axis A.
(c) l=2 axisymmetric solution with an $m = 3$ distortion in the aperiodic regime: the motion is the superposition of the rigid rotation and a pulsation of the legs of the cells.

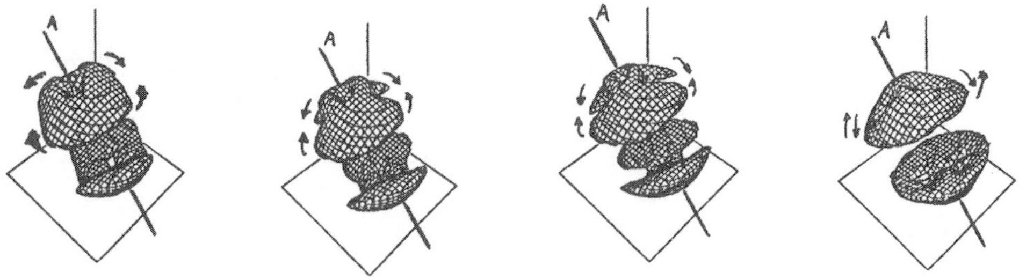

FIGURE 6. $l = 2$ axisymmetric solution with an $m = 2$ distortion in the periodic regime: two counter rotating waves are propagating through the axis A. the time separation between each figure is 1/6 of the period.

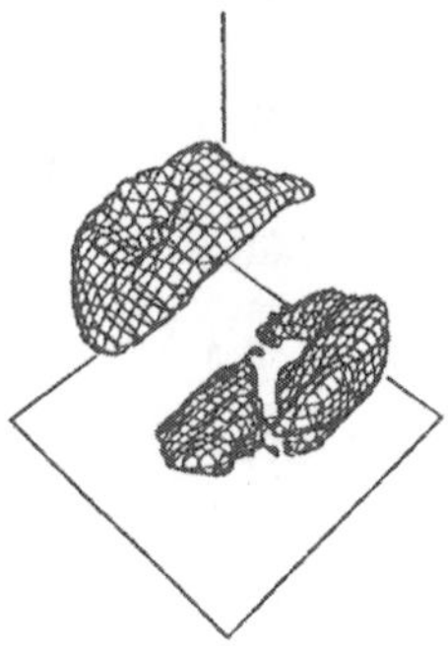

FIGURE 7. l=2 axisymmetric solution with an $m = 2$ distortion in the aperiodic regime: note the break of the equatorial symmetry.

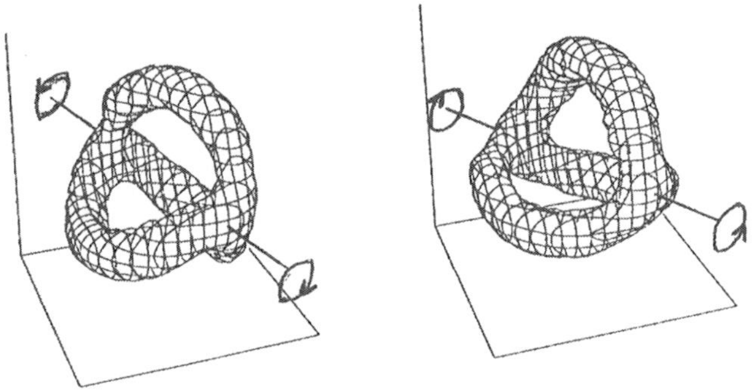

FIGURE 8. "l=3, 3-cells" solution in the periodic regime. The two figures are separated by a time interval of half of the period.

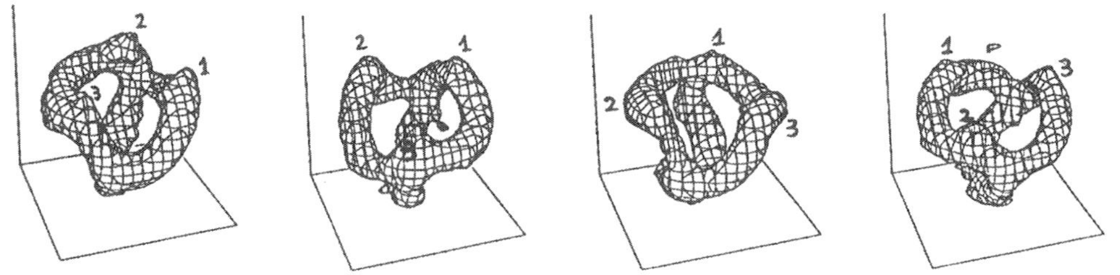

FIGURE 9. "l=3, 3-cells" solution with an $m = 2$ distortion in the aperiodic regime. The time separation between the figures is half of the period corresponding to the first frequency. Note that the waves, labelled 1, 2 and 3, break the phase locking of the 3 branches.

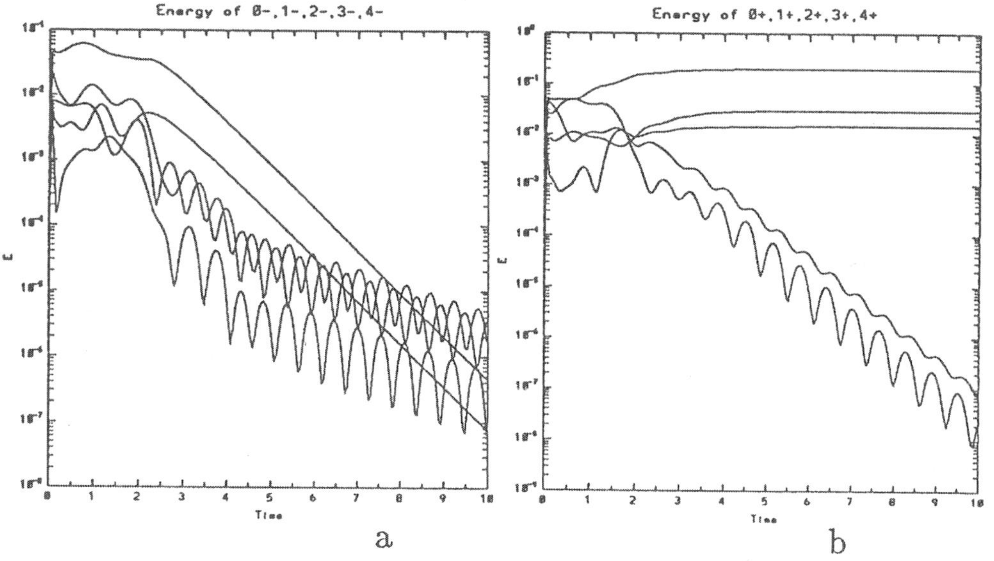

FIGURE 10. (a) Time evolution of the chains $\{(2p)^-\}_{p=0,1,..}$ (oscillating curves) and $\{(2p+1)^-\}_{p=0,1,..}$ (straight curves). $Ra = 1.4Ra_c$ and the time unit is the viscous time.
(b) Time evolution of the chains $\{(2p)^+\}_{p=0,1,..}$ (straight curves) and $\{(2p+1)^+\}_{p=0,1,..}$ (oscillating curves). $Ra = 1.4Ra_c$ and the time unit is the viscous time.

Direct Numerical Simulations of Natural and Controlled Transition of Three-Dimensional Mixing Layers

P. COMTE, M. LESIEUR and E. LAMBALLAIS

Institut de Mécanique de Grenoble*
B.P. 53 X, 38041 Grenoble Cedex, FRANCE

ABSTRACT

We investigate, with the aid of three-dimensional direct-numerical simulations (using pseudo-spectral methods) at high resolution (up to 128^3 grid points in a cubic box containing four fundamental longitudinal wavelengths), the origin and topology of the longitudinal vortex filaments which appear in temporally-growing mixing layers. The basic velocity field is a hyperbolic-tangent profile $U \tanh 2y/\delta_i$, with a Reynolds number $U\delta_i/\nu = 100$. The basic flow is forced initially by two small random perturbations of wide spectrum peaking at the fundamental mode: a three-dimensional one, of kinetic energy $\varepsilon_{3D}\, U^2$, and a two-dimensional one, of kinetic energy $\varepsilon_{2D}\, U^2$. For $\varepsilon_{2D} = 10^{-4}$ and $\varepsilon_{3D} = 10^{-5}$, quasi two-dimensional large coherent Kelvin-Helmholtz rollers are formed. They slightly oscillate in phase, as in the translative instability proposed by Pierrehumbert and Widnall (J. Fluid Mech., **114**, 59, 1982). Between the big rollers, thin hairpin longitudinal vortices are stretched. For $\varepsilon_{2D} = 0$ and $\varepsilon_{3D} = 10^{-4}$ the fundamental rollers which appear have strong spanwise oscillations which are not in phase. Pairings between the primary vortices lead to their reconnection in specific locations of the span, giving rise to a vortex lattice. This is numerical evidence of Pierrehumbert and Widnall's helical-pairing instability. In this case, thin longitudinal vortices are also formed, stretched away from the region of reconnection of the rollers.

1. - INTRODUCTION

Many laboratory or numerical experiments have pointed out the existence of spanwise-organized large-scale vortical structures in the plane mixing layer. These structures have been found to be the result of successive pairings of the primary vortices resulting from the Kelvin-Helmholtz instability, which is basically two-dimensional. However, all

* Unité associée CNRS

mixing layers are known to develop three-dimensionality, both in the small scales and in the large scales: certain laboratory experiments (see [1] and [2], for example) show thin hairpin vortex filaments which are strained downstream between the two-dimensional large rollers. These vortices are also shown in the direct-numerical simulations carried out by Metcalfe *et al.* [3]. Another experiment at a high level of turbulence [4] featured a 'double-helix vortex-structure' after the first pairing of the Kelvin-Helmholtz vortices, this 'helical-pairing' occuring in certain locations of the span only, by contrast with quasi-two-dimensional pairings observed by Bernal and Roshko [2], which occured all along the span.

Among various explanations for the existence of the hairpin vortex filaments observed in the experiments ([1],[2]), let us mention the one associated with translative instability [5], which results from a secondary instability analysis performed on Stuart vortices. It is characterized by a global in-phase spanwise oscillation of the primary rollers. The laboratory experiments of Bernal and Roshko [2] show a spanwise wavelength of the hairpin vortex filament of the order of $2\ \lambda/3$, where λ is the longitudinal wavelength of the Kelvin-Helmholtz vortices, in good agreement with the most amplified spanwise wavelength of the translative instability. However, the thin longitudinal vortices found experimentally are strained between and wrapped around the the big rollers, but are of a different nature. Another mechanism has been proposed ([6],[7]), where vortex filaments (carrying low vorticity) in the stagnation or braid region, perturbed in the spanwise direction, would be strained longitudinally by the basic shear, yielding the hairpin vortex structure. This was confirmed by a three-dimensional numerical simulation using vortex-line methods [8]. An explanation of the 'helical-pairing' has also been proposed in [5], still on the basis of a Stuart-vortex secondary instability analysis. In the latter case, the Stuart vortices were submitted to a spanwise perturbation modulated by a streamwise sub-harmonic mode.

Since all the above-quoted direct-numerical simulations involved both a low number of fundamental Kelvin-Helmholtz vortices and deterministic initial perturbations, we have here carried out a high-resolution calculation (up to 128^3 Fourier modes) with 4 fundamental rollers forming initially: the streamwise length of the computational domain is equal to $4\ \lambda_a$, where $\lambda_a \approx 7\ \delta_i$ is the initial most amplified fundamental longitudinal wavelength. We force initially the mixing layer with a random perturbation, more apt to model the background fluctuations in natural mixing layers. We will show that the vortex structure which develops in the layer depends crucially upon the three-dimensionality of the initial perturbation. In Section 2 the conditions of the numerical simulation will be presented. Section 3 will be devoted to the case of a purely three-dimensional initial forcing, and to the resulting vortex-lattice structure, corresponding to the development of the helical-pairing instability. In Section 4, we will consider a quasi two-dimensional initial forcing.

2. - METHODOLOGY

Using a pseudo-spectral numerical method, the continuity, Navier-Stokes and passive-scalar transport equations are solved in the Fourier space in the form

$$\mathbf{k}.\hat{\mathbf{u}} = 0 \quad , \tag{1}$$

$$\frac{\partial \hat{\mathbf{u}}}{\partial t} = \Pi \left\{ F \left[F^{-1}(\hat{\mathbf{u}}) \times F^{-1}(\hat{\omega}) \right] \right\} - \nu \, \mathbf{k}^2 \, \hat{\mathbf{u}} \tag{2}$$

$$\frac{\partial \hat{\theta}}{\partial t} = -i\mathbf{k}.F \left[F^{-1}(\hat{\theta}).F^{-1}(\hat{\mathbf{u}}) \right] - \kappa \, \mathbf{k}^2 \, \hat{\theta} \quad , \tag{3}$$

where $\hat{\omega} = i \, \mathbf{k} \times \hat{\mathbf{u}}$ is the vorticity in the Fourier space ($i^2 = -1$), F is the discrete Fourier transform operator, Π is the projection on the plane normal to the wave vector $\mathbf{k}$, and $k = \sqrt{\mathbf{k}^2}$.

Periodic boundary conditions are applied in the streamwise x and spanwise z directions. The initial conditions result from the superposition of a small-amplitude random perturbation onto a one-directional basic profile, both for the velocity and the passive scalar fields. This basic velocity profile is defined by

$$u_0(y) = U \tanh \left(2y/\delta_i \right) \quad , \tag{4}$$

δ_i being the initial vorticity thickness. The passive-scalar profile is also proportional to $\tanh \left(2y/\delta_i \right)$. We then use free-slip boundary conditions in the transverse y direction, by means of sine and cosine expansions.

In all cases, the computational domain is cubic, with a side L chosen equal to four times the most amplified streamwise wavelength λ_a predicted by the inviscid linear-stability theory, *viz.*, 7.07 δ_i (Michalke [9]).

The basic velocity and passive-scalar profiles are perturbed initially by two random sets of disturbances whose spectra are broad-banded. The first perturbation is three-dimensional, of kinetic energy $\varepsilon_{3D} \, U^2$. The other one is two-dimensional, of energy $\varepsilon_{2D} \, U^2$.

A good point of comparison is the case $\varepsilon_{3D} = 0$, where the problem is two-dimensional: the corresponding study was done in Lesieur *et al.* [10]. In this study, four primary vortices rolled up at about $t = 15 \, \delta_i/U$, paired at $30 \, \delta_i/U$, and paired again at $70 \, \delta_i/U$.

3. - NATURAL TRANSITION

We present here calculations with $\varepsilon_{2D} = 0$ and $\varepsilon_{3D} = 10^{-4}$, that is, with a purely three-dimensional perturbation. Physically, such a perturbation may be obtained by introducing three-dimensional isotropic residual turbulence into the basic shear layer. *Plate 1*-a, *b* and *c* show top-views of vortex-lines locally coloured by the vorticity magnitude, at $t = 17$, 22 and 30 δ_i/U respectively. *Plate 1-d* and *e* show, at $t = 30 \, \delta_i/U$, the iso-surfaces $||\vec{\omega}|| = 1/3 \, \omega_i$ and $\theta = 0$ respectively, $\omega_i = 2 \, U/\delta_i$ being the initial shear brought about by the basic flow.

At $t = 0$, the initial perturbation superposed onto the basic flow is not visible on visualizations: vortex lines are straight and oriented spanwise, the interface $\theta = 0$ collapses onto the plane $y = 0$ and the locus $\omega = 1/3 \, \omega_i$ consists of two plane sheets $y =$ constant. As from $t = 10 \, \delta_i/U$, the growth of unstable modes becomes visible: one can see the progressive concentration of vortex lines, while spanwise oscillations amplify at various wavelength and scales, with different growth-rates. This mode-selection leads to the progressive formation of a vortex lattice [11], basically made of four Kelvin-Helmholtz vortices; they merge at specific regions of the span, by contrast with the quasi-two-dimensional pairings observed in [2]. In addition to these four Kelvin-Helmholtz rollers, the simulation also features a thin hairpin vortex stretched between them. From *Plate 1-a, b* and *c*, it turns out that this hairpin results from the concentration of vortex lines in the braid region between the Kelvin-Helmholtz vortices. This appears strikingly in *Plate 1-c* where are plotted vortex lines shot from highest vorticity magnitude regions (to show the Kelvin-Helmholtz vortices) and also from points located in the hairpin vortex, in order to materialize the latter. This hairpin vortex is then stretched by the Kelvin-Helmholtz vortices, until it looks like a pair of streamwise counter-rotating vortices analogous to those observed by Bernal and Roshko [2] in a more two-dimensional case.

From *Plate 1*, it seems that the highly three-dimensional lattice structure of the Kelvin-Helmholtz rollers we observe here results from the helical-pairing instability proposed by Pierrehumbert and Widnall [5] (see *Fig. 6* of [5], to be compared with *Plate 1-a*) to explain the *double-helix* vortex structure obtained by Chandrsuda *et al.* [4] in laboratory experiments at a high level of turbulence: as mentioned in the introduction, Pierrehumbert and Widnall [5] not only studied the translative instability, but also the linear de-stabilization of a row of two-dimensional Stuart vortices under the action of a couple of oblique modes $(k_a/2)\vec{\alpha} \pm k_z\vec{\beta}$, where k_z is an arbitrary spanwise wavenumber, and k_a corresponds to the streamwise spacing of the vortices. These modes correspond to a spanwise modulation of the longitudinal sub-harmonic perturbation which causes pairings. Pierrehumbert and Widnall find them unstable, the smaller their k_z the larger their growth-rate (see *Fig. 5* of [5]), which could be the reason why quasi two-dimensional pairings are more often encountered in experiments than helical pairings.

The persistance of the helical-pairing instability beyond the domain of validity of the linear-stability theory, which leads in our case to the formation of a Λ-shaped vortex lattice, might be ascribed to the development of a staggered mode, analogous to the one proposed by Herbert [12], on the basis of a secondary-instability analysis:
in *Fig. 1-a* are sketched vortex lines or tubes of streamwise spacing λ_a, oscillating in yz planes, 180° out of phase from one to another, due to the oblique modes $(k_a/2)\vec{\alpha} \pm k_z\vec{\beta}$, with $k_a = 2\ \pi/\lambda_a$. In first approximation, one can assume that vorticity is stretched by the basic inflectional fluid motion. The vorticity of the peaks P will thus be convected along the upper flow, while that of the valleys V will be convected by the lower flow. This will lead to pairings as indicated in *Fig. 1-b*, resulting in a vortex-lattice structure, which is highly three-dimensional. It seems that the same mechanism is at work in our calculations, with a slight difference: the three-dimensional unstable modes develop as from the beginning of the simulation, together with the fundamental mode $k_a\vec{\alpha}$. Consequently, the helical-pairing process acts while the Kelvin-Helmholtz vortices form and not after this formation is completed, as assumed in [5]. However, since the

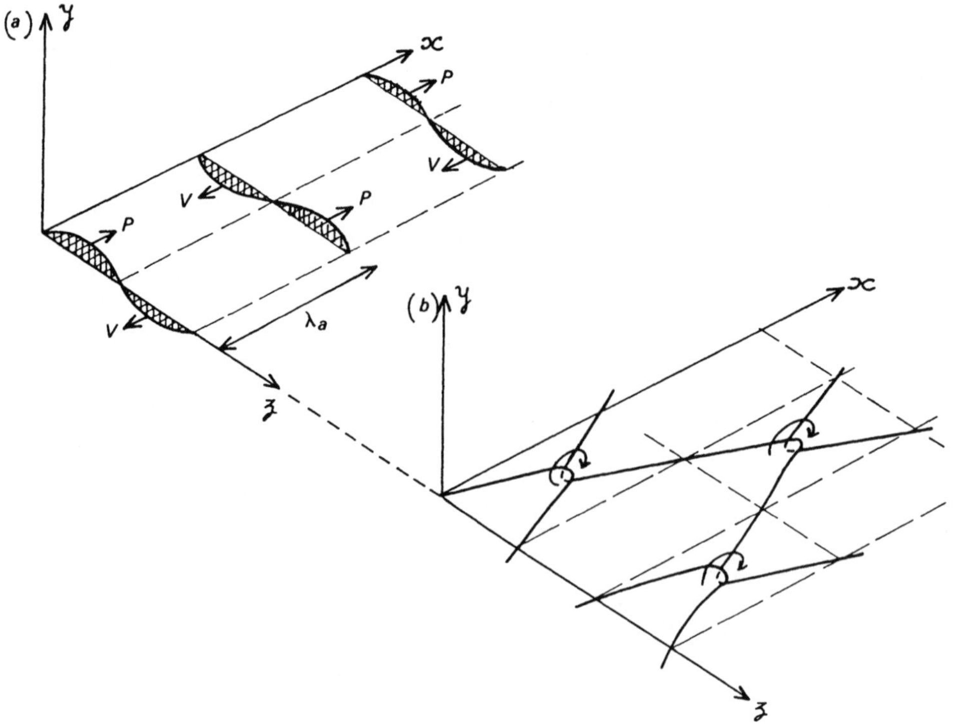

Figure 1. *Schematic representation of the helical-pairing instability, leading to a vortex-lattice structure.*

fundamental mode is more amplified than the other modes, this should come to the same thing.*

The question of the statistical relevance of such helical pairings is of prime interest. In addition to the resemblance with the experimental results of Chandrsuda *et al.* [4] there are several proofs of the existence of helical pairings:

- the present run is not an isolated case; all direct simulations conducted with the present numerical code in a cubic domain with a purely three-dimensional initial forcing, that is, about ten runs, have produced a similar vortex lattice: an example at a resolution of 64^3 (any other parameter being the same as in the present simulation) has been reported in Comte *et al.* ([13],[14]).

- the helical-pairing instability has already been simulated numerically, by means of vortex-line methods [15]. By contrast with the present simulation, the initial perturbation was there deterministic and consisted of the fundamental mode $k_a\vec{\alpha}$ plus a pair of oblique modes, as suggested by Pierrehumbert and Widnall [5]. The calculation stopped at a stage very close to what is obtained here at $t = 17\ \delta_i/U$ (see *Plate 1*-a, to be compared with *Fig. 38* of [15]).

* Notice that the sketch of *Fig.1* also applies to individual vortex filaments. It may be generalized to Herbert's staggered mode in a boundary layer (as already stressed), and to a flow submitted to a constant shear.

- helical pairings have just been found in three-dimensional direct-numerical simulations of a weakly-compressible mixing layer computed by means of finite-difference methods [16]. Helical pairings thus do not result from a numerical artefact of the present code, and are somewhat statistically relevant. One might object that, both in the present study and in [16], periodicity is applied both in the streamwise and the spanwise directions. This imposes a high degree of symmetry to the vortex structure, liable to force helical pairings. However, Rogers and Moser [17] have performed many simulations with such boundary conditions; in all cases, their initial perturbation was purely deterministic, without a pair of oblique modes $(k_a/2)\vec{\alpha} \pm k_z\vec{\beta}$, and no helical pairing was observed. The helical-pairing instability thus corresponds to a real physical phenomenon which occurs only if properly initiated, either by means of explicit forcing as done by Meiburg [15], or random noise as in the present calculations.

Another important issue is the possible existence of a preferential spanwise wavelength. In no cases have we ever found a vortex lattice having a spanwise period different from the box's size: this is in agreement with Pierrehumbert and Widnall's findings [5] concerning the growth rate of the helical-pairing mode, which is, as said above, maximum when $k_z \to 0$. On the other hand, laboratory experiments ([18]-[19]) show branchings of the large rollers, with a spanwise spacing scaling on the vorticity thickness $\delta(t)$, and of the order of twice the fundamental wavelength [20]. This has also been found recently by Fouillet [16] in the case of a weakly-compressible spatially-growing mixing layer at a convective Mach number of 0.3. In our calculations, the longitudinal wavelength at the time when we observe the vortex-lattice structure is 2 λ_a, and the spanwise wavelength is 4 λ_a, which is in agreement with Browand and Troutt's laboratory observations [18]. New calculations in a larger domain (at least in the spanwise direction) are needed in order to see whether this scaling is preserved as time goes on. Spatially-growing calculations could be more relevant for such an investigation.

4. - CONTROLLED TRANSITION

We have repeated the calculation at a resolution of 64^3* with a quasi two-dimensional perturbation, taking $\varepsilon_{2D} = 10^{-4}$ and $\varepsilon_{3D} = 10^{-5}$. *Plate 2* shows perspective views of the passive-scalar interface $\theta = 0$ *(a)* and the vorticity field *(b)*, and a top view of the vortex line *(c)* at $t = 30\ \delta_i/U$. In these snapshots, one can see two quasi two-dimensional rollers which slightly snake in phase, with a spanwise period equal to the box's size 4 λ_a. *Plate 2* also shows hairpin vortex filaments, disposed 180° out of phase with respect to the large rollers and stretched by them. This strongly resembles results of simulations performed by Ashurst and Meiburg using vortex-line methods [8]. It is also reminiscent of the translative instability [5] discussed above, and the associated longitudinal vortex stretching mechanisms. The difference of phase between the spanwise oscillations of the large vortices and that of the hairpin filaments in the braid

* as already stressed above, helical pairings have also been obtained at a resolution of 64^3, but without hairpin vortices ([13],[14]). Simulations of controlled transition at 128^3 are now in progress and will be presented in [21]

region might be ascribed to the emergence of a pair of oblique modes $(k_a/2)\vec{\alpha} \pm k_z\vec{\beta}$, as in the previous case. The main discrepancy lies in the extra amount of two-dimensional energy, which forces the pairing to occur in a quasi two-dimensional manner, instead of helically. Here again, the question of the spanwise spacing and its relation to theoretical [5] or laboratory experiment [2] results needs new temporal and spatial calculations.

5. - CONCLUSION

We have shown, using three-dimensional direct-numerical simulations, that a periodic mixing layer, developing from a hyperbolic-tangent velocity profile, could be subject to a violent three-dimensional instability leading to a vortex-lattice structure if the initial perturbation (of velocity fluctuation $\sim 10^{-2}U$) was random and three-dimensional. This behaviour may be ascribed to the helical-pairing instability discovered by Pierrehumbert and Widnall [5], and is triggered by a spanwise periodic oscillation of the subharmonic perturbation. This could explain the apparent dislocations of the large rollers found in laboratory and numerical experiments ([18],[16]).

On the other hand, calculations with a quasi two-dimensional random perturbation show the development of a weak translative instability of the large rollers, while thinner longitudinal vortices are stretched between them.

It seems therefore that the vortex topology in a mixing layer may depend heavily upon the nature of the residual incoming turbulence.

6. - ACKNOWLEDGEMENTS

The authors are indebted to J. RILEY for numerous discussions during the completion of this study, and to Y. FOUILLET and M.A. GONZE for developing the three-dimensional visualization software. This work was supported by D.R.E.T. under contract 88/150, and by GDR-CNRS *Mécanique des Fluides Numérique*. Part of the calculations were done on a grant of the Centre de Calcul Vectoriel pour la Recherche.

7. - REFERENCES

[1] R. Breidenthal, J. Fluid Mech.,**109**, 1 (1981).

[2] L.P. Bernal and A. Roshko, J. Fluid Mech., **170**, 499 (1986).

[3] R.W. Metcalfe, S.A. Orszag, M.E. Brachet, S. Menon and J. Riley, J. Fluid Mech., **184**, 207 (1987).

[4] C. Chandrsuda, R.D. Mehta, A.D. Weir and P. Bradshaw, J. Fluid Mech., **85**, 693 (1978).

[5] R.T. Pierrehumbert and S.E. Widnall, J. Fluid Mech., **114**, 59 (1982).

[6] G.M. Corcos and S.J. Lin, J. Fluid Mech., **139**, 67 (1984).

[7] J.C. Lasheras and H. Choi, J. Fluid Mech., **189**, 53 (1988).

[8] W.T. Ashurst and E. Meiburg, J. Fluid Mech., **189**, 87 (1988).

[9] A. Michalke, J. Fluid Mech., **19**, 543-556 (1964).

[10] M. Lesieur, C. Staquet, P. Le Roy and P. Comte, J. Fluid Mech., **192**, 511 (1988).

[11] This terminology is due to Chorin (Comm. Pure Appl. Maths., **39**, S47, 1986), who proposes a model of three-dimensional isotropic turbulence based on a vortex-lattice topology.

[12] T. Herbert, Ann. Rev. Fluid Mech., **20**, 487 (1988).

[13] P. Comte and M. Lesieur, in *Separated flows and jets*, V.V. Kozlov ed., Springer-Verlag, in press.

[14] P. Comte, Y. Fouillet, M.A. Gonze, M. Lesieur, O. Métais and X. Normand, in *Turbulence and Coherent structures*, O. Métais and M. Lesieur eds., Kluwer, 45 (1990).

[15] E. Meiburg, *Numerische Simulation der zwei- und dreidimensionalen Strukturbildung in Scherschichten und Nachläufen*, DFVLR Forschungsbericht FB 86-10 (1986).

[16] Y. Fouillet, *Simulation numérique de tourbillons cohérents dans les écoulements turbulents libres compressibles*, Thèse de l'Institut National Polytechnique de Grenoble, en préparation.

[17] R.D. Moser and M.M. Rogers, in *Stirring and Mixing*, special issue of Phys. Fluids A, in press.

[18] F.K. Browand and T.R. Troutt, J. Fluid Mech., **97**, 771 (1980).

[19] F.K. Browand and C.M. Ho, J. Mec. Théor. Appl., Special Issue on *Two-dimensional turbulence*, 99, (1983).

[20] This point has been raised in [5].

[21] P. Comte, M. Lesieur and E. Lamballais, *Large and small-scale stirring of vorticity and a passive scalar in a 3-D temporal mixing layer*, submitted to Phys. Fluids A.

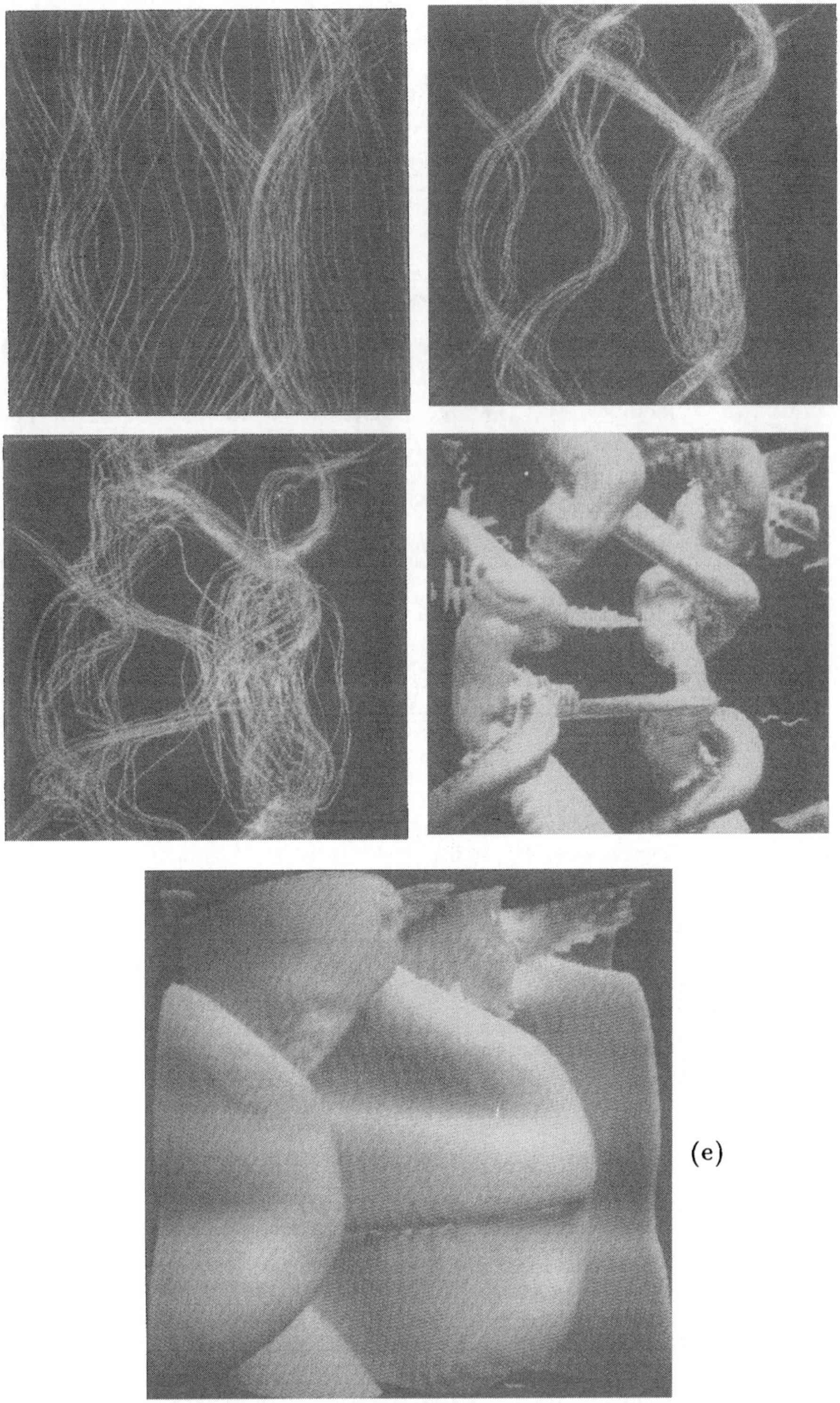

(a) (b) (c) (d) (e)

Plate 1: temporal evolution of a temporally growing mixing layer forced by a three-dimensional perturbation (natural transition)

- *(a), (b)* and *(c)*: top view of vortex lines at $t = 17$, 22 and 30 δ_i/U respectively.
- *(d)*: top view of the surface $\|\vec{\omega}\| = 1/3\ \omega_i$ at $t = 30\ \delta_i/U$.
- *(e)*: top view of the passive scalar interface $\theta = 0$ at $t = 30\ \delta_i/U$.

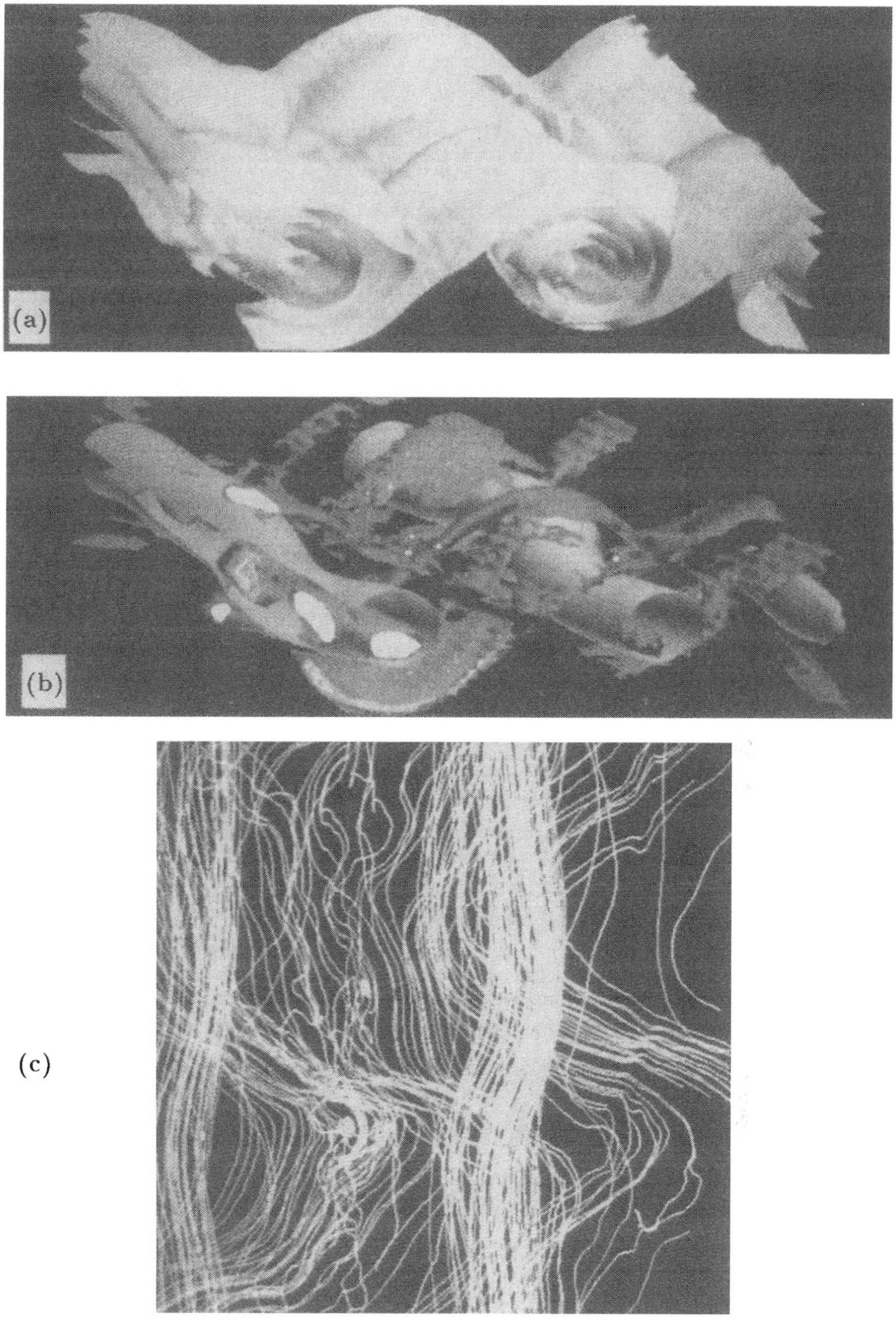

Plate 2: temporal mixing layer forced by a quasi two-dimensional perturbation (controlled transition), at $t = 30\ \delta_i/U$:

- *(a)*: perspective view of the passive scalar interface $\theta = 0$.
- *(b)*: perspective view of the vorticity field [iso-surfaces $|\omega_z| = 1/3\ \omega_i$ (in light blue), $\omega_x = 0.1\ \omega_i$ (dark blue), and $\omega_x = -0.1\ \omega_i$ (green)].
- *(c)*: top view of the vortex lines.

ORGANIZED VORTICES AS MAXIMUM ENTROPY STRUCTURES

Joël Sommeria
Laboratoire de Physique, Ecole Normale Supérieure de Lyon
46 allée d'Italie 69364 Lyon Cedex 07

Abstract: Atmospheric flows often produce organized structures in the presence of strong turbulence. This organization property of two-dimensional turbulence is also observed in laboratory experiments and numerical simulations. After introducing different examples, it is shown that this organization can be explained in terms of equilibrium statistical mechanics on Euler equations.

1. INTRODUCTION

The formation of organized vortices is a striking property of two-dimensional fluid systems in strongly non-linear regimes. In the ordinary middle latitude weather regime, a global flow structure with the shape of a wavy eastward jet propagates eastward, and the succession of low and high pressure at a given location results from the passage of the troughs and crests. By contrast in the so called blocking regime, an anticyclonic vortex can stay over a given area for months without significant evolution and maintains a persistent drought. The switching between these two regimes is relatively fast and difficult to predict. The polar vortex in the stratosphere is another example of organized structure. It persists during the polar night and its core material seems isolated from the outside in spite of strong fluctuations of shape. The ozone hole forms inside this vortex core, and this container effect probably plays a key role in the ozone depletion by preventing mixing with the outside. In the jovian atmosphere, the Great Red Spot has been observed to persist for more than 300 years in a strongly turbulent shear flow, and other similar vortices have been recently discovered on the Giant Planets[1]. The energy of these motions is brought by thermal effects and the formation of vorticity is due to the Coriolis force. However the typical time of forcing and dissipation is often longer than the time of horizontal advection and mixing. Therefore it is reasonable to explain the observed structures as an inertial organization of the flow rather than the direct consequence of driving mechanisms and instabilities.

For such inertial flows (i.e. purely adiabatic motion), a key constraint is the conservation of the local angular momentum, called potential vorticity. Different expressions for the potential vorticity have been proposed, taking into account the planetary curvature, topography and vertical stratification. The structure of the equations is then similar to the Euler equations, with the vorticity replaced by a potential vorticity. In the simplest case, one gets the quasi-geostrophic equations or even the two-dimensional Euler equations, and we shall only discuss here the theory for this latter case. The organisation into inertial structures can also be observed in laboratory experiments in rotating tanks [2,3]. Similar equations are also used to model plasma flows in a magnetic field [4].

Since energy is conserved by the Euler equations, it is tempting to explain the global organization of these strongly stirred flows in terms of equilibrium statistical mechanics (Notice however that this is possible only in two-dimensions, where the regularity of the solutions for all times has been proved. In three dimensions singularities are believed to appear after a fairly short time, as the energy cascade sets up). However this problem is made difficult by the infinite dimension phase space corresponding to a continuous field. The problem was first adressed by Onsager [5] by modelling the continuous field with a finite set of point vortices. Onsager and his successors were able to explain the formation of organized vortices as blobs of point vortices with the same sign, forming for negative temperatures. However this gaz of vortices is not incompressible (although the induced velocity field is divergence free) and this leads to contradiction with the continuous Euler equations. Another way to approximate the Euler equations is to expand the velocity into a Fourier series and truncate the description up to a finite number of Fourier modes [6]. The high wave-number truncation is purely artificial: it is necessary to suppress the ultraviolet divergence of classical statistical field theories. After the truncation, only two constants of the motion remain, energy and enstrophy (integral of the vorticity squared), while in the initial Euler equations the integral of any function of the vorticity is a constant of the motion. This theory deals only with energy spectra, and is not able to predict spatial structures.

A new statistical theory which explains the emergence of organized structures in two-dimensional turbulence has been proposed by Robert [7] and developed in Robert & Sommeria [8]. It is summarized in next section. A similar idea was proposed by Miller [9] but without a less rigorous justification, as discussed by Robert [10].

2. THE MAIN FEATURES OF THE STATISTICAL THEORY:

In its simplest form, the theory applies to an initial condition with piecewise uniform vorticity (which can be an approximation for any regular initial condition). The boundaries of these vorticity patches become in general more and more intricate as time goes on, but the area of each vorticity patch is conserved, as well as the total kinetic energy of the system. This formation of smaller and smaller vorticity scales can be considered as a manifestation of the enstrophy cascade. The goal of the theory is to predict the final state, at the end of this cascade process.

Since the vorticity contours become so intricate, we are not really interested in the exact vorticity field. Indeed the velocity field results from the spatial integration of the vorticity, so that it does not depend on the fine scale fluctuations of the vorticity: it depends only on its local average. In fact, to exploit the whole information given by the constants of the motion, we are led to consider a macroscopic description of the system by introducing the local probability distribution e_i of the different vorticity levels a_i in a small neighborhood. Therefore we define a macroscopic state as the field of these local probabilities, while an individual vorticity field is called here a microscopic state. We consider all the vorticity fields with the same constants of the motion as the initial condition. It was proved [7] that "most" of these accessible microscopic states are very "close" to a well defined macroscopic state. This state is obtained by maximizing an entropy functional, with the constraints due to all the known constants of the motion. Assuming that the system explores fairly uniformly all the accessible microscopic states, taking into account all the constants of the motion (an ergodic hypothesis), we conclude that the system is then most often near this state of maximum entropy.

The entropy density can be obtained by considering partitions of a small surface element into equal area pieces filled with the different vorticity levels a_i, and counting the number of completions. In the limit of a very fine division, physically obtained after a long time of evolution, an expression analogous to Bolztmann mixing entropy for the different vorticity "species" is thus obtained (by integrating over the position vector $\mathbf{x}$ in the whole fluid domain $\mathfrak{D}$)

$$S = -\int_{\mathfrak{D}} \sum_i e_i(\mathbf{x}) \operatorname{Log} e_i(\mathbf{x}) \, d\mathbf{x},$$

The equilibrium states are obtained by maximizing this entropy under the constraints that the total area of each vorticity level and the total energy must correspond to their value in the initial flow condition. The Lagrange parameters, the "chemical potentials" α_i and the "inverse temperature" β are associated with these constraints. The variational problem yields the optimal probability densities

$$e_i(\mathbf{x}) = \frac{\exp(-\alpha_i - \beta\,\alpha_i\,\Psi(\mathbf{x}))}{Z(\Psi(\mathbf{x}))}$$

where Ψ is the stream function associated with the locally averaged vorticity $\overline{\omega}$ by the relation $-\Delta\Psi = \overline{\omega}$ and the partition function Z is given by

$$Z(\Psi) = \sum_{i=1}^{n} \exp(-\alpha_i - \beta\alpha_i\Psi)$$

The total energy of the flow is equal to the integral of $\overline{\omega}\Psi$. The term $a_i\Psi$ can be then interpreted as the energy of the vorticity parcel in the stream function Ψ due to all the other vorticity parcels, therefore the expression corresponds to a mean field theory. The denominator represents a Fermi like exclusion due to the fact that the vorticity parcels form a partition of the space and cannot overlap. The Onsager's theory yields the same mean field relation without the denominator, so that it is analogous to the classical limit of a Fermi distribution. This corresponds to the fact that the point vortices can pack closely together without limitation. This Onsager's relation is recovered here when the vorticity patches are diluted in a sea of irrotational fluid, so that the mutual exclusion becomes marginal. The partition function is then dominated by the zero vorticity level and is close to 1.

The optimal state Ψ is determined as a solution of the differential equation

$$\left\{ \begin{array}{l} -\Delta\Psi = \sum_i a_i e_i(\mathbf{x}) = -\dfrac{1}{\beta}\dfrac{d}{d\Psi}\operatorname{Log} Z(\Psi) \ , \\ \\ \Psi = 0 \quad \text{on } \partial\mathfrak{D} \end{array} \right\}$$

We get a relationship between the locally averaged vorticity and its stream function, which characterizes a steady solution of the Euler equations. The theory selects one state (or several in case of multiple solutions) among all the possible steady solutions of the Euler equations. In the presence of a very small viscosity, the local vorticity fluctuations are smoothed out, and this locally averaged vorticity should become the actual steady flow which emerges from strong turbulent mixing. In the absence of the energy constraint (β=0), a uniform vorticity is obtained, like for the mixing of dyes. For β positive or negative but above a bound given in Robert & Sommeria

[8], the solution for the optimal state is unique, and the vorticity distribution fills fairly evenly the fluid domain. However for sufficiently negative β, a variety of bifurcations can occur and lead to symmetry breaking and isolated vortex structures. Some general properties of the solutions are given in [8], in particular the relationship between $\bar{\omega}$ and Ψ is always monotonous. For a quantitative prediction with a specified initial condition, the determination of the Lagrange parameters α_i and β is a difficulty. Indeed they are only indirectly given by the constraints on the constants of the motion which the solutions of the equation must satisfy.

3. TESTS AND APPLICATIONS:

We have tested the theory by direct numerical computation of the Navier-Stokes equations [11]. A numerical test of the theory, using Euler equations, would be limited by the production of vorticity structures at increasingly fine scales, which become rapidly out of reach of any numerical scheme. Therefore, with any numerical method, a subgrid scale modelling is needed to compute the long time behavior. Such modelling should be able to smear out correctly the local vorticity fluctuations, while conserving the total energy and circulation. In the absence of any known subgrid scale modelling with these properties, an ordinary viscosity was chosen, therefore solving the Navier-Stokes equations. The laplacian term locally averages vorticity, and the total energy decays only moderately when viscosity is small. The theory was tested in the case of a shear layer, in which the initial vorticity is uniform and limited to a band (with smoothed edges). The development of the shear instability and merging processes strongly mix the vorticity with the surrounding irrotational fluid. Then a final vortex is formed (Fig.1), which is nearly a steady solution of the Euler equation (very slowly diffusing by viscosity). Since the boundary conditions are periodic in the direction of the initial band, this final vortex is rather a periodic chain of vortices. This geometry is chosen for its simplicity, but it is similar to an

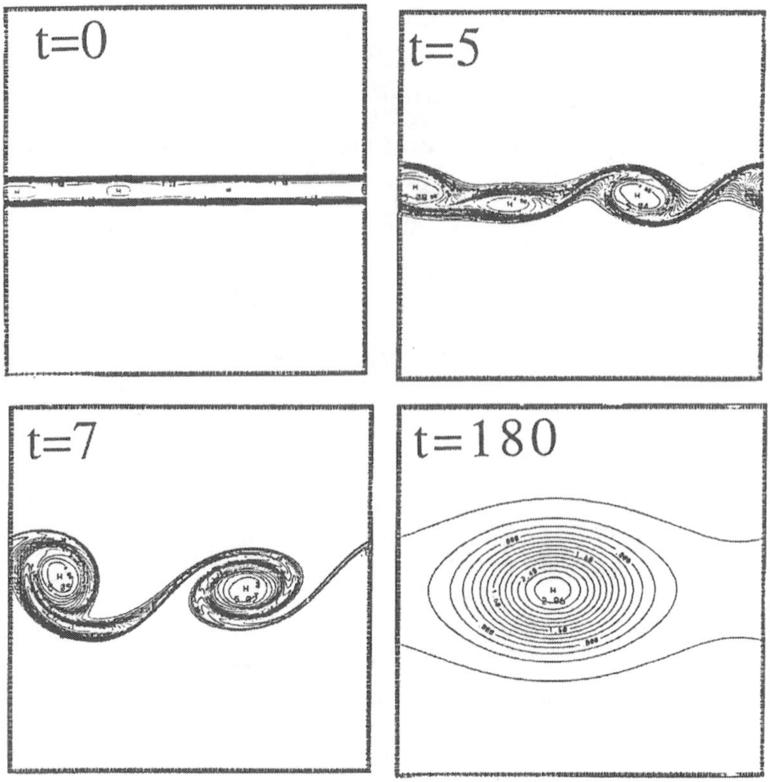

Fig.1: Successive snap shots of the vorticity field in a numerical simulation of a shear layer(from [11]) leading to a steady final vortex. The domain is a channel with periodic boundary conditions in x and the numerical resolution is 256 x 256 grid points. The time unit corresponds to a typical inertial time.

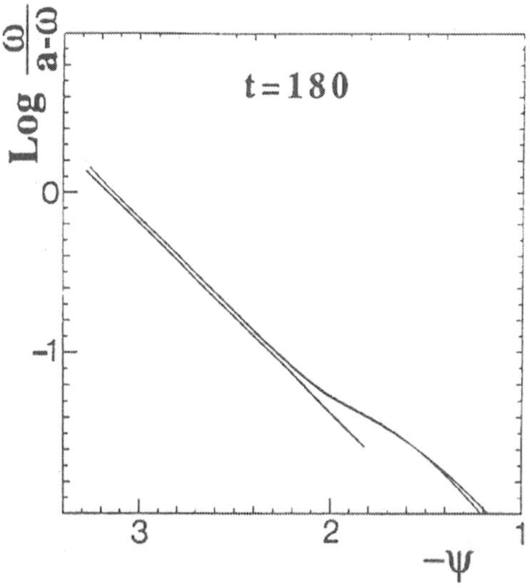

Fig.2 : Scatterplot of Log $\omega/(a-\omega)$ versus stream function for the points of the last field in Fig.1. All the points collapse on a single line, which shows that the flow is very close to a steady solution of the Euler equations. the predicted linear relationship is well verified in the vortex itself.

annular geometry, as found in laboratory experiments [2] and in the atmosphere of the Giant Planets.

Since the initial state has only one non-zero level a, the prediction of the theory can be quantitatively tested by checking that Log $\omega/(a-\omega)$ is a linear function of Ψ (Fig. 2). We find indeed an excellent agreement in the mixed zone. However we observe that the mixing is restricted to an active zone and does not fill the whole domain. We do not know whether the mixing would continue to expand in a longer calculation, and with higher spatial resolution, or whether there is a fundamental limitation for mixing.

The prediction of the theory can be also tested by solving the equation relating ω to Ψ, and a bifurcation leading to a vortex chain is indeed obtained. In the limit of Onsager's theory, and for a wide domain, the equation has an explicit solution and the bifurcated branch of solutions then identifies with the family of Stuart's vortices. These vortices are more and more localized in vorticity as β becomes more negative, but the stream function is always widely spread. By extending the theory to a quasi-geostrophic model suitable for the atmosphere of Jupiter, one can get a vortex which is also isolated in term of stream-lines, as observed for the Great Red Spot of Jupiter [12]. The agreement in structure is remarquable. Therefore the robustness of such an isolated vortex is no more paradoxical. On the contrary it appears as the ineluctable result of complete mixing, when the right constraints are taken into account. Wavy jet structures can also be obtained in other configurations corresponding to atmospheric eastward jets, which would explain the robustness of such jets. The theory is also able to explain experimental flow structures obtained in a circular domain [13].

4. CONCLUSIONS

This theory reconciles organized structures and statistical description for two-dimensional flows. It is in the same spirit as Onsager's theory but it solves its inconsistencies. The conservation of all the known constants of the motion is taken into account. In particular the total enstrophy of the final equilibrium state is equal to the enstrophy of the initial condition. However this enstrophy is contained partly in the mean flow and partly in the local fluctuations. There is thus an irreversible transfer of enstrophy toward the very fine scale fluctuations (where it can be dissipated by a weak viscosity). This result is therefore consistent with the idea of an enstrophy cascade. However the system keeps track of all the conserved quantities, even after the fluctuations are smoothed out, and the result is different from the minimum enstrophy vortices proposed by Leith [14]. While Leith predicts a final structure which depends only on the robust constants of the motion, the energy and angular momentum or circulation, the final state predicted by the present theory depends in addition on the whole distribution of the vorticity levels. However the qualitative behavior does not seem to be sensitive to this vorticity distribution. The application to atmospheric models with one or several vertical layers could provide a conceptual basis for modelling flow regimes on long time scales in climatic systems. Applications to plasma flow structures seems also to be promizing. From the point of view of general physics, the usual ultraviolet divergence of classical field theories is removed by taking into account the local conservation of vorticity, unlike in spectral models. It is remarkable that the statistics is very different from the statistics of an ensemble of atoms, in spite of the fact that the fluid is really made of atoms.

The computations have been performed on the Cray 2 of the CCVR at Palaiseau. The author wishes to thank R. Robert, C. Staquet, M.A. Denoix and T. Dumont for their participation to this work.

REFERENCES:

1- Sommeria J., in Images de la Physique, Courrier du C.N.R.S. (1991)

2- Sommeria J., Meyers S.D. & Swinney H.L. in "Non-Linear Topics in Ocean Physics" A. Osborne ed. Special Issue of Il Nuovo Cimento, Varenna Summer Schol (1988).

3- Sommeria J., Meyers S.D. & Swinney H.L. Nature 337, no 6202, p58-61 (1989)

4- Smith R.A.& O'Neil T.M. Phys. Fluids B2(12) 2961-2975 (1990)
5- Onsager L. Nuovo Cimento Suppl. 6, 279 (1949)
6- Kraichnan R.H. J. Fluid Mech. 67, 155-175. (1975).
7- Robert R., C.R.Acad. Sci. Paris 311 (I), 575-578. (1990)
8- Robert R. & Sommeria J. , J. Fluid Mech. 229, 291-310 (1991)
9- Miller J. Phys. Rev. Letters 65, 17, 2137-2140 (1990)
10- Robert R. J. Stat. Phys., to appear (1991)
11- Sommeria J, Staquet C. & Robert R. J. Fluid Mech. to appear (1991).
12- Sommeria J. , Nore C., Dumont T. & Robert R., C.R. Acad. Sci. Paris, t312, II, 999-1005 (1991)
13- Denoix M.A. & Sommeria J., in preparation.
14- Leith C.E. Phys. Fluids 27, 1388-1395 (1984)

SPECTRAL DEGENERACY AND HYDRODYNAMIC STABILITY

I. Goldhirsch
Dept. of Fluid Mechanics and Heat Transfer
Faculty of Engineering, Tel-Aviv University
Ramat Aviv, Tel-Aviv 69978, Israel

Abstract

It is shown that when the stability spectrum of a system has degeneracies at critical values of a control parameter, below which the system is stable or marginally stable, it may develop instabilities due to the interaction of the degenerate critical eigenmodes. When the spectrum is discrete all instabilities close to criticality are expected to stem from such degeneracies, though not all degeneracies need lead to instabilities. Two examples are briefly reviewed: the instabilities of two dimensional Stokes waves and those of an elliptical vortex.

I. Introduction

It is common knowledge in quantum mechanics [1,2] that when a small off diagonal perturbation is applied to a system having a degenerate spectrum, it leads to the lifting of that degeneracy and to the creation of a gap in the spectrum (also known as an avoided crossing). Consider, for example, the following simple Hamiltonian matrix (as a function of the parameter x):

$$H = \begin{pmatrix} x & \epsilon \\ \epsilon^* & -x \end{pmatrix} \tag{1.1}$$

where ϵ is a "small" perturbation. When $\epsilon=0$, the spectrum of H is: $\lambda_\pm^{(0)} = \pm x$ having a degenerate point at x=0. For finite values of ϵ the spectrum is: $\lambda_\pm = \pm\sqrt{x^2+|\epsilon|^2}$. The degeneracy has been removed and a gap of "forbidden" energies has arisen disallowing values between $-|\epsilon|$ and $|\epsilon|$. This, seemingly trivial, result is of significant consequences in such fields as solid state physics and chemistry. Its importance, when generalized to Hamiltonian systems of arbitrary dimensions, was realized already by Weierstrass [3] and has been investigated by numerous researchers (cf., e.g., ref (4)). It seems that notion of the importance of spectral degeneracies has not percolated as deeply into the field of fluid mechanics as it has into quantum theory. Nevertheless, the problem has been considered in the realm of fluid mechanics. For example, in the study of the stability of Stokes waves it has been noticed that eigenvalues of the problem may cross. Mackay and Saffman [6] and later Kharif and Ramamonjiarisoa [7] have investigated the role of eigenvalue crossings in determining the stability of Stokes waves. They have made heavy use of the Hamiltonian nature of the pertinent dynamics and tested their results using numerical methods.

The aim of the present work is to demonstrate that similar ideas can be relevant to nonhamiltonian stability problems and that, at least in some cases, these same ideas provide useful analytical tools for the investigation of hydrodynamic stability problems. The structure of this paper is as follows. Section (II) provides a formalism, explaining how spectral degeneracies are relevant to stability problems. Section (III) presents a sketch of the derivation of the stability properties of two dimensional Stokes waves. In Section (IV) a brief description of the relevance of our results to the problem of stability of an elliptical vortex is presented. Section (V) provides brief summary.

II. Linear Stability and Degeneracy

II.1. Consider first the following "stability" problem:

$$i\,\dot{u} = \begin{pmatrix} x & \epsilon_1 \\ \epsilon_2 & -x \end{pmatrix} u \tag{2.1}$$

When the perturbations ϵ_1 and ϵ_2 vanish, the solution of eq. (2.1) can be clearly written as: $u = ae^{ixt}+be^{-ixt}$, a and b being the correponding eigenvectors of the stability matrix; the problem is then marginally stable. When nonzero values of ϵ_1, ϵ_2 are considered, the eigenvalues of the matrix in eq. (2.1) are: $\lambda_\pm = \pm\sqrt{x^2+\epsilon_1\epsilon_2}$. In particular when $\epsilon_1\epsilon_2$ is real and negative, i.e., $\lambda_\pm = \pm\sqrt{x^2-|\epsilon_1\epsilon_2|}$ one obtains an instability in the neighbourhood of x=0, i.e., for $|x| < |\epsilon_1\epsilon_2|$. Unlike in the case of a hermitean matrix, mentioned in the Introduction, a nonhermitean perturbation may give rise to instability in the neighbourhood of the crossing point. Notice that as long as $\epsilon_1\epsilon_2<0$, this instability will occur for any absolute value $|\epsilon_1\epsilon_2|$. Consider next a quasidegenerate situation, e.g.:

$$i\,\dot{u} = \begin{bmatrix} \sqrt{x^2+\delta^2} & \epsilon \\ -\epsilon & -\sqrt{x^2+\delta^2} \end{bmatrix} u \tag{2.2}$$

When $\epsilon=0$, the system is only "quasidegenerate" (for $|\delta|<<1$), i.e., no true level crossing occurs. When $\epsilon\neq0$, the spectrum of the matrix in eq. (2.2) is $\lambda_\pm = \pm\sqrt{x^2+\delta^2-\epsilon^2}$, leading to an instability when $|\epsilon|>|\delta|$ and $|x|>\sqrt{\epsilon^2-\delta^2}$. This is an example of a finite amplitude instability occurring when two eigenvalues are close to being degenerate. Then a finite value of the perturbation is needed to trigger the instability. While the case of many levels and many level crossings is technically more complicated, the basic idea sketched here is still the essential mechanism.

Consider a system described by N degrees of freedom (including N=∞) whose linear stability equations can be written as follows:

$$i\,\dot{u} = L_0(x)u + \epsilon L_1(x)u \tag{2.3}$$

Here x is a (control) parameter and ϵL_1 is a small perturbation. Assume that L_0 has a real spectrum $\{\omega_n^{(0)}(x)\}$ and that some pairs $\left\{\omega_{n_1}^{(0)}, \omega_{n_2}^{(0)}\right\}$ can be degenerate for some values of x. The right and left eigenvectors of L_0, corresponding to $\omega_n^{(0)}(x)$ are denoted by $v_n^R(x)$ and $v_n^L(x)$, respectively. Let $(v_n^L\cdot u) \equiv a_n(x,t)$ and $(v_n^L L_1 v_m^R) \equiv T_{nm}$. Assume further that T is a real matrix. It follows that (for a given value of x):

$$i\,\dot{a} = \omega_n^{(0)}\, a_n + \epsilon \sum_m T_{nm}\, a_m \qquad (2.4)$$

Let: $a(x,t) = e^{-i\omega(x)t}\, a(x,0)$. The set of values of ω that solve eq. (2.4) is the spectrum of the stability problem. When $\mathrm{Im}(\omega)>0$, one obtains an instability. Since ϵ is assumed to be small, one expects $\omega_n \sim \omega_n^{(0)}$, with obvious notation. Let $\omega_{n_0}^{(0)}$ be a nondegenerate eigenvalue of eq. (2.4). Then, from eq. (2.4):

$$\left(\omega - \omega_{n_0}^{(0)} - \epsilon T_{n_0 n_0}\right) a_{n_0} = \epsilon \sum_{m \neq n_0} T_{n_0 m}\, a_m \qquad (2.5a)$$

$$n \neq n_0 \qquad \sum_{m \neq n_0} ((\omega - \omega_n^{(0)})\delta_{nm} - \epsilon T_{nm})\, a_m = \epsilon\, T_{n n_0}\, a_{n0}\,. \qquad (2.5b)$$

Let $T'_{nm} = T_{nm}$ for $n,m \neq n_0$, $R'_{nm} = (\omega - \omega_n)\delta_{nm} - \epsilon T'_{nm}$ for $n,m \neq 0$ (i.e., T' is the same as T with row n_0 and column n_0 removed). Then from (2.5b):

$$a_m = \epsilon\, R'^{-1}_{m\ell}\, T_{\ell n_0}\, a_{n_0} \quad ; \quad m \neq n_0 \qquad (2.6)$$

It follows from (2.5a) and (2.6) that:

$$\omega - \omega_{n_0}^{(0)} - \epsilon T_{n_0 n_0} = \epsilon^2 \sum_{m \neq n_0} \sum_{\ell \neq n_0} T_{n_0 m}\, R'^{-1}_{m\ell}\, T_{\ell, n_0} \qquad (2.7)$$

When $\omega \sim \omega_{n_0}^{(0)}$ and ω_{n_0} is not degenerate or quasidegenerate with any other eigenvalue, one can easily deduce that: (1) R' is invertible for small enough ϵ (2) ω is real to all orders in ϵ and eq. (2.7) can be used to generate an appropriate perturbation theory for ω.

Consider next the case of $\omega_{n_1}^{(0)} = \omega_{n_2}^{(0)} \neq \omega_j^{(0)}$ $j \neq n_1, n_2$, i.e., of two degenerate modes. A similar procedure to that outlined above (i.e., by projecting out all modes except for n_1 and n_2) yields:

$$\det \begin{vmatrix} \omega - \omega_{n_1}^{(0)} - \epsilon T_{n_1 n_1} - \epsilon^2 T_{n_1 m'} \tilde{R}^{-1}_{m'k'} T_{k' n_1} & -\epsilon T_{n_1 n_2} - \epsilon^2 T_{n_1 m'} \tilde{R}^{-1}_{m'k'} T_{\ell' n_2} \\ -\epsilon T_{n_2 n_1} - \epsilon^2 T_{n_2 m'} \tilde{R}^{-1}_{m'\ell'} T_{\ell' n_1} & \omega - \omega_{n_2}^{(0)} - \epsilon T_{n_2 n_2} - \epsilon^2 T_{n_2 m'} \tilde{R}^{-1}_{m'n'} T_{n' n_2} \end{vmatrix} = 0$$

(2.8)

Here: $\tilde{R}_{n'm'} = (\omega - \omega_{n'} \delta_{n'm'} - \epsilon T_{n'm'};\ n', m' \neq n_1, n_2)$. $\tilde{R}$ does not have rows and columns n_1 and n_2. The summation convention is assumed in eq. (2.8) and all quantities denoted by ( )′ are assumed to exclude n_1 and n_2. The diagonal terms are degenerate for some value x_0 of x and $\epsilon=0$. Let $f_1(\omega,x,\epsilon)$ and $f_2(\omega,x,\epsilon)$ denote the (1,1) and (2,2) terms of the matrix in eq. (2.8). To linear order in ϵ and $x-x_0$:

$$f_{1,2} = \omega - \omega^{(0)}_{n_{1,2}}(x_0) - (x-x_0)\, \frac{d\omega^{(0)}_{n_{1,2}}}{dx}\Big|_{x_0} - \epsilon T_{n_{1,2}n_{1,2}}(x_0) \tag{2.9}$$

The linear system of (two) equations, $f_{1,2}=0$, can be solved in the neighbourhood of $x=x_0$, $\epsilon=0$, to yield $x=x_c(\epsilon)$, $\omega=\omega_c(\epsilon)$, such that $f_{1,2}(\omega_c(\epsilon), x_c(\epsilon), \epsilon)=0$:

$$x_c(\epsilon) = x_0 + \frac{\epsilon\,(T_{n_1n_1}(x_0) - T_{n_2n_2}(x_0))}{\left.\frac{d\omega_{n_2}}{dx}\right|_{x_0} - \left.\frac{d\omega_{n_1}}{dx}\right|_{x_0}} \tag{2.9a}$$

$$\omega_c(\epsilon) = \omega^{(0)}_{n_1n_2} + \frac{\epsilon\left(\left.\frac{d\omega_{n_2}}{dx}\right|_{x_0} T_{n_1n_1} - \left.\frac{d\omega_{n_1}}{dx}\right|_{x_0} T_{n_2n_2}\right)}{\left.\frac{d\omega_{n_2}}{dx}\right|_{x_0} - \left.\frac{d\omega_{n_1}}{dx}\right|_{x_0}} \tag{2.9b}$$

The condition of solvability is that the lines $\omega_{n_1}(x)$ and $\omega_{n_2}(x)$ cross each other at the point of degeneracy and are not tangent there: a rather mild restriction (else higher orders have to be considered). Clearly, the existence of points x_c and ω_c to all orders in ϵ can be demonstrated for large classes of systems (ϵ being not too large). Thus: $f_{1,2}(\omega,x,\epsilon) = \omega-\omega_c+\epsilon \left.\frac{\partial T_{n_{1,2}n_{1,2}}}{\partial x}\right|_{x_c}(x-x_c)$ to linear order with an obvious generalization to nonlinear orders. Let $\Delta_{1,2}$ and $\Delta_{2,1}$ be the nondiagonal terms in the matrix in eq. (2.8). Then eq. (2.8) implies:

$$\omega = \omega_c - \frac{\alpha_1 + \alpha_2}{2}(x - x_c) \pm \sqrt{\left(\frac{\alpha_1 - \alpha_2}{2}\right)^2 (x - x_c)^2 + \Delta_1\Delta_2} \tag{2.10}$$

where $\alpha_{1,2}= \left.\frac{\partial f_{1,2}}{\partial x}\right|_{x=x_c}$, and only the linear order in $x-x_c$ and $\omega-\omega_c$ in the diagonal terms of the matrix of eq. (2.8) has been taken into account. Eq. (2.10) spells instability when $\Delta_1\Delta_2<0$ and $|x-x_c|$ is small enough. In summary, we have shown that when two eigenvalues cross (at $\omega^{(0)}$) for $\epsilon=0$ and $x=x_0$, their interaction shifts the "effective intersection point" to ω_c and x_c and (if $\Delta_1\Delta_2<0$) a neighbourhood of x_c (of size $2|\Delta_1\Delta_2|$) has an unstable mode, whose growth rate is given by Imω, cf. eg. (2.10). The largest growth rate is for $x=x_c$, i.e., $\sqrt{|\Delta_1\Delta_2|}$. In some cases, $T_{n_1n_2} \propto \epsilon^{|n_1-n_2|}$ (see below). Then the growth rate for modes which are "far apart" ($|n_1-n_2|>>1$) may be negligibly small rendering them essentially stable on time scales of interest.

III. Instabilities of Two Dimensional Stokes Waves (8)

The history of the research of water waves is very curious. Following the pioneering work [4] of Stokes in 1847, a debate of many years, concerning the very existence of propagating water waves of permanent shape ended with the publication of the rigorous proof [10] by Levi-Civita in 1925. The latter has been generalized by Straik [11], Krasovshii [12] and others. In the early sixties, no one questioned the physical and mathematical existence of Stokes Waves.

It therefore came as a shocking surprise, when Benjamin and Feir [13] demonstrated in 1966 that Stokes waves were linearly unstable. Furthermore, it has been found by later investigators [5] that Stokes waves had numerous instabilities, perhaps an infinite number of them (in the absence of viscosity). These results do not imply that not-too-steep waves cannot be observed, since the growth rates of the instabilities are rather small. The fastest growing instability is the one discovered by Benjamin and Feir and it has the effect of disintigrating wave-trains of Stokes waves. Below we present a brief derivation of stability properties of two dimensional Stokes waves, from the point of view described here. In addition to discovering an infinite number of instabilities, we gain another benefit: an ordinary differential equation describing the dynamics of short waves riding on long (Stokes) waves. Consider an irrotational flow of infinite depth, described by a potential ϕ and a free surface η. The equations of motion and $(-\infty < x < \infty)$:

$$\Delta\phi = 0; \quad -\infty \leq y < \eta(x,t) \tag{3.1a}$$

$$\frac{\partial\eta}{\partial t} + \frac{\partial\phi}{\partial x}\frac{\partial\eta}{\partial x} = \frac{\partial\phi}{\partial y}; \; y = \eta(x,t) \tag{3.1b}$$

$$\dot{\phi} + \frac{1}{2}(\nabla\phi)^2 + g\eta = 0 \quad ; \quad y = \eta(x,t) \tag{3.1c}$$

where g is the gravitational instant. When nonlinear terms in (3.1) are dropped - the equations allow for plane wave solutions (linear Stokes waves) whose spectrum is: $\omega = \pm\sqrt{g|k|}$, k being the corresponding wave number. The stability analysis of finite amplitude waves is facilitated when the following transformations are made: (a) The semiifinite domain $-\infty < y < \eta(x,t)$ is conformally mapped onto a half plane $-\infty < v \leq 0$, $-\infty < u < \infty$. (b) small perturbations of the original interface are treated linearly in the amount of change of η. A resulting conformal map of the interface $\eta + \delta\eta$, to linear order in $\delta\eta$, onto a half plane is derived. (c) Since the equations in the transformed half-plane (u,v plane) correspond to a fixed domain, one can linearize arround the solution of intetest (in this case, the Stokes solution). (d) It is convenient (and possible) to cast the resulting stability equations into a single integrodifferential equation. The details of the derivation are too lengthy to reproduce here, the reader is referred to reference [8] for full details. It is convenient to consider the system in a frame of reference in which the basic (Stokes) wave is stationary. In this frame, due to the Doppler shift, the spectrum is: $\omega = ck \pm \sqrt{g|k|}$, where c is the phase speed of the basic wave.

Let T(x,t) be the point in the (u,v) plane corresponding to $(x,\eta(x,t))$ in the physical plane. Then $T(x,t) = f(x,\eta(x,t),t)$, where f is the real part of the analytic function f+ig mapping the physical plane onto the tranformed one. The interface of a Stokes wave, η_0, is mapped to: $(f(\eta_0),0)$ in the (u,v) plane. Hence, a perturbation of η_0 is mapped to a value of v: $\Delta v = g(x,\eta_0 + \delta\eta,t) \approx \frac{\partial g}{\partial v}\delta\eta = \frac{\partial f}{\partial x}\delta\eta$. The function δv can be shown [8] to satisfy the following interodifferential equation (assuming a time dependence $e^{i\omega t}$):

$$\frac{ic^2\omega}{2g\pi} I + \frac{c}{\pi} \frac{d}{du} (\eta_0 I) = c \frac{d}{du} \left\{ \left(\frac{1}{2} - \frac{\eta_0 \eta_0''}{\gamma^2} \right) \frac{\delta v}{T'} \right\} \tag{3.2a}$$

where: $\gamma = \sqrt{1+\eta_0'^2}$ and

$$I = P \int_{-\infty}^{\infty} \frac{du'}{u'-u} \left[- \frac{i\omega}{c} \frac{\gamma^2}{T'^2} \delta v(u') + \frac{d}{du'} \delta v(u') \right] \tag{3.2b}$$

In spite of its somewhat cumbersome appearance, eq. (3.2) is easy to analyze. Notice that this equation can be written as follows:

$$(\omega^2 \hat{L}_2 + \omega \hat{L}_1 + \hat{L}_0)\delta v = 0 \tag{3.3}$$

where $\hat{L}_0$, $\hat{L}_1$ anmd $\hat{L}_2$ are linear operators, independent of ω and periodic in u (since they all depend on the periodic shape of the Stokes wave). In other words, eq. (3.3) is a Floquet problem. That the stability equation for Stokes waves should be of the Floquet type is a-priori clear: one linearizes around a periodic solution (both in the physical plane and the transformed one), rendering the coefficients periodic functions of space. Let $k_0 a = \frac{2\pi}{\lambda_0}$ a be the steepness of the basic Stokes wave, λ being the wavelength and a - the amplitude. Denote: $\epsilon = k_0 a$. When $\epsilon=0$, the problem is reduced that that of the stability of a zero amplitude Stokes wave (a semiplane). The solutions are: $\delta v \propto e^{iku}$ and $\omega = ck \pm \sqrt{g|k|}$, as expected. Taking into account the Floquet nature of the $\epsilon \neq 0$ problem, it is convenient to write the $\epsilon=0$ limit as $\delta v = e^{iqu} e^{ink_0 u}$, where $0 \le q < k_0$. Following Floquet's theorem the eigenmodes of the stability problem can be written as follows:

$$\delta v_n^{(\sigma)}(q) = e^{iqu} \sum_n A_{n,m}^{(\sigma)}(q,\epsilon) e^{imk_0 u} \tag{3.4}$$

As $\epsilon \to 0$, $A_m^n \to \delta_{n,m}$. The corresponding eigenvalues, as $\epsilon \to 0$, are: $\omega_n^{(\sigma)} = C(nk_0+q) + \sigma\sqrt{|nk_0+q|g}$, where $\sigma =$ ± 1 (corresponding to left or right moving perturbations). The full integrodifferential equation (3.2) does not couple solutions having different values of q. Thus one can consider the spectrum to depend on 4 parameters:

1. The value of q. 2. The value of n (determined by the $\epsilon \to 0$ limit). 3. The value of ϵ. 4. The sign σ (determined by the $\epsilon \to 0$ limit).

It is convenient to regard $0 \le q < 0$ as a variable parameter (equivalent to x in the simple example above) and the various eigenvalues as functions of q. The parameter ϵ is to be regarded as the perturbation. From this point of view it is clear that (since the zeroth order, in ϵ, spectrum is real and discrete) one expects (linear) instabilities only at crossing points of the zeroth order spectrum and finite amplitude instabilities at values of (q,σ) for which two eigenvalues are close enough. In this paper we shall deal only with the former case. A preliminary step would thus be to find all possible degeneracies of the zeroth order spectrum. Defining k_0 as the scale of wavenumbers and $ck_0 = \sqrt{gk_0}$ as the frequency scale, all degeneracies can be found by solving:

$$n_1 + \tilde{q} + \sigma_1 \sqrt{\left| n_1+\tilde{q} \right|} = n_2 + \tilde{q} + \sigma_2 \sqrt{\left| n_2+\tilde{q} \right|} \tag{3.5}$$

where n_1, n_2 are integers, $\tilde{q} = q/k_0$ and $\sigma_1^2 = \sigma_2^2 = 1$.

All solutions of eq. (3.5) appear in the following list (ℓ is any possible integer):

1) $0 + \sqrt{0} = 0 - \sqrt{0} = 1 - \sqrt{1} = -1 + \sqrt{1}$

2) $\ell^2 - \ell + \frac{1}{4} + \sqrt{\ell^2 - \ell + \frac{1}{4}} = \ell^2 + \ell + \frac{1}{4} - \sqrt{\ell^2 + \ell + \frac{1}{4}}$

3) $\ell^2 + \sqrt{\ell^2} = (\ell+1)^2 - \sqrt{(\ell+1)^2}$

4) $-(\ell^2 + \ell + 1) + \frac{3}{4} + \sqrt{\ell^2 + \ell + 1 + \frac{3}{4}} = -(\ell^2 - \ell + 1) + \frac{3}{4} - \sqrt{\ell^2 - \ell + 1 - \frac{3}{4}}$

5) $-(\ell^2 + 2\ell + 1) + \sqrt{\ell^2 + 2\ell + 1} = -\ell^2 - \sqrt{\ell^2}$

In cases (1), (3) and (5): q=0. In case (2): $q = \frac{1}{4}$. In case (4): $q = \frac{3}{4}$. It can be shown that case (1) corresponds to the Benjamin-Feir instability whereas the other cases correspond to high order (in ϵ) instabilities, (to leading order in ϵ, only modes $n = \pm 1$ contribute). The spectrum corresponding to $n_1 = 1$, $n_2 = -1$ is:

$$\omega = \frac{g}{2} \pm \sqrt{\left(\frac{q^2}{8} - \frac{\epsilon^2}{2}\right)^2 - \frac{\epsilon^2}{4}} \tag{3.6}$$

and it is in full agreement with known results (13). Instead of dealing with the high order instabilities directly we wish to show here how high wavenumber instabilities can be dealt with by transforming eq. (3.2) into an ODE.

It follows from general considerations that the amplitude $A_{n,m}$ scales as $\epsilon^{|n-m|}$ for not too large values of ϵ. Similarly, if we write $\eta_0(u) = \sum_n b_n \cos(k_0 n u)$ then $b_n \propto \epsilon^{|n|}$. The same holds for quantities such as $\frac{\gamma^2}{T'^2}$ appearing in the integrand of eq. (3.2b). Hence if a mode having a large value of n (cf. eq. (3.4)) is considered and if, for example, n is positive, then the Fourier expansion of the term in the square brackets in eq. (3.2b) can be considered to consist of a sum of the type $e^{iqu} \sum_{m>0} d_m e^{ik_0 m u}$ the error being at most $O(\epsilon^{n+1})$.

Since $P \int \frac{e^{iku'}}{u'-u} du' = i\pi \,\mathrm{sgn}(k) e^{iku}$, it follows that up to an error $O(\epsilon^{n+1})$ the integral I in eq. (3.2b) can be replaced by the integrand multiplied by $i\pi$. Upon doing so eq. (3.2a) becomes an ODE for δv:

$$\eta \delta v'' + (2\pi a + \eta_0' + i\varsigma)\delta v' + \left(\frac{\pi^2 a^2}{\eta} + i\varsigma\right)\delta v = 0 \tag{3.7}$$

where: $a = \frac{i\omega c}{2\pi g}$ and $\varsigma = \left(\frac{1}{2} - \frac{\eta_0 \eta_0''}{\gamma^2}\right)\frac{1}{T'}$.

Upon defining a new variable τ:

$$\frac{d\tau}{du} = - \frac{1}{2\eta_0} \tag{3.8}$$

and noting that in our convention (cf. eq. 3.1c) with $\dot{\phi} = 0$) $\eta_0 < 0$, we obtain, in the new variable,

$$\left[\left(\frac{\partial}{\partial t} - i\Omega\right)^2 - 2i\bar{\zeta}\frac{\partial}{\partial \tau}\right]\delta v = 2i\frac{d}{d\tau}(\zeta_1 \delta v) \tag{3.9}$$

where $\Omega = \frac{\omega}{\omega_0}$, $\omega_0 = \sqrt{gk}$, $\bar{\zeta}$ is the average of ζ over a wavelength $\frac{2\pi}{k_0}$ and $\zeta_1 = \zeta - \bar{\zeta}$. The l.h.s. of eq. (3.9) has no τ dependence. The r.h.s. depends on τ through ζ_1, which can be shown to be $O(\epsilon^3)$. Thus, to order ϵ^3 the high wavenumber spectrum is of the form:

$$\frac{\omega}{\omega_0} = Q \pm \sqrt{2\bar{\zeta}|Q|} \tag{3.10}$$

where $|Q| >> 1$ is a real number and: $2\bar{\zeta} = 1 + \epsilon^2 + O(\epsilon^3)$. The instabilities come about due to the r.h.s. of eq. (3.9).

Using a procedure similar to that explained in section II, one can show that any pair of degenerate eigenmodes corresponding to (n_1, n_2) leads to an instability, whose maximal growth rate scales as: $\epsilon^{|n_1 - n_2|}$. The value of q for which the maximal instability occurs is shifted by $O(\epsilon^2)$ with respect to the value of q corresponding to the zeroth order level crossing and the real part of the eigenvalue is also shifted with respect to its $\epsilon=0$ value by $O(\epsilon^2)$. The range of unstable values q, around q_c (corresponding to the maximal growth rate) is $O(\epsilon^2)$. Furthermore, the corresponding eigenmodes have their highest amplitudes at the forward side of the basic wave, near its peak (i.e., in the direction in which the basic wave moves).

The theory outlined here is readily generalizable to include surface tension and viscous effects. A three dimensional generalization is also feasible.

IV. The Instability of an Elliptical Vortex

While a full understanding of the transition to turbulence is still lacking, there has been significant progress in identifying instability mechanisms, which are responsible for the destabilization of laminar flows, such as parallel shear flows [15]. Following the establishment of the existence of linearly unstable modes in such flows, it has been shown that (at lower Reynolds numbers) the system had finite amplitude nonlinear, two dimensional travelling wave solutions [16]. The latter are unstable to three dimensional perturbations, which could be precursors of a fully developed turbulent state. It has been noted, in particular, that elliptical streamlines give rise to strong three dimensional instabilities.

A model consisting of a single (infinite) vortex having elliptical streamlines has been shown [17-19] to possess three dimensional instabilities on all scales, limited only by molecular viscosity. In our own work [14] we have performed a full stability analysis of the elliptical vortex (including the viscous terms). Our findings are as follows: (a) The stability problem can be reduced to a Floquet type equation (like in the Stokes' waves case). (b) the eigenmodes of the problem form a complete set (though the equation is not self adjoint). (c) In the limit of zero eccentricity the spectrum is doubly degenerate and neutrally stable. (d) All unstable modes for small values of eccentricity stem from degenerate eigenmodes at zero eccentricity, the maximal growth rate being proportional to the eccentricity. (e) At finite or high values of eccentricity, the growth rate of unstable modes (in the inviscid limit) is logarithmic in the eccentricity. (f) The Greens function corresponding to the linear (stability) analysis equations is localized on cones of elliptical cross section in the inviscid case and strongly localized (for finite times) in the viscous case. The latter result is of importance since it justifies the use of a single vortex for the stability analysis: since instability is shown to be a local property of the (elliptical) flow, the fact that the far field is not necessarily elliptical or does not "belong" to the same vortex is immaterial. Due to space limitations we shall not dwell here on the technical details of the analysis, and we refer the reader to ref. [14].

Summary

The interaction of degenerate modes has been shown to be an important source of instabilities and the search for such modes - a useful technique to find instabilities. In some cases all linear instabilities can be traced to this mechanism. Moreover, the method is not limited to Hamiltonian or nondissipative systems. An obvious generalization would be a nonlinear stability analysis based on the linearly unstable modes. To conclude, and perhaps further demonstrate the power of the method we wish to consider a case in which instabilities follow from a well known parametric resonance, i.e., the Mathieu equation. Consider: $u_{xx} + (a^2 + \epsilon\cos x)u = 0$. When $\epsilon=0$, the spectrum is the set of all negative numbers $(-a^2)$. Since the problem is of the Floquet type, one can write a general solution (for $\epsilon=0$) as $e^{inx+iqx}$ where n is an integer (e^{inx} has the periodicity of the equation) and $0 \leq q < 1$ is a "Floquet exponent". For a given q (as before, different values of q are uncoupled). The condition of degeneracy is: $(n+q)^2 = (n'+q)^2$, where n, n' are integers. A solution with $n \neq n'$ is possible only for $q=0$ (i.e., $n=-n'$) or $q = \frac{1}{2}$ (then $n' = -n-1$) and, as is well known, this accounts for all small ϵ instabilities (the beginning of the known tongues).

References

1. e.g. L.I. Schiff, "Quantum Mechanics", (McGraw Hill, N.Y., 1955), Second edition, pp. 156-160.

2. e.g. N.W. Ashcroft and N.D. Mermin, "Solid State Physics", (Holt, Rinehart and Winston, N.Y., 1976), pp. 156-159.

3. K. Weirstrass, "Mathematishe Werke", Vol. 1, pp. 233-246 (1984).

4. V.I. Arnold, "Mathematical Methods of Classical Mechanics", (Springer Verlag, Heidelberg and Berlin, 1978).

5. M.S. Longuet Higgins, Proc. R. Soc. Lond. A 360, 471-488 (1978). ibid A 360, 489-505.
J.W. Mclean,. J.F.M. 114, 315-330 (1982).

6. R.S. Mackay and P.G. Saffman, Proc. R. Soc. Lond. A406, 115-125 (1986).

7. C. Kharif and A. Ramamomjiurisoa, J.F.M. 218, 163-170 (1990).

8. B. Spivak and I. Goldhirsch, to be published.

9. G.G. Stokes, "On the Theory of Oscillatory Waves", Trans. Camb. Phil. Soc. 8, 441-473 (1847).

10. T. Levi-Civita, "Determination rigoureuse des ondes permanentes d'ampleur finie", Math. Ann. 93, 264 (1925).

11. D.J. Struik, Math. Ann. 95, 595 (1926).

12. Y.P. Krasovskii, "On the Theory of Permanent Waves of Finite Amplitude", (in Russian), Zh. Vychise Mat. Mat. Fiz. 1, 836 (1961).

13. T. Brooke Benjamin and J.E. Feir, "The Disintegration of Wave Trains on Deep Water", J.F.M. 27, 417-430 (1967).
For a review see H.C. Yuen and B.M. Lake, Advances in Appl. Mech. 22, 67 (1982).

14. N. Stein and I. Goldhirsch, to be published.

15. For a recent review, see e.g.: B.J. Bayley, S.A. Orszag and T. Herbert, Ann. Rev. Fluid Mech. 20, 359 (1988).

16. S.A. Orszag and A.T. Patera, J.F.M. 128, 347, 1983.

17. R.T. Pierrehumbert, Phys. Rev. Lett. 57, 2157 (1986).

18. B.J. Bayly, Phys. Rev. Lett. 57, 2160 (1986).

19. Further analysis of the elliptical instability is found in: F. Waleffe, Phys. Fluids A2, 76-80, (1990), and tests of this instability are described in M.J. Landman and P.G. Saffman, Phys. Fluids 30, 2339 (1987).

Non-linear stability of plane Couette flow

Bérengère DUBRULLE[1,2]

[1]C.E.R.F.A.C.S., 42 avenue Coriolis, 31057 Toulouse, France
[2]Observatoire Midi Pyrénées, 14 avenue Belin, 31400 Toulouse, France

1 Historical background

Most geophysical and astrophysical shear flows can be approximated *locally* by the plane Couette flow, a plane parallel stream of constant shear (i.e. of constant vorticity). No one could have ever dreamt about a simpler shear flow and yet, the determination of its stability characteristics turns out to be a hard nut to crack. As early as 1936, Taylor proved in a series of experiments the unstable nature of the plane Couette flow. In 1956, Reichardt undertook a new set of experiments and was able to exhibit a critical Reynolds number characterising the transition between laminar and turbulent state. Using the channel width as the characteristic length scale, this critical Reynolds number was found to be of the order of 1500.

On the theoretical side, progress came more slowly and it is only in the early seventies (e.g. Romanov, 1973) that the linear stability of plane Couette flow was unambiguously proved. The confrontation of such a result with the previous experiments led to the suspicion that instability in plane Couette flow was of non-linear nature and generated by finite-amplitude perturbations. For a long time, no rigorous proof a such a conjecture nor complete characterisation of the instability (e.g. dependence of the threshold amplitude with the Reynolds number) have been available. The main problem lies in the intrinsic "infinite nature"[1]of the bifurcation occurring in the plane Couette flow, while most of the methods to explore the non-linear stability are only deviced to handle finite bifurcations: one uses for example expansions in the neighbourhood of the critical point, or linearly unstable modes as the starting point of the non-linear exploration.

The study of the non-linear stability of plane Couette flow remained therefore confined to the numerical domain. Such a study was pioneered in 1980 by Orszag and Kells using a pseudo-spectral numerical code. Their linear exploration was inspired of the method described above, the finite amplitude perturbation being constructed from the least stable linear mode. The main results of their exploration are connected with the onset of the instability: the finite amplitude instability was shown to be confined to three-dimensional regimes and to Reynolds numbers larger than about 2000. Moreover, the developing of

[1]The linear critical Reynolds number being infinite.

the instability was found to coexist with the appearance of inflection points in the mean profile, which suggests a causal relation between the two phenomena. This would not be too surprising, since those inflection points would create a local maximum of vorticity and the resultant flow would then satisfy the Rayleigh-Fjørtøft necessary condition for instability.

A major break-through was made recently by Nagata (1990) who displayed the first theoretical evidence of the existence of three-dimensional finite amplitude solutions in plane Couette flow. These solutions are obtained numerically by extending the bifurcation problem of a circular Couette system between co-rotating cylinders with a narrow gap to the case with zero average rotation case. The critical Reynolds number for appearance of those finite amplitude solutions is found to be of the order of 1000. A striking feature of the solutions is that they tend to create an inflection point in the mean flow, confirming thereby the results obtained by Orszag and Kells.

While much progress has been made regarding the proof of the finite amplitude nature of instability in plane Couette flow, very little is still known about the characteristics of such instability. This is a major source of unhapppiness for both astro and geophysicists, who would like to know more about the thresholds for instability and the main feature of the possibly resulting turbulence, in order to better understand and parametrise the objects they are dealing with. This is the main motivation of the work which is described in this report.

The starting point of our approach is the investigation of the importance of the inflection points appearing in the mean profile and which have been observed by Orszag and Kells and Nagata. As already said, such inflection points are a major source of instability via the local maximum of vorticity they generate. They are therefore likely to play an important role, both on the onset and on the developing of the instability into a self-sustained equilibrium ("turbulent") state. The study of the influence of inflection points in a Couette flow was originated by Lerner and Knobloch (1988), who restricted their investigation to the inviscid case. However, it is clear that viscosity has an important stabilising effect by smoothing the inflection points. Our work was therefore to extend Lerner and Knobloch's work by including viscous dissipation. A more detailed account of the results presented here can be found in Dubrulle and Zahn (1991) and Dubrulle (1990).

2 Analytical results

2.1 The model

By the Rayleigh-Fjørtøft theorem, any inflection point will give rise to an instability on a certain time scale τ_I. Under the assumption-to be checked afterwards- that the evolution of the inflection point occurs on time scales long compared with τ_I, the instability can be studied via a plain linear stability analysis performed on the total flow (Couette plus the inflection point). To enable analytical computation, the total profile was chosen piece-wise linear (cf. figure 1.). ε is the non-dimensional amplitude of the inflection point (defect), while d is the width of the defect. ε is taken to be small (less than unity) but

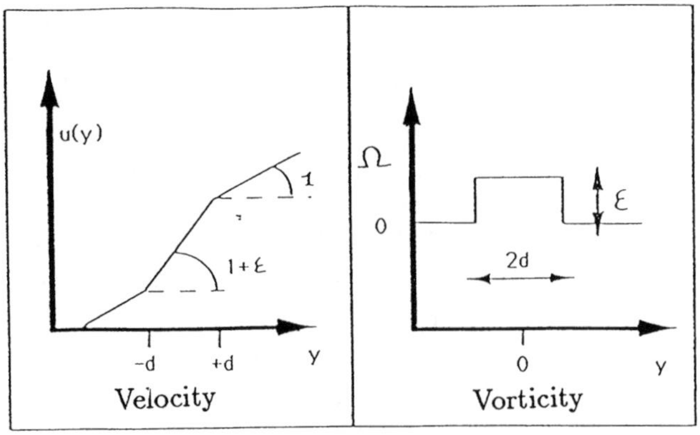

Figure 1: The profile used for analytical computation

not infinitesimally small (finite amplitude perturbation). The computations can therefore be performed using ε as the expansion parameter. To avoid any effect of the boundary on the instability, the flow is assumed to be unbounded in the y-direction, but can be bounded in the x-direction.

2.2 Results

The instability generated by the defect was found to be confined to small wavenumber: $k < \varepsilon$. Moreover, the growthrate was found to be very little sensitive to the Reynolds number, keeping a value of the order of $\varepsilon/2$. When the flow is bounded in the x-direction, two necessary conditions for instability arise: first, the magnitude of the defect has to be larger than $R_L^{-1/3}$, R_L being the Reynolds number of the mean flow based on the downstream scale L. Second, for a given ε satisfying this condition, the profile defect which leads to instability must have a relative width in the interval:

$$(\varepsilon R_L)^{-1/2} < \frac{d}{L} < \varepsilon.$$

2.3 Discussion

One may wonder whether the results obtained with a physically insane (discontinuous) profile can be trusted. To counter such a perfidious questioning, we solved numerically the Orr-Sommerfeld equation using a smooth, continuous version of our discontinuous profile. For computational convenience, the plane Couette flow was considered to be bounded as well in the y-direction, which differs noticeably from the case studied in 2.1. However, for defect with small relative width d/L, the boundary effects can be thought to be negligible. In table 1., we display a comparison between the critical wavenumber computed

εR	$(k/\varepsilon)_{\mathrm{free}}$	$(k/\varepsilon)_{\mathrm{rig}}$	$(k/\varepsilon)_{\mathrm{unbd}}$
11.53	0.92	0.92	0.94
2.8	0.85	0.85	0.82
1.5	0.76	0.76	0.75
0.85	0.70	0.70	0.69
0.45	0.52	0.52	0.61
0.3	0.4	0.4	0.55

Table 1: Comparison between bounded and unbounded Couette flow

numerically using both free-slip and rigid boundary conditions, and the theoretical one, derived analytically. As can be seen, the agreement is rather satisfying.

The basis of our analytical study was the assumption that the defect evolves on time scales longer than the growth of the instability. This assumption can be checked noticing that a localised defect evolves with time as $t^{-1/2}\exp(-y^2/4\nu t)$ according to the one-dimensional Orr-Sommerfeld equation. Therefore, the amplitude of the defect varies as $t^{-1/2}$ and its width as $t^{1/2}$. The viscous damping of the defect thus decreases with time as t^{-1}, which is faster than the $t^{-1/2}$ decline of the growth rate, which validates our assumption.

The conditions derived in section 2.2 contains already a great whole of information about the existence of a self-sustained equilibrium state. We can now suspect that no turbulence can be expected from finite-amplitude solutions which do not satisfy the necessary conditions. To complete our analysis and try to obtain sufficient conditions, we have now to resort to numerical simulations.

3 Numerical simulations

The numerical simulation presented in this section were done using a pseudo-spectral code using Fourier series in the downstream (x-)direction and Tchebychev polynomia series in the crosstream (y-)direction. The basic profile was taken as a pure plane Couette flow. At the beginning of the simulations, we add to the basic flow a perturbation consisting of a finite amplitude defect and a white noise of infinitesimal amplitude. The defect is chosen so that it creates an inflection point in the mean profile. Its amplitude ε and its width d (see figure 1.), along with the Reynolds number based on the channel width are the free parameters of the problem. The total flow (Couette plus perturbation) is then let freely evolving, and the evolution is followed for periods varying between 50 and 100 turn-over times.

Two sets of boundary conditions (in the y-direction) were tried: rigid and free-slip. With the former, boundary layers are expected to form. This may induce high dissipation

2D Numerical Simulations

Fig. 2.: Rigid boundary conditions

Evolution of the defect amplitude and total kinetic energy at a Reynolds number 5000 and for initial defect amplitude 0.02.

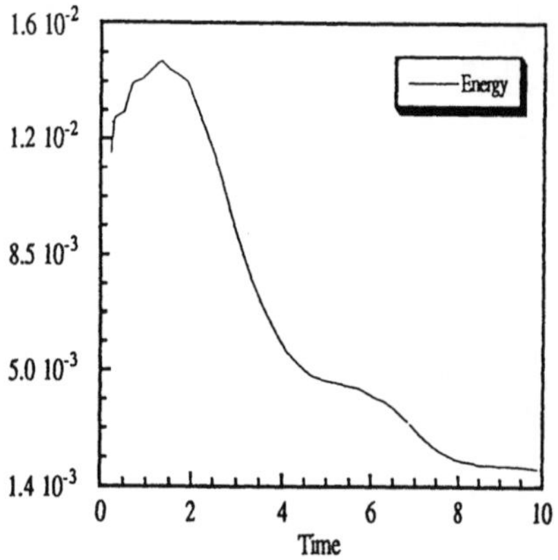

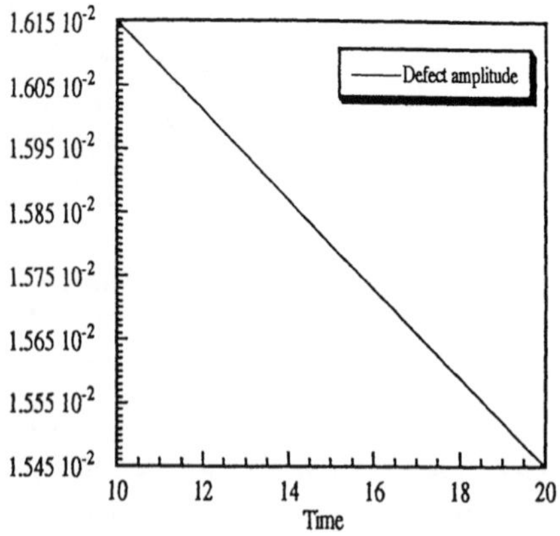

Fig. 3.: Free-slip boundary conditions

Evolution of the defect amplitude and total kinetic energy at a Reynolds number 5000 and for initial defect amplitude 0.02.

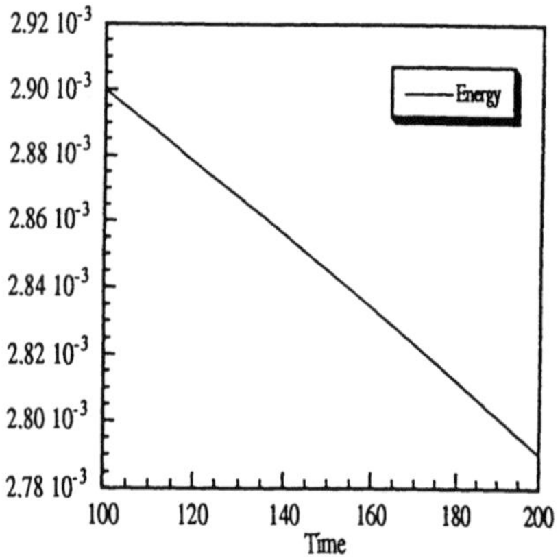

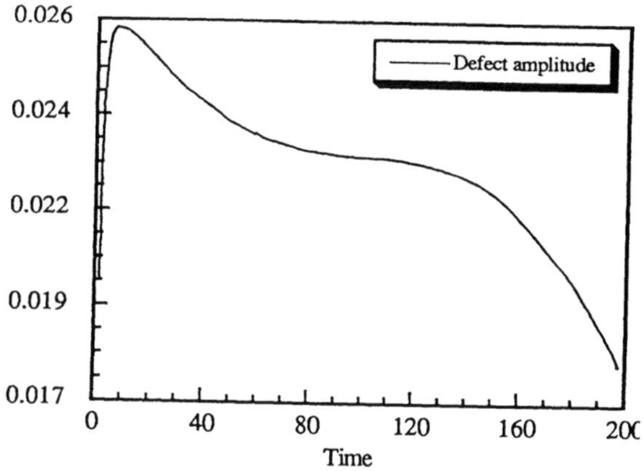

at the wall and then influence the total dynamics which would not be concentrated in the neighbourhood of the defect, as implicitly assumed the previous section. With free-slip boundary conditions, however, one expects the boundary conditions to be of lesser influence on the dynamics of the instability. This case is therefore expected to fit better the unbounded Couette flow treated in the section 2, and therefore, more relevant for describing astrophysical situations.

3.1 Two-dimensional simulations

3.1.1 Rigid boundary conditions

A typical run, performed with a Reynolds number of 5×10^3 and a defect amplitude of $\varepsilon = 0.02$, is shown in figure 2. During the early stages of the instability, the energy starts increasing, at the expense of the defect whose amplitude decreases with time. Then, the energy experiences a strong decrease, followed by many "plateau", until the defect has given away all its "energy" and disappears. We were not able to find any regime in which the non-linear development of the instability would produce any back-reaction on the mean flow and therefore, "sustain the defect" against viscous dissipation. This seems to rule out the possibility of getting any self-sustained equilibrium state with this set of boundary conditions.

3.1.2 Free-slip boundary conditions

This case appears to be rather different from the previous case, as expected. We were indeed able to find regimes in which the amplitude of the defect would be increasing and then sustained for period as long as 100 turn-over times. One example is given in figure 3. using the same parameters as in figure 2. After a sufficiently long time, however, the defect is decreasing again and the flow goes back to the original plane Couette flow.

Those two simulations confirm the previous findings of Orszag and Kells, according to which no turbulence can be expected in incompressible two-dimensional plane Couette flow. The situation is reminiscent for example of the MHD case, where anti-dynamos theorems exists in two dimensions, although regimes which temporarily increasing magnetic energy (due to magnetic flux ejection) can be found. In our case, the equivalent of such anti-dynamo theorem is not yet available. Due to the analogy between magnetic field and vorticity evolution equation, it can however be suspected that such a theorem should be centered on vorticity properties. Indeed, in two-dimensional shear flows, vorticity is advected like a passive scalar and does not experience the stretching that might be essential in the transition to turbulence.

3.1.3 Influence of compressibility

We then investigated whether compressibility could help sustaining the instability in two-dimensions. Indeed, in compressible flows, sharpening of edges through shock formation is possible and could therefore help holding the defect against viscous damping. However, we limited ourselves to rather low Mach numbers, for it is known from the work of Glatzel

3D Numerical Simulations

Fig. 4.: Early evolution

Evolution of the amplitude of the mode m=5 and the x-component of the perturbed velocity at the early stage of the computation (8 turnover times). Rigid boundary conditions are used. The Reynolds number is 2000. The beginning of the exponential growth of the mode coincides with a symmetry breaking occuring in the profile.

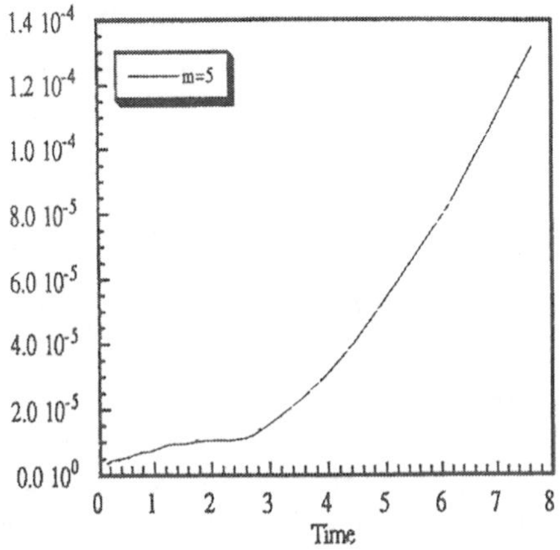

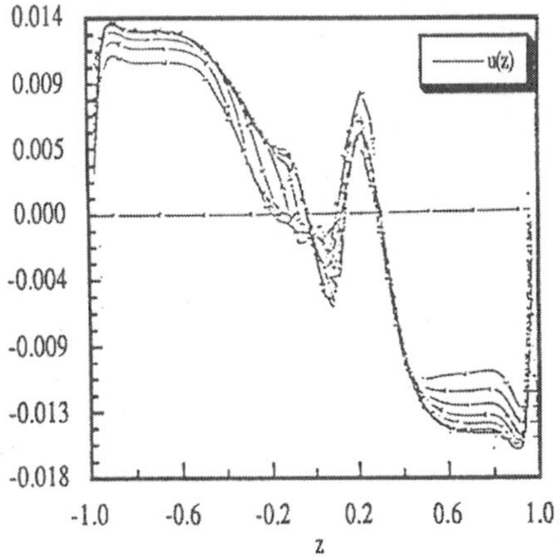

Fig. 5.:

Evolution of the amplitude of the mode m=5 after 53 turnover times.

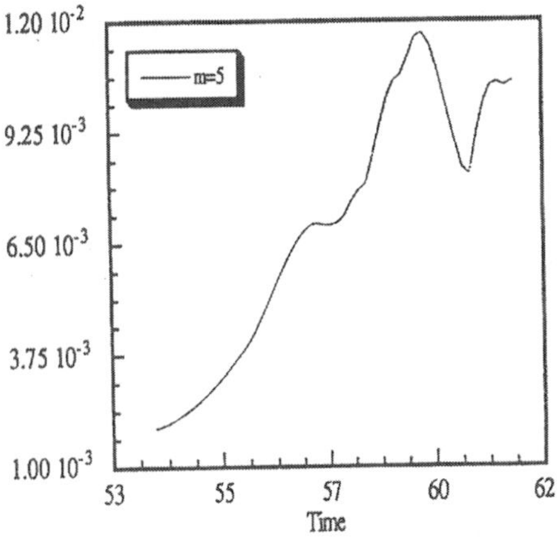

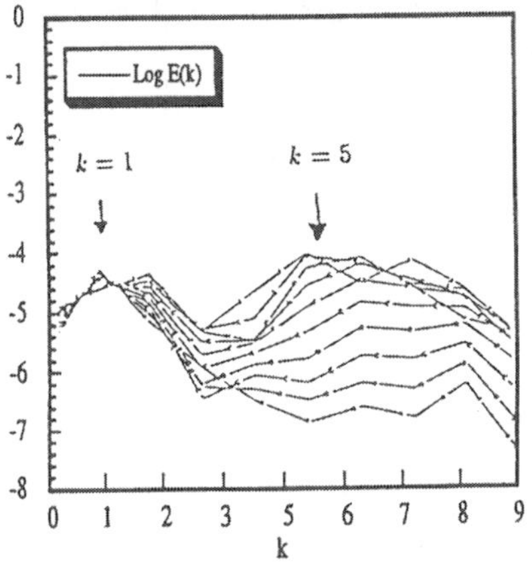

Fig. 6.:

Energy spectrum after 53 turnover times.

(1989) that at large Mach numbers, the plane Couette flow becomes linearly unstable. Mach numbers up to 0.4 were then tried but the results obtained differed very little with the incompressible case. This is not too surprising, since shock formation is rather limited in such cases. In realistic objects, where the boundaries are sufficiently remote so that they do not influence local instabilities, it may well be that compressibility effects are important, triggering supersonic instability via resonance at large scale (cf Glatzel) and local instabilities at smaller scales. Explicit verification of such a conjecture by numerical or experimental simulations are however clearly beyond the limits of our current technology and might stay so for ever.

3.2 Preliminary three-dimensional simulation

We started to investigate the situation in three dimensions. A preliminary incompressible simulation, using rigid boundary conditions, performed at a Reynolds number of 2000 seems to bring more hope into the possibility of turbulence. The evolution of the perturbation was found to be divided in two periods. During the first stages of the evolution (about 3 turnover times, see fig 4.), the perturbation follows closely the two-dimensional scenario (decay of the amplitude of the defect and of the energy, increase of the other modes). Then, a spontaneous symmetry-breaking occurs (reflectional symmetry with respect to the midplane $y = 0$), immediately followed by an increase of the total kinetic energy and an exponential increase of all the modes. This effect seems to be related with the dominance of the mode $k = 3$ which, in two dimensional simulation, was never exceeding the others. The instability was then followed for about 50 turnover times, before the lack of resolution would prevent further integration. The flow has already evolved into a state of strong non-linearity, the amplitude of the modes reaching $1/100$ of the maximum velocity (fig 5.). The energy spectrum (fig 6) exhibits clear peaks on two different modes, $k = 1$ and $k = 5$. The former, being the largest available scale, could be linked with an inverse cascade of energy often observed in shear flows (e.g. in Kolmogorov flow, see She 1987). The second one, associated with a scale corresponding to $1/5$ of the width of the channel, is of great interest if one bears in mind the possible astro or geophysical applications of our work. It would then represent the typical scale associated with turbulent transport of momentum. Such a result is not at odd with previous investigations on that subject. In experiments performed in a rotating tank, using an incompressible fluid and in the limit of zero rotation rate (which corresponds to the plane Couette flow), Wendt (1933) measured viscous torques corresponding to a typical scale of $1/100$ of the gap between the two-cylinders. More recently, Brandenburg et al (1990) computed numerically the turbulent viscosity in a convective plane Couette flow in a channel. The characteristic scale associated was here found to be of the order of $1/5$ of the channel width, which is surprisingly close to our result.

4 Conclusion

The results presented here are clearly to be developed in the future. The way how turbulence sets up and organise itself to lead to the characteristics scales quoted above is still a deep mystery and would need further experimental and numerical support. It might be also worthwhile to look for an equivalent of the anti-dynamo theorem to understand the peculiarities of the two-dimensional case. This might help understanding the importance of breaking of the parity invariance in the transition to turbulence, by analogy with the MHD situation where this condition is known to be necessary for the alpha-effect. The exciting possibilities of applications of this kinds of studies (e.g. for accretion discs in astrophysics, or the atmosphere in meteorology) should provide the necessary motivations to resume intensive studies on the plane Couette flow, which was for sometime treated as a rather interesting but very academic case.

5 Acknowledgements

Part of this work was done in collaboration with J-P. Zahn and N. Dolez. I thank S. Orszag for his Orr-Sommerfeld code and A. Brandenburg for a critical reading of the manuscript. The numerical simulations were conducted on the Cray2 of the CCVR, at Palaiseau.

6 References

Brandenburg, A. A., Nordlund, A., Pulkkinen, P., Stein, R.F. & Tuominen, I. (1990) Turbulent diffusivities derived from simulations, in *Proceedings of the Finnish Astronomical Society 1990*, eds. K. Muinonen, M. Kokko, S. Pohjolainen, and P. Hakala, Helsinki 1990, p. 1-4

Dubrulle, B. (1990) Instability, turbulence and transport in accretion discs via asymptotics, Ph. D. Thesis, University of Toulouse III.

Dubrulle, B. & Zahn, J-P. (1991) Non-linear stability of plane Couette flow. I. Analytical approach to a necessary condition, to appear in *J. Fluid Mech.*

Glatzel, W. (1989) The linear stability of viscous compressible plane Couette flow", *J. Fluid Mech.*, **202**, 515-542..

Nagata, M. (1990) Three-dimensional finite amplitude solutions in plane Couette flow: bifurcation from infinity, *J. Fluid Mech.*, **217**, 519-527.

Lerner, J. & Knobloch, E. (1988) The long wave instability of a defect in a uniform parallel shear. *J. Fluid Mech.*, **189**, 117.

Orszag, S.A. & Kells, L.C. (1980) Transition to turbulence in plane Poiseuille and plane Couette flow. *J. Fluid Mech.* **96**, 159.

CONTINUED-FUNCTION SOLUTIONS AND EIKONAL APPROXIMATIONS†

H. M. Fried

Physics Department, Brown University

Providence, RI, 02912

Abstract

Formal solutions to a class of initial-value fluid problems are converted to explicit solutions given in terms of "continued functions". Simplifications analogous to eikonal approximations of particle scattering theory are suggested, and an exact but implicit solution is constructed for a special situation of two-dimensional Euler flow.

1. Introduction

These remarks are an outgrowth of the "Infrared (IR) Method" devised for approximating large-scale solutions to ODEs, together with the "rescaling group" method of reinserting previously suppressed, high-frequency components into the output of the IR method[1]. When generalized to the PDEs of the Euler or Navier-Stokes equations, it turns out that "rescaling" in time, t, depends on the position, $\mathbf{r}$, and one requires an appropriate formalism to describe that non-trivial complexity.

This presentation consists of such a formalism (Section II) defined in terms of ordered exponentials, and its more explicit representation in terms of "continued-function" solutions. Eikonal approximations to the latter may be constructed (Section III) in analogy with those used for particle scattering amplitudes of quantum field theory,[2] while exact solutions corresponding to a subclass of continued functions can be exhibited (Section IV).

For simplicity, we restrict attention to two-dimensional flow, with velocity $\mathbf{v}$ in the (x, y) or (r, θ) plane, and with vorticity $\mathbf{w} = \nabla \times \mathbf{v}$ in the z-direction. The class of problems under consideration here are those defined by an initial condition, with $\mathbf{v}(\mathbf{r}, t = 0) = \mathbf{v}^{(0)}(\mathbf{r})$ and $\mathbf{w}(\mathbf{r}, t = 0) = \mathbf{w}^{(0)}(\mathbf{r})$ specified vector functions. That is, at $t = 0$ one inserts a given velocity distribution into a quiescent background, and watches it develop and change as time increases. An example of such behavior is the experimentally observed[3] and theoretically understood[4,5] question of vortex-pair fusion. The formalism defined here may include viscosity, although the special example presented has assumed inviscid flow.

† Supported in part by the U.S. Department of Energy Contract No. DE-AC02-76ER03130, A022-Task A.

Reichardt, H. (1956) Uber die Geschwindigkeitsverteilung in einer geradlinigen turbulenten Couetteströmung, *Z. angew. Math. Mech.*, 26

She, Z-S. (1988) Ph. D. Thesis, Université de Nice.

Wendt, F. (1933) Turbulente Strömungen zwischen zwei rotierenden konaxialen Zylindern, *Ing. Arch.*, **4**, 577, 595.

2. Formalism

Consider, first, the N-S equations for $\mathbf{w}(\mathbf{r}, t)$ and $\mathbf{v}(\mathbf{r}, t)$, where for simplicity, we restrict attention to two dimensions,

$$\partial \mathbf{v}/\partial t + (\mathbf{v}\cdot\nabla)\mathbf{v} = -p + \nu\nabla^2\mathbf{v} \tag{1}$$

and

$$\partial \mathbf{w}/\partial t + (\mathbf{v}\cdot\nabla)\mathbf{w} = \nu\nabla^2\mathbf{w}\,, \tag{2}$$

and for the moment neglect viscous effects, so that the RHS of (2) and the last RHS term of (1) vanishes. Look for a solution of form $\mathbf{w}(\mathbf{r}, t) = \mathbf{w}^{(0)}\left(\mathbf{R}(\mathbf{r}, t)\right)$, where $\mathbf{R}(\mathbf{r}, t)$ satisfies the same, homogeneous equation

$$\partial \mathbf{R}/\partial t + (\mathbf{v}\cdot\nabla)\mathbf{R} = 0 \tag{3}$$

with the initial condition

$$\mathbf{R}(\mathbf{r}, t = 0) = \mathbf{r}\,. \tag{4}$$

A formal solution to (3) and (4) may be written in terms of an "ordered exponential" (OE),

$$\mathbf{R}(\mathbf{r}, t) = \left(e^{-\int_0^t ds(\mathbf{v}(\mathbf{r}, s)\cdot\nabla)}\right)_+ \vec{\mathbf{r}} \tag{5}$$

where $(\ )_+$ is ordered such that terms with larger s values stand to the left. For example,

$$\mathbf{R} \simeq \left[1 - \int_0^t ds(\mathbf{v}\cdot\nabla) + \int_0^t ds_1 \int_0^{s_1} ds_2 \left(\mathbf{v}(\mathbf{r}, s_1)\cdot\nabla\right)\ \left(\mathbf{v}(\mathbf{r}, s_2)\cdot\nabla\right) + \ldots\right]\mathbf{r}$$

or

$$\mathbf{R} \simeq \mathbf{r} - \int_0^t ds\, \mathbf{v}(\mathbf{r}, s) + \int_0^t ds_1 \int_0^{s_1} ds_2 \left(\mathbf{v}(\mathbf{r}, s_1)\cdot\nabla\right)\ \vec{\mathbf{v}}(\mathbf{r}, s_2) + \ldots .$$

To construct an explicit solution, divide the range of s-integration into small intervals, and expand sequentially. One finds a result equivalent to the iteration of

$$\mathbf{R}(\mathbf{r}, t) = \mathbf{r} - \int_0^t ds\, \mathbf{u}(\mathbf{r}, s|t)\,, \tag{6}$$

where

$$\mathbf{u}(\mathbf{r}, s|t) = \mathbf{v}(\mathbf{r} - \int_s^t ds'\, \mathbf{u}(\mathbf{r}, s'|t), s)\,. \tag{7}$$

If the velocity and its first time and spatial derivatives are bounded in magnitude, one can prove that (6) and (7) generate a solution to the Euler equation. Note that the solution of (7) may be given as a "continued function",

$$\mathbf{u}(\mathbf{r}, s|t) = \mathbf{v}(\mathbf{r} - \int_s^t ds_1\, \mathbf{v}(\mathbf{r} - \int_{s_1}^t ds_2\, \mathbf{v}(\mathbf{r} - \ldots, s_1), s))\,, \tag{8}$$

in which one can easily imagine multiple bifurcations and eventual chaos.

To be of practical use, one needs a method of upgrading the knowledge of the velocity $\mathbf{v}(\mathbf{r}, t)$ which was used to calculate $\mathbf{u}$ or $\mathbf{R}$. For this, return to (1) and assume that div $v = 0$, so that the density satisfies (3), and the specific pressure p may be calculated as

$$p(\mathbf{r}, t) = \int d^2r'\, G(\mathbf{r} - \mathbf{r}')\, \nabla' \cdot \left[\left(\mathbf{v}(\mathbf{r}', t) \cdot \nabla'\right) \mathbf{v}(\mathbf{r}', t)\right], \tag{9}$$

where the Green's function $G(\mathbf{r} - \mathbf{r}')$ contains all specification of boundary conditions. If the pressure P may be considered as given by an equation of state, $P \sim \rho^\gamma$, then p will also satisfy (3). One need therefore calculate (9) but once - at $t = 0$, using the initial fields $\mathbf{v}^{(0)}(\mathbf{r})$ - and one can then upgrade the $p_0(\mathbf{r})$ so calculated, or its gradient: $p(\mathbf{r}, t) = p_0(\mathbf{R}(\mathbf{r}, t))$.

The formal solution to (1) can be expressed as

$$\mathbf{v}(\mathbf{r}, t) = \mathbf{v}^{(0)}\left(\mathbf{R}(\mathbf{r}, t)\right) - \int_0^t dt' \left(e^{-\int_{t'}^{t} ds\, \mathbf{v}(\mathbf{r}, s)\cdot\nabla}\right)_+ \vec{\nabla} p(\mathbf{r}, t'). \tag{10}$$

For this one requires a slight generalization of $\mathbf{R}(\mathbf{r}, t)$ to

$$\mathbf{R}(\mathbf{r}, t, t_1) = \left(e^{-\int_{t_1}^{t} ds\, \mathbf{v}(\mathbf{r}, s)\cdot\nabla}\right)_+ \mathbf{r}, \tag{11}$$

which can be expressed in terms of the same function $\mathbf{u}$, and is given by:

$$\mathbf{R}(\mathbf{r}, t, t_1) = \mathbf{r} - \int_{t_1}^{t} ds\, \mathbf{u}(\mathbf{r}, s|t).$$

In this way, $\mathbf{v}(\mathbf{r}, t)$ may be calculated as

$$\mathbf{v}(\mathbf{r}, t) = \mathbf{v}^{(0)}\left(\mathbf{R}(\mathbf{r}, t)\right) - \int_0^t dt'\, \nabla p\left(\mathbf{R}(\mathbf{r}, t, t'), t'\right). \tag{12}$$

Given $\mathbf{u}$, or $\mathbf{R}$, (13) provides the relation $\mathbf{v} = \mathbf{v}\{\mathbf{R}\}$, while for a given $\mathbf{v}$, (7) provides the necessary $\mathbf{R} = \mathbf{R}\{\mathbf{v}\}$. One begins at $t = 0$ with the specification of the initial velocity, calculates $\mathbf{R}$ at a small time-step, and then uses that function to calculate $\mathbf{v}(\mathbf{r}, t)$; and then the process is repeated at the next time-step, etc. In this way, one has a complete, if sequential, method of calculating $\mathbf{v}(\mathbf{r}, t)$ in terms of an initial $\mathbf{v}^{(0)}(\mathbf{r})$. If viscosity is to be included, that term can easily be inserted on the RHS of (11), while in three dimensions, a vortex-stretching term can be inserted on the RHS of (2).

3. Eikonal Approximations

The sequential steps described above can be defined in essentially two distinct ways, approximately and exactly, with the first category far simpler than the second. What is involved here is the in-principle, infinite number of iterations needed in solving the continued function equations at a fixed t. If one "breaks the chain", and uses a finite number of iterations, the forms generated resemble those of the eikonal approximation long familiar in particle scattering theory.[2].

For example, the exact (7) might be approximated by

$$\mathbf{u}(\mathbf{r},\, s|t) \simeq \mathbf{v}(\mathbf{r},\, s)\,,$$

or by

$$\mathbf{u}(\mathbf{r},\, s|t) \simeq \mathbf{v}(\mathbf{r} - \int_s^t ds'\, \mathbf{v}(\mathbf{r},\, s'),\, s)\,.$$

The implicit assumption here is that one need treat correctly only sufficiently small frequency components of the spatial Fourier transform of $\mathbf{v}(\mathbf{r},\, t)$, frequencies $\mathbf{k}$ for which $0(|\mathbf{k}|t|\mathbf{v}|) < 1$.

In quantum field theory, one has exact representations[6] for Green's functions in the presence of an external, or background field of form

$$G_c[A] \sim \int_0^\infty ds\, e^{-ism^2} \int d[\phi]\, e^{i\int_0^s ds'\, \phi_\mu^2(s')}.$$
$$\cdot\, e^{ig\int_0^s ds'\, \phi_\mu(s')\, A_\mu\left(x - \int_0^{s'} ds''\, \phi(s'')\right)} \cdots.$$

There, the eikonal approximation replaces the last factor by

$$\exp\left[ig\int_0^s ds'\, \phi_\mu^{(0)}\, A_\mu\left(x - s'\,\phi^{(0)}\right)\right]\,,$$

where $\phi_\mu^{(0)}$ denotes an asymptotic four-velocity of an incident or scattered particle. Such an approximation neglects variations of $\phi_\mu(s')$ and is only viable in the limit of soft quanta exchanged.

In the present context, one would expect that such an eikonal approximation would be useful when not much change in the spatial form of $\mathbf{v}$ is expected, e.g., during time intervals well before or well after a vortex-fusion process takes place. It would be very interesting to use a computer to perform such iterations, and to see explicitly what sort of convergence properties are displayed as the number of iterations are increased.

4. Exact Iterations

For sufficiently simple flows, it is possible to construct implicit solutions to the equations corresponding to an infinite number of iterations, although the passage from implicit to explicit solutions is not clear. In contrast to the approximations described above, one would expect such exact results to be important when the flow changes substantially in a short time interval.

The simplest illustration is in the small-t limit, where the infinite number of iterations of (7) are given in terms of an initial velocity $\mathbf{v}^{(0)}(\mathbf{r},\, \theta) = \hat{\theta}\, \beta(r)$, where β is a specified function of the radial coordinate. Let the solution to (7) for $\mathbf{u}(\mathbf{r},\, s|t) \to \mathbf{u}(\mathbf{r},\, t - s)$ be expressed in the form $\mathbf{u} = \hat{\mathbf{r}}\, a(r|\tau) + \hat{\theta}\, b(r|\tau)$, where $\tau = t - s$. The argument of $\mathbf{v}^{(0)}$ on the RHS of (7) may be written as $\rho = \hat{\mathbf{r}}\,(r - A) + \hat{\theta}\, B$, where A and B are the τ'-integrals (between 0 and τ) of a and b, respectively. A straightforward analysis shows that an implicit solution for $B(r|\tau)$ may be obtained in the form

$$\tau = \int_0^B \frac{dB'}{\beta\left(\sqrt{r^2+2B'^2}\right)} \cdot \sqrt{\frac{r^2+2B'^2}{r^2+B'^2}}, \tag{13}$$

in the sense that the result of the integration of (14) must be inverted to give $B(r|\tau)$. If this can be done, then $r - A = \left[B^2 + r^2\right]^{1/2}$, and ρ is known. Finally, $\mathbf{u}(r|\tau) = \mathbf{v}^{(0)}(\rho) = \hat{\theta}(r - A)\,\beta(\rho)/\rho - \hat{\mathbf{r}}\,B\,\beta(\rho)/\rho$, and this small part of the problem has been solved exactly. Similar but even more implicit remarks may be made when the initial $\mathbf{v}^{(0)}$ distribution has both radial and angular components, depending on both variables.

5. Summary

The above remarks have presented a method of approach to Euler and Navier-Stokes flows which may be of some usefulness in obtaining a physical understanding of at least the qualitative aspects of problems which can otherwise only be handled by extensive numerical calculations. The methods described are new to the author, although that certainly does not guarantee that they have never before appeared in the voluminous literature on these subjects.

A variety of other problems can be investigated using this formalism, including those of random forcing, and studies of possible "local coherence" in turbulent flows. Whether or not the results are worth the effort remains to be seen; but, at the very least, it is of non-trivial interest to explore in this subject those low-frequency, "eikonal" approximations familiar in particle scattering theory.

The author is indebted to J-D Fournier and B. Legras for several most useful and informative discussions.

6. References

1. J-D Fournier and H. M. Fried, *Phys. Rev.* D (to be published).
2. See, for example, H. M. Fried, *Functional Methods and Eikonal Models*, Édition Frontières (1990), Gif-sur-Yvette, France.
3. M. Rabaud and Y. Couder, *J. Fluid Mech.* **136** (1983) 291.
4. N. J. Zabusky, M. H. Hughes and K. V. Roberts, *J. Comput. Phys.* **30** (1979) 96.
5. B. Legras and D. G. Dritschel, Physics of Fluids (to be published).
6. E. S. Fradkin, *Nucl. Phys.* **76** (1966) 588. Application of Fradkin's lovely representation to problems of quantum physics and classical fluids may be found in reference 2.

QUANTUM CHAOS AND SABINE'S LAW OF REVERBERATION IN ERGODIC ROOMS

Olivier LEGRAND and Didier SORNETTE
Laboratoire de Physique de la Matière Condensée, CNRS URA 190,
Université de Nice-Sophia Antipolis, Parc Valrose, 06034 Nice Cedex, France

Abstract : We investigate the standard acoustical problem of sound decay in a room, due to a small absorption at the walls, both in the geometrical approximation (1) and for the full wave problem (2). The classical universal Sabine's law of reverberation is shown to rely on ergodic properties of both geometrical billiard-like trajectories (1) and eigenmodes (2). A paradigm of an ergodic auditorium is used to test numerically these ideas: a two-dimensional (2-D) room with the shape of a Stadium. In both approaches, Sabine's law for the characteristic reverberation time is verified with good accuracy.

INTRODUCTION

Almost a century ago, Sabine [1] studied experimentally the decay of sound in concert halls due to absorption. In order to account for his observations, Sabine made the assumption that the pressure field in the room is uniform and isotropic. He then arrived at the famous formula for the decay time

$$T = \text{const}\frac{V}{\Sigma \alpha_j S_j} \tag{1}$$

in which V is the volume of the room and the summation is over different types of absorbing materials with an absorption coefficient α_j and an area of exposed surface S_j. The constant factor is universal (const $= 4/c$ with c being the sound velocity) and does not depend on details of the room. In the acoustics literature, the reverberation time T_R is defined as the time necessary for a 60-dB decrease of the acoustic pressure and is thus simply related to the decay time T through $T_R = (6 \ln 10)T \simeq 13.8T$. In the literature, it has been mainly argued that uniformity and isotropy may be validated by assuming sufficiently irregular walls or diffuse surfaces so as to ensure an efficient isotropic scattering of sound [2-9]. Nevertheless, in the high frequency limit one should consider that all the characteristic lengths of the room become larger than the wavelength. In this regime where the diffuse sound hypothesis does not apply, is the Sabine's law still relevant? Following the general discussion given by Joyce [8] on Sabine's law in the geometric acoustics limit, we presented in [10] a test for its validity, focusing on a pure ray approach in the case of a 2-D room with the shape of a stadium. The problem of sound decay in a room then becomes equivalent to an escape problem for particles bouncing off the enclosure of a billiard and being captured by a trap in the wall. This is a very general problem [11] which has been considered in other contexts such as transient chaos for connecting the (internal) dynamical properties of chaotic billiards to the (external) problem of chaotic scattering.

In what follows, we shall briefly summarize our results concerning the decay law for particles and the associated reverberation time for sound in a chaotic billiard with mirror reflection on the boundaries as a function of billard surface and effective absorption length. We shall then resort to the validation of Sabine's law in the limit of small absorption through a perturbative wave approach relying on the ergodic hypothesis introduced by Berry [12] and Voros [13]. This hypothesis concerns the correspondance between characteristic modes and classical phase space in nonintegrable ray systems. Restrictions will rise about this derivation due to the existence of localized enhancements of the field coined "scars" by E.J. Heller [14] and further described by E.B. Bogomolny [15]. We shall finally present numerical results on the morphology of modes in the Stadium billiard which clearly exemplify the role of "scars" in the decay of sound in an ergodic auditorium.

I. SABINE'S LAW WITHIN GEOMETRIC ACOUSTICS

Consider the following problem. Within the enclosure of a Stadium billiard (made of two half-circles of unit radius connected by two parallel segments of length ϵ), launch, from a given source point, a large amount of pointlike particles in all directions. Now define a disk D of radius r taken as a perfect absorber i.e. such that when the trajectory of a particle hits it, this particle is lost. Each particle is reflected following Snell's law on the boundary of the Stadium. Due to the ergodic property of the billiard each particle will almost surely hit the disk D after a length L. In previous papers [10, 16], we studied the probability $P(L)$ for a particle of having "survived" after a finite length L. We numerically found (see Fig. 1) and also justified theoretically that $P(L)$ is given by

$$P(L) = P_0 \exp[-\beta(\epsilon, r)L], \tag{2}$$

with

$$\beta(\epsilon, r) = \frac{r}{\frac{\pi}{2} + \epsilon}. \tag{3}$$

These results are founded on the Markovian character of the chaotic trajectories in the Stadium, which is met in the limit of small r or equivalently for long trajectories. The mean length of capture $[\beta(\epsilon, r)]^{-1}$ is proportional to the effective length of absorption, namely the perimeter $2\pi r$ of the capturing disk. The quantity $[\frac{\pi}{2}+\epsilon]$ is half the surface of the Stadium and approximately weighs (for a small enough capturing disk) the fraction of surviving particles.

These results are nothing but Sabine's law for which, in the case of a 2-D room, the expression of the characteristic decay time may be rewritten as

$$T = \frac{1}{\beta c} = \frac{\pi S}{cl}, \tag{4}$$

where $S = \pi + 2\epsilon$ is the total area of the Stadium and $l = 2\pi r$ is the effective absorption length. Note that the universal constant $= 4/c$ in three dimensions (3-D) becomes π/c in 2-D.

In fact, we have shown that, in the case of the Stadium (for $\epsilon = 1$ or 2), the decay law is more complicated than a pure exponential when the absorption perimeter becomes larger than 0.3 due to the existence of finite time correlations which stem from the finite mean rate of exponential divergence between nearby trajectories, the so-called Kolmogorov entropy. Indeed, special trajectories will play a crucial role in the breakdown of Sabine's law, namely ***periodic*** orbits which appear to be of two kinds in the Stadium: (1) a continuous set of the so-called marginally stable "bouncing-ball" orbits which bounce perpendicular to the rectilinear portions of the boundary, (2) an infinite countable set of isolated unstable periodic orbits from which any neighboring trajectory diverges exponentially with a local rate given by the Lyapunov exponant of the periodic orbit. Among those periodic orbits, "bouncing-ball" ones will be responsible for substancial echoes between the flat parts of the Stadium whereas unstable ones will induce significant anisotropy of the flow of particles at least if their time rate of instability is longer than the characteristic reverberation time. Those effects are evidently particularly strong when all particles start at initial time from a single point source. On the contrary, for particles that have randomly sampled initial positions (corresponding to a diffuse acoustic source), we have numerically checked that the decay law departs from a pure exponential for an absorption perimeter only larger than 0.6. Obviously, this perimeter must remain small enough so as to ensure that the dynamics is truly Markovian such that Sabine's law still holds.

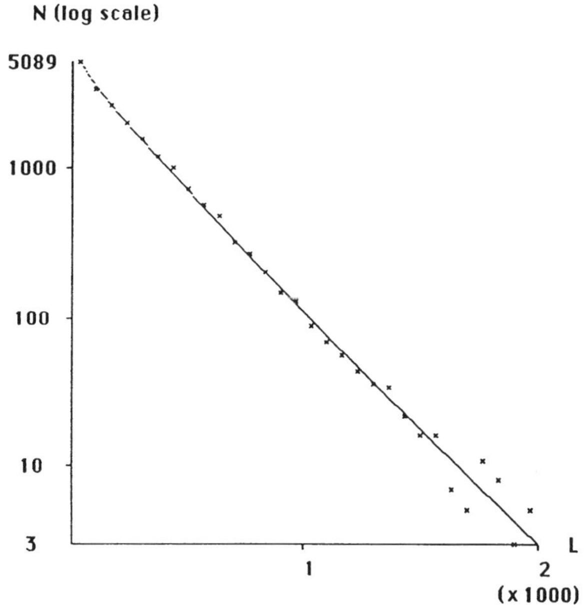

Fig. 1. Semi-log representation of the number $N(L)$ of orbits captured by the disk at length L for 10^4 initial orbits in the Stadium with end caps of unit radius and $\epsilon = 1$. The radius of the absorbing disk is $r = 10^{-2}$. N(L) is well fitted by an exponential $N(L) = N_0 \exp(-\beta L)$ with $\beta \simeq 3.9 \times 10^{-3}$. Crosses are numerical results and the continuous line corresponds to fit by an exponential with a powerlaw correction (see ref.[16]).

II. A WAVE APPROACH : HOW TO TEST ERGODICITY

Geometrical acoustics is the frame within which statistical room acoustics was developed [8]. Beyond this approximation, one might expect that the essential feature of waves, namely the phase leading to interferences, could lead to departures from the homogeneous statistical distribution of sound energy assumed to establish formula (1) and based on geometric chaos.

In rooms with simple shapes such as parallelepipedic ones (where the wave problem is exactly integrable due to separability) with absorbing materials evenly distributed over walls, floor and ceiling, Schuster and Waetzmann [17] were the first to present general expressions for the time decay constants of eigenmodes. Using the concept of the absorption exponent α' defined as $\alpha' = -\ln(1-\alpha)$ where α is the absorption coefficient, the decay constant δ for a characteristic mode with complex frequency $\omega - i\delta$ reads :

$$\delta = \frac{c}{4}\left[2\alpha'_x\frac{\cos\theta_x}{L_x} + 2\alpha'_y\frac{\cos\theta_y}{L_y} + 2\alpha'_z\frac{\cos\theta_z}{L_z}\right], \tag{5}$$

where L_x, L_y and L_z are the side lengths of the parallelopiped, θ_x (resp. θ_y and θ_z) being the angle of incidence of the given mode on both walls perpendicular to the x axis (resp. y and z axes).

For the example at hand, a reverberation process generally results from the decay of all eigenmodes in a given frequency band. But, since different modes correspond to different angles of incidence and thus to different δ's, one cannot obtain a pure exponential decay. Rather, the decay will start with a mean decay constant $\langle\delta\rangle = \sum_i \delta_i I_i$ where the sum is over all modes of intensity I_i in the band and will eventually be determined by the decay constant of the least damped mode. In fact, a pure exponential Sabine-like decay as in Eq.(2) would necessitate quite a specific incidence dependence of the absorption exponent, namely

$$\alpha'_{x,y,z} = \frac{\alpha'_0}{\cos\theta_{x,y,z}}. \tag{6}$$

This behavior may be approximately recovered if one assumes nongrazing incidence and a large enough ***specific acoustic impedance*** ζ defined as

$$\zeta = \frac{1}{\rho c} \times \left(\frac{\text{air pressure at wall}}{\text{normal air velocity at wall}}\right) = |\zeta|\exp(i\varphi), \tag{7}$$

where ρ is the density of air. Under those assumptions, the expression of the absorption coefficient as a function of ζ and of the angle of incidence [18]

$$\alpha(\theta) = 1 - \left|\frac{\zeta\cos\theta - 1}{\zeta\cos\theta + 1}\right|^2 \tag{8}$$

can be expanded in powers of $\frac{1}{|\zeta|\cos\theta}$ to yield the following expression of the absorption exponent to leading order

$$\alpha'(\theta) = \frac{4\cos\varphi}{|\zeta|\cos\theta}. \tag{9}$$

Thus, using the *specific acoustic admittance* $\beta = \zeta^{-1}$ whose real part is denoted β_R, this allows to re-write equation (5) in a form closely related to Sabine's formula :

$$\delta = \frac{c}{V} \sum_j \beta_{R_j} S_j \tag{10}$$

where the sum is over the walls of rectangular areas S_j and V is the volume. This formula was first derived by van den Dungen [19] from energy balance arguments. Of course, if the wall is not hard (impedance not large enough) or if the wave is not oblique (near grazing incidence), the *wall coefficient* $\alpha' \cos\theta$ significantly differs from the so-called *normal wall coefficient* $\alpha_N = 4\beta_R$ as shown in the extensive work of Morse and Bolt [20]. Note that formula (10) would yield Sabine's law if its validity were not restricted to particular types of eigenmodes. This particular case of a room with an "integrable" enclosure clearly exemplifies the fact that, in highly regular and symmetric rooms, noticeable differences in the decay rate of various modes occur due to the existence of so-called "axial" or "tangential" modes [20].

It has long been asserted [20] that sufficiently irregularly shaped rooms would average out those differences between decay rates which would all approach the one predicted by statistical geometrical acoustics. Here we wish to emphasize that, in a 2-D room as regularly shaped as the Stadium, ergodic wave motion is obtained at high frequencies preluding to a validation of Sabine's law in a full wave treatment.

In a first stage, consider the following unperturbed eigenvalue problem in a room with an arbitrary shape, in which characteristic functions ϕ_n satisfy a Neumann boundary condition (perfectly rigid walls):

$$\nabla^2 \phi_n + \frac{{\omega_n^0}^2}{c^2} \phi_n = 0, \tag{11a}$$

$$\text{with } \partial_n \phi_n = 0 \text{ on the boundary.} \tag{11b}$$

Now, with a small homogeneous specific acoustic admittance β ("hard walls"), the perturbed characteristic functions ψ_n satisfy

$$\nabla^2 \psi_n + \frac{{\eta_n}^2}{c^2} \psi_n = 0, \tag{12a}$$

$$\text{with } \partial_n \psi_n = i\frac{\omega}{c} \beta \psi_n \equiv i\frac{\omega}{c}(\sigma + i\gamma)\psi_n \text{ on the boundary.} \tag{12b}$$

A standard perturbation calculation [20] then yields (to first order) :

$$\eta_n \equiv \omega_n - i\delta_n \simeq \omega_n^0 - \frac{c}{2\int\int\int \phi_n^2 dV} \int\int (\sigma + i\gamma)\phi_n^2 dS. \tag{13}$$

We next resort to the results of Schnirelman, Zelditch, and Colin de Verdière [21], and also Voros [13] (hereafter referred to as the "ergodic theorem") which roughly state that, if the geometrical limit is ergodic, then the eigenmodes are also ergodic (save possibly for a set of measure zero), in the sense that, on a coarse-grained scale (i.e. averaged over many wavelengths), the intensity is uniform and the spatial correlations are isotropic. Thus, for almost all modes in an ergodic room, averaging the absorption coefficient on the walls amounts to averaging it over all directions of incidences. In terms of the specific acoustic admittance, the absorption coefficient given in Eq.(8) depends on angle of incidence θ as

$$\alpha(\theta) = \frac{4\gamma\cos\theta}{(\cos\theta+\gamma)^2+\sigma^2} \tag{14}$$

and approximately averages to (using $\sigma \ll \gamma \ll 1$ since $\mathrm{Re}\zeta \ll 1 \ll \mathrm{Im}\zeta$ for "hard walls")

$$\langle\alpha\rangle \simeq 8\gamma. \tag{15}$$

Note that this results is obtained in 3-D and that for a 2-D room, the corresponding formula is

$$\langle\alpha\rangle \simeq 2\pi\gamma. \tag{16}$$

From equations (13) and (15) one finally gets a decay constant which reads (in the 3-D case):

$$\delta_n = \frac{c}{2} \times \frac{\langle\alpha\rangle}{8\int\int\int \phi_n^2 dV}\int\int \phi_n^2 dS. \tag{17}$$

This decay constant precisely yields Sabine's formula, since $T = 1/(2\delta)$, provided that the ratio of the squared characteristic function integrated over the walls to the squared function integrated over the total volume of the room be twice the ratio of the total surface of the walls to the total volume:

$$\text{ratio} \equiv \frac{\int\int \phi_n^2 dS}{\int\int\int \phi_n^2 dV} = 2\frac{S}{V}. \tag{18}$$

Such a requirement is, following the "ergodic theorem" mentioned above, expected to be fulfilled in an ergodic room. Indeed, as Berry suggested [12], one may view a typical characteristic function, in an ergodic room, as a random superposition of plane waves, with different phases and directions, but with, locally, the same wavelength. Having this picture in mind, one can then argue that, in the limit of small absorption, for each plane wave incident on a wall with a given amplitude A, RA with $R \simeq 1$ is reflected back therefore leading to the desired ratio $\frac{|1+R|^2}{1+|R|^2} \simeq 2$. Here, we would like to stress that no mode-mixing processes as those proposed by Morse and Bolt [20] are involved to achieve ergodicity. The room is geometrically ergodic and diffusive reflection is not necessary.

We checked how equation (18) is verified for the true modes in the Stadium. In this 2-D room, the corresponding ratio must equal $2L/S$ in order to recover the statistical expression (4) of the decay time. We have numerically evaluated this ratio for high frequency odd-odd modes with Neumann boundary conditions (Eqs.(11a,b))(i.e. for rigid walls) in the case of a Stadium with end caps of unit radius and a rectilinear portion ϵ equal to the radius. As a rule, eigenmodes may be roughly classified among three categories : a) the so-called "bouncing-ball" modes [22] localized between the flat parts of the boundary, b) modes "scarred" by one or more unstable periodic [14] orbits or perhaps also homoclinic orbits [23] and c) apparently random modes. As can be seen in figure 2 (where the square of the eigenfunction is represented in only a quarter of the Stadium due to the odd-odd symmetry), of the three represented eigenmodes, the "bouncing-ball" mode at $k \equiv \omega/c = 80.259$ (b) and the mode at $k = 80.099$ (a) with a "whispering gallery" (due to a grazing orbit along the wall) exhibit the largest departures from the expected value of the ratio given in equation (18) whereas the mode at $k = 80.278$ (c) displays no remarkable feature and the value of the ratio is very close to $2L/S$. O'Connor and Heller [22] have shown that the existence of "bouncing-ball" modes persists up to infinite frequency but this is, by no means, in contradiction with the above mentioned "ergodic theorem", for the proportion of those highly localized modes vanishes as the frequency tends to infinity. We have checked over the range $k = 30 - 90$ that the overwhelming majority of modes have a ratio approximately ranging from $1.7L/S$ to $2.3L/S$, thereby qualifying numerically the "ergodic theorem". Note that the ratio defined in Eq.(18) provides a novel physical measure for the "ergodicity" of modes.

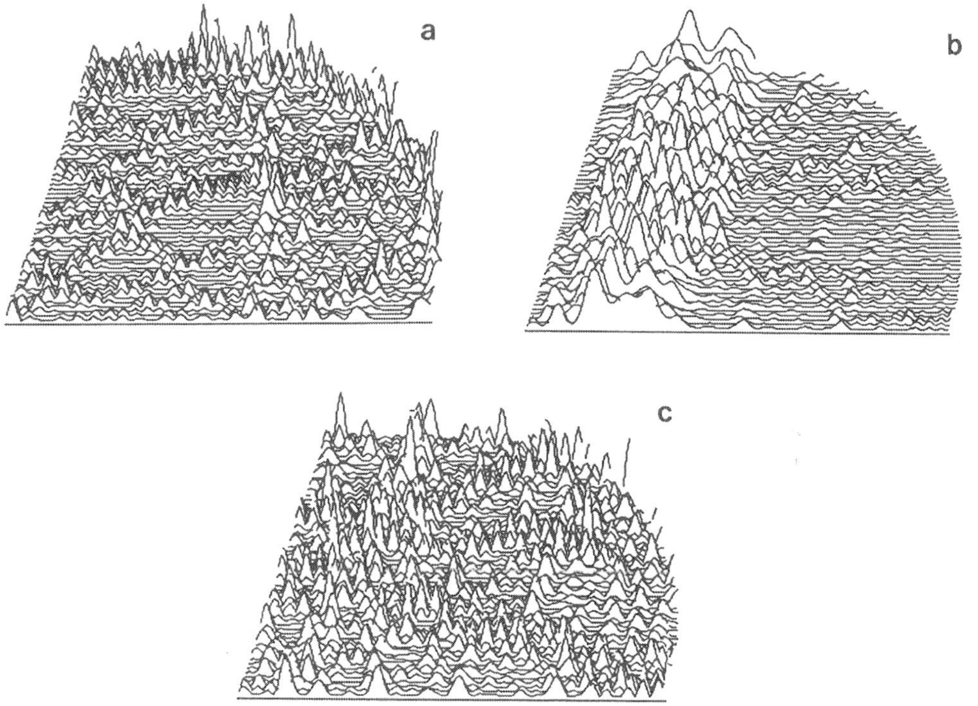

Fig. 2. Three odd-odd eigenmodes for the Neumann problem (see Eqs. (11a,b)) in the Stadium with circular end caps of unit radius and with $\epsilon = 1$. The square of the eigenfunction is represented only in a quarter of the Stadium due to odd-odd symetry. For the three modes, the wave number defined by $k \equiv \omega/c$ and the ratio defined in Eq.(18) are given: a) $k = 80.099$, ratio $= 4.45$; b) $k = 80.259$, ratio $= 1.27$; and c) $k = 80.278$, ratio $= 2.17$.

CONCLUSION

We have checked that Sabine's law of reverberation is verified with good accuracy for almost all modes in a 2-D chaotic room with the shape of a Stadium in the high frequency limit and for small enough absorption ("hard walls"). Besides the global problem of reverberation, wave acoustics in ergodic rooms offer a true richness of phenomena which can be tackled with the help of tools developed in "Quantum Chaos". Among those, one can cite the phenomenon of early recurrence in the transient response of an ergodic room related to periodic orbits in the associated ray problem and also the signature of those same orbits in the fluctuations of the spectral density. This will be pursued in a future publication.

ACKNOWLEDGEMENT

We are very grateful to M. Berry for suggesting us the wave approach to Sabine's law through the ergodic hypothesis for modes.

REFERENCES

[1] W. C. Sabine, *Collected Papers on Acoustics* (Harvard University Press, Cambridge, 1923), reprinted by Dover, New York, 1964).
[2] M. R. Schroeder and D. Hackman, Acustica **45**, 269-273 (1980).
[3] E.N. Gilbert, J. Acoust. Soc. Am. **69**, 178-184 (1981).
[4] E.N. Gilbert, Acustica **66**, 275-280 and 290-292 (1988).
[5] E.N. Gilbert, J. Acoust. Soc. Am. **83**, 1804-1808 (1988).
[6] H. Kuttruff, Acustica **23**, 238 (1970).
[7] H. Kuttruff, Acustica **24**, 356 (1971).
[8] W. B. Joyce, J. Acoust. Soc. Am. **58**, 643 (1975).
[9] W. B. Joyce, J. Acoust. Soc. Am. **64**, 1429 (1978).
[10] O. Legrand and D. Sornette, J. Acoust. Soc. Am. **88** 865 (1990).
[11] O. Legrand and D. Sornette, Europhys. Lett. **11**, 583 (1990); Phys. Rev. Lett. **66**, 2172 (1991).
[12] M. V. Berry, J. Phys. A **10**, 2083 (1977)
[13] A. Voros, in *Stochastic Behavior in Classical and Quantum Hamiltonian Systems*, edited by G. Casati and J. Ford, Lecture notes in Physics **93** (Springer-Verlag, Berlin and New York, 1979).
[14] E. J. Heller, Phys. Rev. Lett. **53**, 1515 (1984).
[15] E. B. Bogomolny, Physica (Amsterdam) **31D**, 169 (1988).
[16] O. Legrand andd D. Sornette, Physica (Amsterdam) **44D**, 229 (1990).
[17] K. Schuster and E. Waetzmann, Ann. Phys., V, **1** 671 (1929).
[18] E.T. Paris, Philos. Mag. **V**, 489 (1928).
[19] F. H. van den Dungen, Acoustique des Salles, Paris (1934).
[20] P. M. Morse and R. H. Bolt, Rev. Mod. Phys. **16**, 69-150 (1944).
[21] A.I. Schnirelman, Usp. Mat. Nauk **29**, 181 (1974); S. Zelditch, Duke Math. J. **55**, 919 (1987); Y. Colin de Verdière, Commun. Math. Phys. **102**, 497 (1985).
[22] P. W. O'Connor and E. J. Heller, Phys. Rev. Lett. **61**, 2288 (1989).
[23] A. M. Ozorio de Almeida, *Hamiltonian Systems: Chaos and Quantization* (Cambridge University Press, 1988).

SELF-ORGANIZED CRITICALITY, EARTHQUAKES AND PLATE TECTONICS

Didier SORNETTE, Anne SORNETTE and Christian VANNESTE
Laboratoire de Physique de la Matière Condensée, CNRS URA 190,
Université de Nice-Sophia Antipolis, Parc Valrose, 06034 Nice Cedex, France

One of the most important breakthroughs in Geology is the theory of Plate Tectonics according to which collisions between continents lead to deformations within the plates. Many important questions, notably in connection to the generation of earthquakes, are related to the ways the interior of the plates accomodate and suffer from these collisions : does the deformation remain localized in the neighborhood of the plate boundaries and if not, why? What is the cause of the large deformations observed thousands of kilometers within the Asian plate, for instance? What is the relative role of large faults with respect to smaller ones in the accomodation of the deformation? Where does the heterogeneity of the fault patterns, observed within a plate at the earth surface, come from? How can one explain the observation of large undeformed regions imbedded within strongly deformed ones? These are some of the key questions connected to the physics of the deformations of the earth lithosphere. For the physicist, the creation of a wealth of complex geological structures, such as fractal fault patterns, self-similar or self-affine mountain ranges, etc, is very stimulating in relation to the modern trends of research aiming at understanding the general feature of pattern formations. This subject of the mechanics of plate tectonics is also fascinating because it is intermediate between the standard problem of the mechanics of deformation in solids (regime of small deformations) and fluid mechanics (which corresponds to a regime of infinitely large deformations).

In this work, we propose a general unifying approach to these and related questions. We suggest that the problem of plate deformations can be tackled with the general concept of fractal growth phenomena. This enables one to identify the dominant features of the problems, namely, screening, enhancement and non-local effects. In order to demonstrate the relation between the plate tectonics problem and growth processes, we have carried out laboratory experiments of a continental plate collision, which are scaled down for gravity. These experiments focus on the formation of faults in a laboratory model of the earth crust. A novel general quasi-two dimensional mechanical model of crack pattern formation is introduced which exhibits both fractal and non-fractal rupture patterns, depending upon the nature of the applied strain and of the imposed boundary conditions. It consists into a brittle layer (dry sand) lying on top of a ductile layer (silicon putty) which are submitted to a simple shear or to more complex strain fields resulting from the indentation of a wedge penetrating steadily inside the system. This model is inspired from the structure of the earth crust and mantle and its kinematics is thought to represent a simplified version of plate tectonics. We argue that this system can be seen as a generalized version of a dual problem of dielectric breakdown, similar to a dual diffusion limited aggregation problem, well-known to develop fractal growth structures. We have measured the fractal dimension and the multifractal spectrum of generalized dimensions of the fault pattern. The fractal dimension is found close to 1.7 and independent of the rheologies of the samples. We also measured the distribution of fault lengths which is found to be a powerlaw, in agreement with the assumption that fault formation is akin to a self-similar process.

Finally, we propose that the concept of self-organized criticality (SOC) is relevant for understanding the processes underlying the occurence of earthquakes and the mechanics of plate tectonics. The SOC concept here implies that earthquakes organize the earth crust, both at the spatial and temporal levels. This idea allows to rationalize many observations on occurences and magnitude of earthquakes. Building on the SOC concept, we have attempted to clarify the basics mechanisms responsible for the organization of the crust within a continental plate. An order parameter is introduced which is the local fluctuating strain tensor. The mechanical equilibrium of the continental plate showing faults over its whole surface and submitted to the boundary conditions imposed by the motion of the neighboring plates implies the existence of a global conservation law acting on the stress and strain fields. SOC is intimately related to the existence of such a global constraint, which means that local fluctuations are relaxed via a generalized diffusion equation. This global conservation law can be shown to be related to the fact the earth crust is almost everywhere at the rupture threshold, in a way similar to the sand-pile at its critical slope. A method is proposed for computing the long-range space and time correlations in the strain fluctuations building up within the crust as a result of the SOC kinematics. We show how to compute from this theory the Gutenberg-Richter exponent of the distribution of earthquake magnitudes. Recent developments building of the SOC concept and on the mechanics of large deformations can be found in the references indicated below.

REFERENCES

[1] A. Sornette and D. Sornette, *Self-organized criticality and earthquakes*, Europhys. Lett. **9**, 197 (1989).
[2] A. Sornette and D. Sornette, *Earthquake rupture as a critical point : Consequences for telluric precursors*,Tectonophysics **179**, 327 (1990).
[3] D. Sornette, P. Davy and A. Sornette,*Structuration of the lithosphere in plate tectonics as a self-organized critical phenomenon*,J.Geophys.Res. **95**, 17353 (1990).
[4] P. Davy, A. Sornette and D. Sornette,*Some consequences of a proposed fractal nature of continental faulting*,Nature **348**, 56 (1990).
[5] A. Sornette, P. Davy and D. Sornette,*Growth of fractal fault patterns*, Phys.Rev.Lett. **65**, 2266 (1990).
[6] D. Sornette, C. Vanneste and A. Sornette,*Dispersion of b-values in the Gutenberg-Richter law as a consequence of a proposed fractal nature of continental faulting*, Geophys.Res.Lett. (1991) in press.
[7] D. Sornette and P. Davy,*Fault growth model and the universal fault length distribution*, Geophys.Res.Lett. (1991) in press.
[8] D. Sornette and A. Sornette,*Les systmes critiques auto-organises : un concept plutot qu'un phenomene*, Pour la Science (Mars) **161**, 62 (1991).
[9] A. Sornette, P. Davy and D. Sornette,*Fault growth in brittle-ductile experiments and the mechanics of continental collisions*, J.Geophys. Res.(1991), in press.

[10] P. Evesque and D. Sornette,*A dynamical system theory of large deformations and patterns in non-cohesive solids*, Phys.Rev.Lett. (submitted).
[11] D. Sornette,*Self-organized criticality in plate tectonics*, In the Proceedings of the NATO ASI *Spontaneous formation of space-time structures and criticality*, Geilo, Norway 2-12 april 1991, edited by T. Riste and D. Sherrington, (Kluwer Academic Press, 1991).

Psi-series and their Summability
for Nonintegrable Dynamical Systems

J.-D. Fournier
C.N.R.S.
Observatoire, BP 139,
06003 Nice Cedex, France

and

M. Tabor
Division of Applied Mathematics,
Department of Applied Physics,
Columbia University,
New York, NY 10027, USA.

1. Introduction

The role of complex time singularities in determining the real time behavior of dynamical systems has become a topic of growing interest over the last decade. The nature and distribution of these singularities is relevant to the behavior of both integrable and nonintegrable systems.

In the case of the former there is much interest in the idea of the "Painlevé Property" as a test for integrability. Here, as is by now well known, there seems to be an intimate connection between the ability of a system to exhibit only poles - and possibly essential singularities - as movable (i.e. initial condition dependent) singularities and its integrability. However, the very notion of what one means by "integrability" is, with the exception of Hamiltonian systems, by no means clear or universally accepted - this issue will not concern us here. An introduction to some of the general principles of singularity analysis and associated ideas of integrability can be found in Clarkson and Kruskal(1), Kruskal et al. (2), Ramani et al. (3) and Tabor (4). At this point, however, it is useful to recall some of the standard terminology. In a local expansion about some arbitrary singularity position t0 the power of (t - t0) at which an arbitrary coefficient appears is called a resonance and the conditions that have to be satisfied for the coefficient to indeed be arbitrary are the compatibility conditions. If the leading order is integer and all the resonances occur at integer powers and their associated compatibility conditions satisfied, the local expansion is a Laurent series, i.e. the solution has the Painlevé Property. In the cases where compatibility conditions for integer resonances are not satisfied or the resonances occur at noninteger powers of (t - t0) the solution has to be represented by a more complicated local expansion known as a psi-series.

Our primary interest here is the singularity structure of nonintegrable systems - especially the nature of the distribution of the singularities and the way these distributions can influence the real time behavior. Apart from intrinsic mathematical interest these issues may have real physical consequences. For example, early work by Frisch and Morf (5), to understand aspects of intermittency in turbulence, investigated the singularity distributions of randomly forced systems and showed how the high frequency portion of the spectrum is determined by those distributions. In their study of the Lorenz system Tabor and Weiss (6) showed how intermittent bursts (in the dynamical systems sense) in the real time dynamics are correlated to the approach of singularities towards the real axis. At a superficial level it might be said that regular trajectories are characterized by more regular distributions of singularities above the real axis (an extreme and obvious example being the pole lattice of elliptic functions) whereas chaotic orbits seem to have an associated singularity distribution that is more random looking. An early illustration of this idea is given in Chang et. al. (7).

Just as the association of "pure pole" behaviour (i.e. the Painlevé property) with integrability is still not fully understood (it seems to be a sufficient but not necessary condition) the type of singularity structure that might guarantee non-integrability, or be required for a given orbit to be chaotic, is also not clear-cut. In this context a number of ideas have been discussed; in particular the idea that nonintegrable systems might be characterized by some sort of natural boundary. This idea was first mooted in the work of Chang et. al. (8) in their study of the Henon-Heiles system. In the system parameter regimes studied the local expansions have complex resonances and this leads to a consideration (to be enlarged on in the next section) of the asymptotic (in the sense of t -->t0) form

$$x(t) = (1/t)\, f(z)$$

where $z = t^{p}$ and p the complex resonance (t stands for t-t0). Mapping the singularities of f(z) back onto the t-plane reveals a remarkable pattern of self-similar spirals of singularities - these theoretical predictions being well matched by numerical experiment. At first sight it might appear that the ever-tightening spirals lead to a natural boundary through which analytic continuation is impossible. However, as pointed out by Bessis and Chafee (9), great care must be exercised in reaching this conclusion since the spirals lie on different Riemann sheets and it may be that it is only their projection onto the plane that appears impenetrable. In addition, a fundamental and, as yet, unresolved question remains: since the singularity distribution, in this case spiraling clusters, is a global property, what analytic structure feature (if any) distinguishes the regular orbits from the chaotic ones? This may well be associated with the radius of convergence of the associated series - the pathologies of which make this a most difficult question.

2. Psi-series for the Duffing oscillator

Here we reviev some of our results concerning the analysis of psi-series for the Duffing equation (Fournier et.al.(10), henceforth referred to as I) and the Lorenz (Levine and Tabor (11), henceforth referred to as II) system. We work with the former in the form

$$\ddot{x} + \lambda\dot{x} + \frac{1}{2}x^3 = \varepsilon\ F(t) \tag{2.1}$$

where F(t) represents some external driving force (the traditional linear term does not affect our analysis and is therefore dropped). In the case of $\lambda = \varepsilon = 0$ the equation reduces to one that is easily integrated in terms of elliptic functions and indeed the local expansion takes the form of the simple Laurent series

$$x(t) = \Sigma\, a_j\,(t - t_0)^{j-1} \tag{2.2}$$

in which a_4 is the arbitrary coefficient (resonance). In the presence of damping and/or driving the arbitrariness of a_4 is lost and can only be restored by resorting to the formal psi-series expansion of the form

$$x(t) = (1/t)\, \Sigma\Sigma\, a_{jk}\, t^j\, (t^4 \ln t)^k \tag{2.3}$$

where t stands for $(t-t_0)$. It is on such series, about which so little seems to be known, that we concentrate our attention.

At first glance, in examining the properties of (2.3) in the complex plane deprived of the logarithmic cut, one cannot argue that any one set of terms, such as the higher power ones, can be neglected in favour of others in the t-->0 limit. In this limit indeed

$$|t|^j\ |t^4 \ln t|^k \ll |t^4 \ln t|^k\ ,\quad \forall\ k, j > 0,$$

but for $0 < j < 4(k-\ell)$ the opposite ordering

$$|t^4 \ln t|^k \ll |t|^j\ |t^4 \ln t|^\ell$$

holds. If one however retains their multivaluation the logarithms

$$\ln t = \ln r + i\theta + 2\, i\, \pi\, k \tag{2.4}$$

can still take on complex values with arbitrarily large modulus by going to sufficiently high sheets k. This suggests ways of ordering terms in the series and, in particular by summing the double series in vertical strips we obtain the resummed series

$$x(t) = (1/t)\Sigma\, t^{n}\,\theta_{n}(z) \tag{2.5}$$

where the θ_n are certain functions of $z = t^4 \ln t$. Formally these can be interpreted as the generating functions for the coefficient sets ajn and are found to satisfy certain linear differential equations (except for n = 0). The leading function θ_0 plays a particularly significant role. Setting formally

$$x(t) = (1/t)\,\theta_0(z) \tag{2.6}$$

substituting into eqn (2.1) and taking the above ordering in the limit t--> 0 gives the equation

$$16\, z^2 \ddot{\theta}_0(z) + 4\, z\, \dot{\theta}_0(z) + 2\, \theta_0(z) + (1/2)\, \theta_0^3(z) = 0 \tag{2.7}$$

which on making the substitution $\theta_0 = z^{1/4} f(z^{1/4})$ transforms to

$$\ddot{f} + \frac{1}{2} f^3 = 0 \tag{2.8}$$

which is precisely the integrable part of the original equation. Thus we have locally mapped the nonintegrable equation onto an integrable one. In fact, what the rescaling at this order of t does is to map the problem onto that part of the equation which exhibits scaling invariance (the dominant balance part, in fact) and which, in this case, is also exactly integrable. The exact solution to (2.8) reveals an elliptic function lattice of poles in the complex $y = z^{1/4}$ plane. Under the inverse mapping $z = t^4 \ln t$ each member of the lattice produces a multisheeted four arm pattern of singularities in the complex t-domain. In most cases excellent agreement is found between this theoretical prediction and numerical studies. However the results are not perfect and in fig 7 of reference I certain arms of the lattice seem to be missing near the real axis. We have as yet not been able to explain this satisfactorily.

Substitution of the entire series (2.5) into eqn. (2.1) yields a whole hierarchy of coupled linear equations for the θ_n's . In our first investigation of this hierarchy it appeared that the θ_n's might all be sufficiently behaved to suggest that the resummed series may actually converge. Subsequent work has, however, revealed significant pathologies in these functions and it these results that we report below.

3. Pathologies in the Psi-series

In continuing our studies of the psi-series for the Duffing oscillator we have chosen to work with a slightly different form of the equation, namely

$$\ddot{x} + \lambda\dot{x} + \mu x + \frac{\varepsilon}{2}x^3 = 0 \qquad (3.1)$$

The introduction of the linear term μx gives the important special case of $\mu = 2\lambda^2/9$ for which (3.1) has the Painlevé property. Indeed one may also show that here a nontrivial (point) Lie symmetry may be found and the equation transformed into a form that can be integrated in terms of elliptic functions. For $\mu \neq 2\lambda^2/9$, even with no driving, the Painlevé property is lost and the local expansion becomes a logarithmic psi-series exactly like the one described in I and discussed above. The special choice $\mu = 2\lambda^2/9$ provides a useful limit case in the analysis. For reference purposes we note that if we substitute the Laurent series

$$x(t) = \sum_{j=0}^{\infty} a_j\,(t-t_0)^{j-1}$$

into (3.1) the first few coefficients are found to be

$$a_0 = 2i\sigma/\varepsilon^{1/2}. \quad (\sigma = \pm 1) \qquad (3.2a)$$

$$a_1 = -i\lambda\sigma/3\,\varepsilon^{1/2} \qquad (3.2b)$$

$$a_2 = i\sigma\,(\mu - \lambda^2/6)/3\varepsilon^{1/2} \qquad (3.2c)$$

$$a_3 = -i\lambda\sigma\,(\frac{\lambda^2}{27} - \frac{\mu}{6})/\varepsilon^{1/2} \qquad (3.2d)$$

$$0.\,a_4 = -i\sigma\lambda^2(\frac{2}{3})(\mu - \frac{2\lambda^2}{9})/\varepsilon^{1/2} \qquad (3.2e)$$

Thus one sees that a_4 will not be arbitrary unless $\mu = 2\lambda^2/9$ (or $\lambda = 0$ conservative integrable case).

Working with the resummed psi-series (2.5) for the general case leads to a hierarchy of equations for the θ_k of the general form (here $'$ denotes d/dz):

$$16\,z^2\ddot{\theta}_k + 4(2k+1)\,z\,\dot{\theta}_k + (k-1)(k-2)\,\theta_k + \frac{3\varepsilon}{2}\theta_0^2\,\theta_k$$

$$= -[\,8z\,\ddot{\theta}_{k-4} + (2k-3)\,\dot{\theta}_{k-4} + \ddot{\theta}_{k-8}\,] - \lambda[\,4\lambda\,\dot{\theta}_{k-1} + (k-2)\,\theta_{k-1} + \dot{\theta}_{k-5}]$$

$$- \mu\,\theta_{k-2} - \frac{\varepsilon}{2}\sum_{\ell m}\theta_{k-m-\ell}\theta_m\theta_\ell \qquad (3.3)$$

The system (3.3) is then subjected to the transformation

$$\theta_k(z) = (z^{1/4})^{1-k}\,\psi_k(z^{1/4}) \qquad (3.4)$$

and then the rescaling

$$\psi_k(y) = 2i\sigma\, C\, X_k(Cy) \quad , \tag{3.5}$$

where C is a certain arbitrary constant. The net result is the system

$$\ddot{X}_k - 6 X_0^2 X_k = R_k \quad , \qquad k \geqslant 1 \tag{3.6}$$

where ' denotes differentiation to the independent variable $\xi = Cy$ and the $R_k = R_k\,(X_\ell, \ell < k)$ are the inhomogeneous terms of general form

$$R_k = -\frac{1}{2}\ddot{X}_{k-4} - \frac{1}{16}\ddot{X}_{k-8} + \frac{1}{4}(2k-11)\frac{1}{\xi}\dot{X}_{k-4} + \frac{1}{16}(2k-15)\frac{1}{\xi}\dot{X}_{k-8} - \frac{1}{4}(k-5)\frac{1}{\xi^2}X_{k-4}$$
$$-\frac{1}{16}(k-9)(k-5)\frac{1}{\xi^2}X_{k-8} \; -\lambda\left(\frac{1}{C}\right) \; [\dot{X}_{k-11} + \frac{1}{4}X_{k-5} + \frac{1}{4}(6-k)\frac{1}{\xi}X_{k-5}\,]$$
$$-\,\mu\left(\frac{1}{C}\right)X_{k-2} + 2\,\varepsilon \sum_{\ell m} X_{k-\ell-m}X_\ell X_m \tag{3.7}$$

We will mainly be interested in just the first few equations, namely

$$\ddot{X}_0 - 6X_0^3 = 0 \tag{3.8a}$$

$$\ddot{X}_1 - 6X_0^2\,X_1 = -\frac{\lambda}{C}\dot{X}_0 \tag{3.8b}$$

$$\ddot{X}_2 - 6X_0^2\,X_2 = 6\varepsilon X_1^2\,X_0 - \frac{\lambda}{C}\dot{X}_1 - \mu\left(\frac{1}{C^2}\right)X_0 \tag{3.8c}$$

$$\ddot{X}_3 - 6X_0^2\,X_3 = 12\,\varepsilon\,X_2 X_1 X_0 + 2X_1^3 - \frac{\lambda}{C}\dot{X}_2 - \frac{\mu}{C^2}X_1 \tag{3.8d}$$

The homogeneous equation (3.8a) has the two independent solutions,

$$X_I = \dot{ds}(\xi) \;\text{ and }\; X_{II} = \dot{\overline{\xi\, ds(\xi)}} \quad ,$$

where ds is a standard elliptic function, which can be used to write the general inhomogeneous solution

$$X_k = 2X_I\int^{\xi}X_{II}(\omega)\,R_k(\omega)\,d\omega - 2X_{II}\int^{\xi}X_I(\omega)\,R_k(\omega)\,d\omega \tag{3.9}$$

The original variable x(t) thus has the formal expansion

$$x(t) = 2i\sigma C\sum_{k=0}^{\infty}(t-t_0)^{k-1}\,C^{k-1}X_k(\xi)/\xi^{k-1} \tag{3.10}$$

To maintain the ordering in powers of $t-t_0$ we impose (see I for details)

$$\lim_{\xi\to 0} C^{k-1}\,X_k(\xi)/\xi^{k-1} = 0 \text{ or finite limit}. \tag{3.11}$$

In analyzing the functions X_k it is important to distinguish between the behaviour at $\xi = 0$, which corresponds to the original singularity at $t = t_0$ from the other singularities which may arise at some $\xi = \xi_c$. It is these that introduce the pathologies into the psi-series.

The behaviour around $\xi = 0$ enables us to determine the arbitrary constant C introduced in (3.5). From (3.11) we set

$$X_k \simeq \bar{X}_k \xi^{k-1} \tag{3.12}$$

Direct substitution of this into the equations (3.8) and explicit use of the expansion of $\overset{\prime}{ds}(\xi)$ about $\xi = 0$ gives the following

$$\bar{X}_1 = -\lambda/6C$$

$$\bar{X}_2 = -(1/6C^2)(\frac{\lambda^2}{6} - \mu)$$

$$\bar{X}_3 = -(\lambda/12C^3)(\frac{2\lambda^2}{39} - \mu)$$

which should be compared to the a_j given in (3.2). At $k = 4$ we find

$$0 = -2\frac{\lambda}{C}\bar{X}_3 - \frac{\mu}{C^2}\bar{X}_2 + 12\bar{X}_3\bar{X}_1 + 6\bar{X}_2^2 + 6\bar{X}_2\bar{X}_1^2 - \frac{1}{8}$$

from which C is determined to be

$$C = \sqrt[4]{8\lambda^2(2\lambda^2 - 9\mu)/27} \tag{3.13}$$

In the case of $X_1(\xi)$ it is straight forward to perform local analysis about both $\xi = 0$ and $\xi = \xi_c$. About the former one finds a well behaved Taylor series of the form

$$X_1 = \sum_0^\infty b_n \xi^n$$

whith $b_0 = -\lambda/6C$ ($\equiv \bar{X}_1$), $b_1 = b_2 = 0$, b_3 arbitrary, $b_4 = \lambda/48C$, etc. About ξ_c one finds the Laurent expansion

$$X_1 = \sum_{n=0}^\infty C_n (\xi - \xi_c)^{n-2}$$

with C_0 arbitrary, $C_1 = 0$, $C_2 = -\lambda/6C$, $C_3 = C_4 = 0$, C_5 arbitrary. Note that since X_1 satisfies a linear equation the pole position ξ_c is not arbitrary. Note also that X_1 exhibits second order poles. In fact one can integrate the equation for X_1 explicitly using (3.9) to obtain

$$X_1 = -\frac{1}{3}\frac{\lambda}{C}\xi ds - \frac{1}{6}\frac{\lambda}{C}\xi^2 \overset{\prime}{ds} + \mu_1 \overset{\prime}{ds} + \mu_2 \overline{\xi ds}^{\,\prime}$$

where μ_1 and μ_2 are arbitrary coefficients multiplying the homogeneous contributions to the general solution. Judicious choice of these coefficients lead to the result.

$$X_1(\xi) = -\frac{\lambda}{6C}\xi\,[\,2\,ds + \xi d\dot{s}\,] \tag{3.14}$$

Explicit expansion of this exact solution about $\xi = 0$ and $\xi = \xi_c$ confirms the coefficient b_n, C_n found by the local analysis.

Our main interest is in the pole behaviour of the X_k. The pole order increases with increasing k. This can be seen by looking at the inhomogeneous term R_k. The triple sum part of it (see (3.7)) will generate the most singular terms. For example using the fact that X_0 has first order poles and X_1 second order poles we can see that

$$R_2 = 6\,X_1^2 X_0 - \frac{\lambda'}{C}X_1 - \frac{\mu}{C^2}X_0$$

has a term (the triple product one) that can blow up as $(\xi - \xi_c)^{-5}$. For this to be balanced by the homogeneous part of the equation, i.e.

$$\ddot{X}_2 - 6\,X_0^2 X_2 \quad ,$$

we see that X_2 must exhibit a $(\xi - \xi_c)^{-3}$ behaviour. Using this type of argument we find the leading order behaviour for each X_k to go as

$$X_k \sim \frac{r_1^k}{(\xi - \xi_c)^{k+1}} \tag{3.15}$$

where $r_1 = \lambda\,\xi_c / 6\,C$. Since we know that x(t) itself can only support first order poles this result looks rather alarming! However, these singularities can be resummed in the following sense. Consider summing up the most singular terms of each X_k, i.e.

$$\begin{aligned} S &= \sum_{k=0}^{\infty} X_k \sim \sum_{k=0}^{\infty} r_1^k/(\xi - \xi_c)^{k+1} \\ &= \frac{1}{(\xi - \xi_c)}\sum_{k=0}^{\infty}\left(\frac{r_1}{\xi - \xi_c}\right)^k = \frac{1}{(\xi - \xi_c)}\ \frac{1}{\left(1 - \frac{r_1}{\xi - \xi_c}\right)} \\ &= \frac{1}{\xi - (\xi_c + r_1)} \end{aligned}$$

Thus the singularities (that is their leading orders) sum up to give back a first order, shifted, pole. Transforming all variables back to their dependency on t and the various $\theta_k(z)$ we find at $\xi = \xi_c$

$$x(t) = \sum_{k=0}^{\infty}(\,t_c - t_0\,)^{k-1}\ \ 2i\gamma\,(\lambda/\,6C\,)^k/(\,\xi - \xi_c\,)^{k+1}$$

$$= \frac{2i\sigma}{(t_c - t_0)} \quad \frac{1}{[\xi - (\xi_c + \lambda/6C)]} \tag{3.16}$$

where t_c is determined from the relation

$$\xi_c = C(t_c - t_0)[\ln[t_c - t_0]]^{1/4}$$

Of course the above results are only at leading order and in no way implies that other problems are not lurking in the expansion. Indeed these can be found in X_2 . Using the exact solution for X_1 the inhomogeneous term R_2 can be shown to have the form

$$R_2 = \frac{\lambda^2}{6C^2}\{\xi^4 ds\,(d\dot{s})^2 + 4\xi^3 ds^2 d\dot{s} + 6\xi^2 ds^3 + 4\xi d\dot{s} + (2 - \frac{6\mu}{\lambda^2})\, ds\}$$

Although we have not been able to explicitly integrate (3.9) using this form of R_2 we can now at least examine the expansion of X_2, term by term. Thus, for example, in expanding about $\xi = 0$ the previously found form of $\bar{X}_2$ is easily confirmed. In expanding about $\xi = \xi_c$ the integrands in (3.9) are seen to have nonvanishing terms proportional to $(\xi - \xi_c)^{-1}$ which thus integrate to give logarithmic terms, the first one in X_2 behaving as

$$X_2 \sim (\xi - \xi_c)^3 \ln(\xi - \xi_c).$$

In fact we show that X_2 itself has a (formal) psi-series of the form

$$X_2(\xi) = \sum_{nm} a_{nm} (\xi - \xi_c)^{n-3} [(\xi - \xi_c)^6 \ln(\xi - \xi_c)]^m \tag{3.17}$$

We can now start to see the pathological nature of the original psi-series for x(t) : terms in that series generating logarithmic psi-series of their own!

4. Conclusion

The present knowledge of the analytic structure of nonintegrable dynamical systems has two sources. One is the delicate numerical singularities search, whose interpretation requires much care. The other is the study of formal double " ψ-series" and their partial resummations. We have shown that the actual understanding of the local and global phenomena exhibited by these systems is much more limited than usually thought. Specifically, assumptions often implicitly made on the analytic behaviour of partial resummations have been found too much optimistic – this includes a recurence rule conjectured for the full resummation in a first exploration in I. We believe that progress on a firm ground will require a conclusive study of the interaction between local and global analyticity features of nonintegrable systems.

We thank D. Bessis for discussions on this work.

5. References

1. Clarkson P.A. and Kruskal M.D. (1989) J. Math. Phys. 30, 2201.

2. Kruskal M.D., Ramani A., Grammaticos B. in Partially Integrable Evolution Equations in Physics Les Houches 1989 ed. Conte R. and Boccara N., Kluwer Publish. Nato ASI Series C 310 (1990), 321.

3. Ramani A., Grammaticos B. and Bountis T. (1989) Phys. Rep. 180, 159.

4. Tabor M., Chaos and Integrability in Nonlinear Dynamics, An Introduction, Wiley-Interscience New York (1989).

5. Frisch U. and Morf R.H. (1981) Phys. Rev. A. 23, 2673.

6. Tabor M. and Weiss J. (1981) Phys. Rev. A. 24, 2157.

7. Chang Y., Tabor M., Weiss J. and Corliss G., (1981) Phys. Lett. 85, 211.

8. Chang Y., Greene J.M.,Tabor M. and Weiss J., (1983) Physica 8 D, 183.

9. Bessis D. and Chafee N. (1986) in Chaotic Dynamics and Fractals, ed. Barnsley M.F. and Demko S., Academic Press Publ., New York (1986), 69.

10. Fournier J.-D., Levine G. and Tabor M. (1988) J. Phys. A 21, 33.

11. Levine G. and Tabor M. (1989), Physica D 33, 189.

BOUND SOLITONS IN THE NONLINEAR SCHRÖDINGER/GINZBURG-LANDAU EQUATION

Boris A. Malomed
P.P. Shirshov Institute for Oceanology, Moscow, 117259, USSR

Interaction of slightly overlapping solitary pulses (SP's) is considered in the cubic nonlinear Schrödinger equation with small pumping and dissipation terms, and in the quintic Ginzburg-Landau equation with small dispersion terms. In both cases, the small perturbing terms render the asymptotic wave form of a SP spatially oscillating. Using the description of the interaction of SP's in terms of an effective potential, it is demonstrated that this fact may give way to formation of two-pulse and multi-pulse bound states, which are weakly stable.

The subject of this work is the perturbed nonlinear Schrödinger (NS) equation with pumping and damping terms:

$$iu_t + u_{xx} + 2|u|^2 u = i\gamma_0 u + i\gamma_1 u_{xx} - i\gamma_2 |u|^2 u \qquad (1)$$

$(\gamma_0, \gamma_1, \gamma_2 > 0)$, which has attracted attention as a dynamical model of plasma physics and hydrodynamics.[1-3] Recently, Eq.(1) has also found an application in the theory of optical solitons in fibers.[4,5] Eq.(1) describes a situation when the trivial solution u = 0 is unstable against small disturbances. In various physical problems, there occurs situation, then the trivial state is stable against small disturbances, but can be triggered into a nontrivial state by a finite disturbance. The simplest model describing this situation is based upon the quintic perturbed NS equation:[6]

$$iu_t + u_{xx} + 2|u|^2 u = -i\gamma_0 u + i\gamma_1 u \quad + i\gamma_2 |u|^2 - i\gamma_3 |u|^4 u \qquad (2)$$

where the damping $(\gamma_0, \gamma_1, \gamma_3)$ and pumping (γ_2) coefficients are all positive.

One treats Eqs.(1) and (2) as perturbed NS equations if the dimensionless parameters γ_1, γ_2, and γ_0, γ_3 are small. In the opposite case, the same equations can be regarded as the Ginzburg-Landau (GL) equations, which also have applications in plasma physics[3] and in hydrodynamics,[7,8] and attract a great attention as general models for pattern formation and onset of chaos.[9]

An important object governed by Eqs.(1) and (2) is a solitary pulse (SP). It is known[1] that Eq.(1) with $\gamma_2 = 0$ has an exact SP solution; if $\gamma_2 \neq 0$ but γ_1 and γ_2 are small, the SP can be found approximately as a solution close to the soliton of the unperturbed NS equation with a fixed amplitude:

$$u = 2i\eta \ \operatorname{sech}(2\eta(x - z_0)) \exp(4i\eta^2 t - ik|x - z_0| + i\Phi_0), \tag{3}$$

$$\eta^2 = (3/4)\gamma_0(\gamma_1 + 2\gamma_2)^{-1}, \quad k/\eta = (2/3)(2\gamma_1 + \gamma_2), \tag{4}$$

z_0 and Φ_0 being arbitrary constants. The presence of the small wave number k produced by the perturbing terms implies that the asymptotic wave field of the soliton is oscillating in x, unlike that in the absence of perturbations. As for Eq.(2), in the near-NS regime $(\gamma_1, \gamma_2, \gamma_0\gamma_3 \ll 1)$ it has the soliton solution in the form (3) with $k/\eta = \gamma_1 - \gamma_0/4\eta^2$, where

$$\eta^2 = (64\gamma_2)^{-1}\left(5(2\gamma_2 - \gamma_1) + \sqrt{5[5(2\gamma_2 - \gamma_1)^2 - 96\gamma_0\gamma_3]}\right). \tag{5}$$

In the opposite (near-GL) regime, Eq.(2) has a stable solution in the form of a broad SP.[11,12]

The aim of this work is to demonstrate that, in both regimes, slightly overlapping SP's can form stable bound states (BS's). These can be two-pulse states, multi-pulse ones, and periodic arrays of SP's. This result may be important in applications. For instance, a casual formation of a two-soliton BS is detrimental for operation of fiber communication lines, therefore it is necessary to know how this can happen.

Note that the soliton solution of Eq.(1), given be Eqs.(3) and (4), is unstable as, at $|x| = \infty$, it coincides with the trivial unstable solution $u \equiv 0$. However, this circumstance is not so important, at least in application to the optical solitons in fibers.[4,5] Anyway, the soliton solution of Eq.(2) given by Eqs.(3) and (5) is stable, and the general results obtained below apply as well to these stable solitons.

The interaction of the slightly overlapped solitons in the unperturbed NS equation was analyzed by means of the perturbation theory in Ref.13. To obtain an effective potential of the soliton-solution interaction, it is sufficient to insert the linear superposition of the two unperturbed solitons into an exact expression for the energy of the system, and calculate a term produced by overlapping of each soliton with the "tail" of another one. It has been found[13] that the interaction potential in the unperturbed NS equation has no local minimum, so that it cannot give rise to a stable bound state of the two solitons. This inference accords with the well-known fact that the exact solution of the NS equation admits only unstable two-soliton and multi-soliton states with zero energy.

The circumstance that drastically alters the situation for the slightly perturbed equation is that the tail of the soliton (3) contains the oscillating factor $\exp(-ik|x-z_0|)$. Let us reproduce the calculation of the effective potential, taking account of this factor. We will consider the interaction of two solitons (3) with equal amplitudes η. The solitons are separated by a large distance $z = z_0^{(1)} - z_0^{(2)}$ $(z\eta \gg 1)$ and have a phase shift $\Phi = \Phi_0^{(1)} - \Phi_0^{(2)}$. The soliton-soliton interaction is accounted for by the term

$$H_{int} = -\int_{-\infty}^{+\infty} |u(x)|^4 dx \tag{6}$$

in the full Hamiltonian of the unperturbed NS equation. Following Ref.13. one inserts the linear superposition of the two slightly overlapping solitons, $u = u_1 + u_2$, into Eq.(6). It is easy to see that in the first approximation the effective potential U of the soliton-soliton interaction following from Eq.(6) after this substitution is the sum of two symmetric terms,

$$U = -4\int_{-\infty}^{+\infty} |u_1(x)|^2 \mathrm{Re}(u_1(x)u_2^*(x))\,dx \;+\; (1 \leftrightarrow 2), \tag{7}$$

where u_1 (x) is realized as the soliton wave form (3) with $z_0 = 0, \Phi_0 = 0$, and u_2 (x) is the tail of the second soliton which can be taken in the form

$$u_2(x) = 2i\eta \ \exp\ (4i\eta^2 t - 2\eta|x-z| - ik|x-z| + i\Phi). \tag{8}$$

Note that, when inserting the tail of the first soliton into the second term of the potential (7), one must change signs in front of z and Φ in Eq.(8). Subsequent straightforward calculations yield the expression

$$U = -256\eta^3 \exp(-2\eta z)\cos\Phi \cdot \cos(kz). \tag{9}$$

which coincides with the unperturbed effective potential if $K = 0$.

The potential (9) has two sets of the stationary points:

$$\cos\Phi = 0, \quad \cos(kz) = 0; \tag{10}$$

$$\sin\phi = 0, \quad \cos(kz) + (k/2\eta)\sin(kz) = 0. \tag{11}$$

The stationary states (10) are unstable (saddles) as their binding energy is exactly equal to zero, see Eq.(9). However, the states (11) are stable, provided

$\cos\Phi \cdot \cos(kz) > 0$. Thus the oscillating potential (9), unlike the unperturbed one with k = 0, gives rise to the set of stable two-soliton BS's with the distances between the solitons

$$z_n \approx (2n+1)\pi/2|k|, n = 0, 1, 2, \ldots \quad . \tag{12}$$

Note that the underlying assumption $\gamma_1, \gamma_2, \gamma_0\gamma_3 << 1$ implies $|k|/\eta << 1$ [see, e.g., Eq.(4)], so that the bound solitons are indeed slightly overlapped, as it was presumed: $\eta z_n \gg 1$ By the same reason, the binding energy E of the BS is exponentially small:

$$E_n \equiv -U(z = z_n) \approx 128|k|\eta^2 \exp[-(2n+1)\pi\eta/|k|] \tag{13}$$

Note that the potential energy was analyzed in the system which, strictly speaking, had no potential at all as it contained small dissipative terms. However, analysis of full equations of motion for the solitons' parameters, which is an exercise on the perturbation theory, leads to an effective equation of motion for a particle in the potential (9) in the presence of friction, so that the minima of the potential are stable indeed.

Alongside the two-soliton BS's, there may as well exist multi-soliton ones, and also the BS's in the form of periodic or irregular arrays of solitons, with distances z_n between neighboring solitons. The same mechanism implies a possibility of self-trapping of one soliton in a system with periodic boundary conditions due to the interaction with its own tail.

Let us proceed to the GL regime. In this case it is convenient to rewrite Eq.(2) in the form of the GL equation proper:

$$v_t = -(1-\epsilon)v + (1+i\beta)v_{xx} + (1+i\alpha)|v|^2 v - (3/16)(1+i\delta)|v|^4 v, \tag{14}$$

where α and β are small dispersive parameters, the additional one δ has been added for generality (it is implied $\alpha \sim \beta \sim \delta$), and ϵ is assumed small too. At $\alpha = \beta = \delta = \epsilon = 0$, Eq.(14) has the exact kink solution,

$$v_0(x) = 2\exp(i\Phi_0)[1 + \exp(-2\sigma(x - z_0))]^{-1/2}, \sigma = \pm 1, \tag{15}$$

where Φ_0 and z_0 are arbitrary constants, cf. Eq.(3). As it has been demonstrated in Ref.12 (see also Ref.11), when the perturbing parameters in Eq.(14) are different from zero the kink($\sigma = +1$) and antikink ($\sigma = -1$) can form a large-size SP, provided $\epsilon \sim \alpha^2$. The frequency of the SP is $\omega = 4\alpha - 3\delta$, and the local wave number in a vicinity of each kink is

$$k(x)=\sigma[A+B|v_0(x)|^2], A=2\alpha-(\beta+3\delta)/2, B=3(\beta-\delta)/16. \tag{16}$$

Thus the kink has the SP on one side of it, and on another side the wave field falls off exponentially, cf. Eg.(8) :

$$v(x)=2\exp(i\Phi_0)\exp[i\omega t-(1+iA)|x-z_0|]. \tag{17}$$

Again, Eq.(17) tells us that the small perturbing terms render the tail of the SP oscillating.

Let us now recollect that, in the case $\alpha=\beta=\delta=0$, Eq.(14) can be presented in the gradient form, $v_t=-\delta L/\delta v$, where the Lyapunov functional is

$$L=\int_{-\infty}^{+\infty} dx[(1-\epsilon)|v|^2+|v_x|^2-(1/2)|v|^4+(1/16)|v|^6]. \tag{18}$$

Stable configurations correspond to minima of L. We will consider the interaction of two SP's with a relative phase Φ , Which are separated by a large distance $z >> 1$. One can again insert the linear superposition of the two slightly overlapping SP's into Eq.(18). The lowest-order term that accounts for their interaction is [cf. Eq.(7)]

$$U=2\int_{-\infty}^{-\infty}[(3/16)|v_1(x)|^2-1]|v_1(x)|^2\mathrm{Re}(V_1(x)v_2^*(x))dx+(\leftrightarrow 2), \tag{19}$$

where $v_1(x)$ is the kink (15) with $z_0=0, \Phi_0=0$ times exp $(i\omega t)$, and $v_2(x)$ is the tail (17) of the adjacent antikink (belonging to the other SP), with $z_0=z, \Phi_0=\Phi$. The eventual result is [cf. Eq.(9)]

$$U=-32\exp(-z)\cos\Phi\cdot\cos(Az), \tag{20}$$

where A is defined by Eq.(16). Evidently, the effective pseudopotential (20) has the set of stable equilibria with $\sin\Phi=0$, $\cos(Az) + A\sin(Az) = 0$ [cf. Eq.(11)], i.e., with

$$z_n=(2n+1)\pi/2|A|, \quad n=0,1,2,... \tag{21}$$

[cf. Eq.(12)]. The "margin of stability" of the bound states, characterized by the value of the pseudopotential (20) at $z = z_n$, is again exponentially small [cf. Eq.(13)]:

$$E_n = -U(z=z_n) \approx 32|A|\exp[-(2n+1)\pi/2|A|]. \tag{22}$$

Like the solitons in the NS regime, in the GL regime the SP's may form multi-pulse BS's and periodic or irregular arrays alongside the pair SP's.
To analyze the interaction of two SP's, it was assumed z >> 1, but z need not be large as compared to the proper size 1(el) of the SP. (el)1 is large too, but it is governed by another small parameter and it is uniquely determined, 12 unlike the distance z_n which depends on the arbitrary integer n.

The analytical treatment made it possible to reveal the BS's whose "margin of stability" was exponentially narrow, see Eqs.(13) and (22). The fundamental reason for this was that the dimensionless parameters γ_1, γ_2, and $\gamma_0\gamma_3$ in the underlying Eqs.(1) and (2) had to be assumed either small or large. However, in the intermediate case $\gamma_1, \gamma_2, \gamma_0\gamma_3 \sim 1$ (when the SP's can only be investigated numerically 15) the BS's, if any, could be more robust.

In many cases, the generalized NS equation must include additional terms which account for higher dispersion. For instance, the nonlinear optical fibers usually operate in a spectral range near the zero of the second dispersion; in this case, the third dispersion must be taken into account, i.e., the term $i\zeta u_{xxx}$ with real ζ must be added to the r.h.s. of Eq.(1) 16. The term gives rise to the additional oscillating factor $\exp(iq(x-z))$, $q = 2\zeta\eta^2$, in the soliton's asymptotic (8). The crucial difference from the previous factor $\exp(ik|x-z|$ is that we have (x-z) instead of |x-z|. After straightforward calculations, one can see that all the difference introduced by the new factor is the change of $\cos\Phi$ in the potential (9) to $\cos(\Phi + qz)$. Eventually, this amounts to the fact that the value of Φ in the stationary states is determined by the equation $\sin(\Phi + qz) = 0$ instead of $\sin\Phi = 0$, see Eq.(11). The higher dispersion does not influence the stability and binding energies of the BS's; in particular, the BS's are absent if $q \neq 0$ but k = 0. The same pertains to the "skew" terms, like $i\zeta u_{xxx}$, added to the basic (this time - dissipative) part of the GL equation (14).

In conclusion, let us briefly discuss feasible experimental manifestations of the effect revealed. A plausible object that could be interpreted as a soliton in a nonlinear system combining the dispersion and dissipation is the quasi-one-dimensional (strongly stretched) localized spot of convection in a layer of a liquid crystal heated from below, discovered in Ref.17. One might try to interpret a stationary pattern of the spots observed in Ref.17 as a multi-pulse BS. Another interesting object is the localized convection pulse observed in a binary liquid filling a narrow annular channel. 18. Interaction of two pulses in this system was recently studied in Ref.19. It was demonstrated that, when the two pulses are not far from each other, they suffer a slow fusion into one pulse. It remains to understand if this interaction can be described within the framework of the approach developed in the present paper.

REFERENCES

1. N.R. Pereira and L. Stenflo, Phys. Fluids 20, 1733 (1977).
2. N.R. Pereira and F.Y.F. Chu, Phys. Fluids 22, 874 (1979).
3. A.L. Fabrikant, Zh. Eksp. Teor. Fiz. 86, 470 (1984) [JETP 59, 274 (1984)].
4. K.J. Blow, N.J. Doran and D. Wood, Opt. Lett. 12, 1011 (1987).
5. A. Höök, D. Anderson and M. Lisak, Opt. Lett. 13, 1114 (1988).
6. V.I. Petviashvili and A.M. Sergeev, Dokl. AN SSSR 276, 1380 (1984) [Sov. Phys. Doklady 29, 493 (1984)].
7. K. Stuartson and J.T. Stuart, J. Fluid Mech. 48, 529 (1971).
8. P.C. DiPrima, W. Eckhaus and L.A. Segel, J. Fluid Mech. 48, 705 (1971).
9. Y. Kuramoto and T. Tsuzuki, Progr. Theor. Phys. 55, 356 (1976); H.T. Moon, P. Huerre and L.P. Redekopp, Physica D7, 135 (1983); C.D. Doering, J.D. Gibbon, D.D. Holm and B. Nicolaenko, Nonlinearity 1, 279 (1988); B.A. Malomed and A.A. Nepomnyashchy, Phys. Rev. A 42, 6238 (1990).
10. B.A. Malomed, Physica D29, 155 (1987); S. Fauve and O. Thual, Phys. Rev. Lett. 64, 282 (1990).
11. W. van Saarlos and P. Hohenberg, Phys. Rev. Lett. 64, 749 (1990); V. Hakim, P. Jacobsen and Y. Pomeau, Europhys. Lett. 11, 19 (1990).
12. B.A. Malomed and A.A. Nepomnyashchy, Phys. Rev. A 42, 6009 (1990).
13. V.I. Karpman and S.S. Solov'ev, Physica D3, 487 (1981).
14. V.E. Zakharov and A.B. Shabat, Zh. Eksp. Teor. Fiz. 61, 118 (1971) [JETP 34, 62 (1972)].
15. O. Thual and S. Fauve, J. Phys. (Paris) 49, 1829 (1988).
16. P.K. Wai, C.R. Menyuk, Y.C. Lee and H.H. Chen, Optics Lett. 11, 464 (1986).
17. R. Ribotta and A. Joets, Phys. Rev. Lett. 60, 2164 (1988).
18. J.J. Niemela, G. Ahlers and D.S. Cannell, Phys. Rev. Lett. 64, 1365 (1990); K.E. Anderson and R.P. Behringer, Phys. Lett. A145, 323 (1990).
19. J.A. Glazier and P. Kolodner, Phys. Rev. A, to be published.

The Colour of the Force in the Renormalized Navier-Stokes Equation : A Free Parameter?

D. Carati,

Physique Statistique, Plasmas, et Optique non Linéaire, CP 231,
Association Euratom-Etat Belge,
Campus Plaine, Université Libre de Bruxelles. 1050 Bruxelles, Belgium.

Abstract

The renormalization group is applied to the Navier-Stokes equation for randomly forced media. Contrary to the previous works on the subject, the stochastic forcing is not assumed to be a white noise (the Galilean invariance can then be broken). The influence of deviations from the white noise on the renormalized hydrodynamic equations is estimated.

1. Introduction

When the renormalization group (RG) is used to describe fully developed turbulence, it is usual to introduce a stochastic noise in the Navier-Stokes equation.[1,2] The assumption of similarity between the experimentally observed turbulence and the turbulence generated by the hydrodynamic equations including a noise represents then a basic ingredient of the present application of RG.[3,4] Even if it is not supported by any theoretical demonstration, this correspondance principle has been numerically tested with a certain success.[5] Many other arguments are put forward to justify this method. For example, the stochastic force is considered as the source of energy necessary to entertain turbulence. Moreover, the introduction of suitable stochastic forcings does not break the symmetries of turbulent flows. On the contrary, deterministic sources (pressure gradient, gravity field, ...) can break large scale properties such as isotropy and homogeneity. Nevertheless, the main justification to introduce such a stochastic noise is much more simple: at the present time, no way has been found to apply RG techniques to the deterministically forced Navier-Stokes equation. It is then very important to evaluate the influence of the properties of the noise on the results obtained by the RG approach of fully developed turbulence. In this work we investigate the relation between the temporal behaviour of the noise correlations and the value of some typical constants of turbulence such as the Kolmogorov constant and the Prandtl number.

2. The RG technique

Let us start by a brief presentation of the RG approach of turbulence. As many papers have been devoted to this subject,[1-4] we only present the main steps of the method. In Fourier space, the Navier-Stokes equation reads:

$$v_l(\hat{\mathbf{k}}) = G(\hat{\mathbf{k}})\; f_l(\hat{\mathbf{k}}) - \frac{i\lambda}{2}\; G(\hat{\mathbf{k}})\; P_{lmn}(\vec{\mathbf{k}}) \int \frac{d\hat{\mathbf{q}}}{(2\pi)^{d+1}}\; v_m(\hat{\mathbf{k}} - \hat{\mathbf{q}})\; v_n(\hat{\mathbf{q}}) \tag{1}$$

where

$$P_{imn}(\vec{k}) = k_m \; P_{in}(\vec{k}) + k_n \; P_{im}(\vec{k})$$
$$P_{ij}(\vec{k}) = \delta_{ij} - k_i \; k_j/k^2$$
$$\hat{k} = (\omega, \vec{k})$$
$$G = (-i\omega + \nu \; k^2)^{-1} \; .$$

The stirring force $\mathbf{f}$ is usually assumed to be Gaussian. The parameter λ is introduced for further convenience and will be used as a perturbation parameter. Using the properties of the system such as isotropy, homogeneity and incompressibility, the correlations of $\mathbf{f}$ are nearly completely specified:

$$\left\langle \; f_i(\hat{k}) \; f_j(\hat{k}') \; \right\rangle = \frac{2D_0(2\pi)^{2d+1}}{S_d} \; k^{-y} \left(\frac{|\omega| \; k^{(y-d-2)/3}}{D_0^{1/3}} \right)^{\alpha} \delta(\hat{k} + \hat{k}') \; P_{ij}(\vec{k}) \tag{2}$$

where D_0 is an undetermined constant and both y and α remain free parameters at this stage. The numerical constant $(2\pi)^{2d+1}/S_d$ (S_d = the area of a d-dimensional unit sphere) is introduced to simplify further notations. The parameter α characterizes the temporal dependence of the noise correlations. It is usually assumed to be equal to zero. In this case $\mathbf{f}$ is δ-correlated in time and the stochastic force is a white noise. Our goal is to investigate the influence of deviations from white noise characterized by non-zero values of the parameter α.

The main steps of the RG method are:

First, the wavenumbers $e^{-r}\Lambda < k < \Lambda$ corresponding to the more fluctuating phenomena are eliminated from the equation. The cutoff Λ represents the larger wavenumber which has to be taken into account in the system. The parameter r characterizes the fraction of eliminated wavenumbers. Mathematically, this elimination procedure is performed by splitting the velocity field into two different components $v(\hat{k}) = v^{>}(\hat{k}) + v^{<}(\hat{k})$. Following the range of wavenumbers, we define: $v^{>}(\hat{k})$ if $e^{-r}\Lambda < k < \Lambda$ and $v^{<}(\hat{k})$ if $k < e^{-r}\Lambda$. By introducing these new variables into the Navier-Stokes equation, we obtain two similar and coupled equations for $v^{>}(\hat{k})$ and $v^{<}(\hat{k})$. The goal of the procedure is to remove $v^{>}(\hat{k})$ from the equation for $v^{<}(\hat{k})$. This is done by using a perturbation scheme where λ plays the role of the expansion parameter. As a matter of fact, the effective expansion parameter (usually noted $\overline{\lambda}$) is proportional to λ and is given by:

$$\overline{\lambda}^2 = \lambda^2 \left(\frac{D_0 \; \Lambda^{d-y-4}}{\nu^3} \right)^{1-\alpha/3} \tag{3}$$

Secondly, the lengths are rescaled so that the initial value of the cutoff is restored. This is done in order to allow the comparison between the equation obtained after the small scales elimination and the initial Navier-Stokes equation. Indeed, following the arguments of Ma,[6] different equations can only be compared if they describe the phenomena with the same degree of precision. In the present work, the degree of precision is clearly given by the cutoff.

The rescaling of the cutoff leads in turn to the rescaling of all the variables:

$$\begin{aligned} \omega(e^{-r}\Lambda) &\rightarrow e^{-zr}\omega(\Lambda) \\ v(e^{-r}\Lambda) &\rightarrow e^{\gamma r} v(\Lambda) \end{aligned} \tag{4}$$

At this stage, the rescaling dimensions for the frequency and the velocity (z and γ) remain undetermined parameters. The rescaling dimensions for the expansion parameter, the viscosity and the forcing are determined in order to preserve the structure of the Navier-Stokes equation:

$$\begin{aligned}
&\lambda(e^{-r}\Lambda) = e^{(3-\alpha)\ (z+(y-d-2)/3)r/2}\ \lambda(\Lambda) \\
&\nu(e^{-r}\Lambda) \to e^{(-z+2)r}\ \nu(\Lambda) \qquad (5)\\
&f(e^{-r}\Lambda) \to e^{(\gamma-z)r} f(\Lambda)
\end{aligned}$$

Remarkably, up to the second order in the $\overline{\lambda}$ expansion, these operations (elimination/rescaling) do not change the structure of the equation in the limit of small wavenumbers. In this limit, the equation for $v^{<}(\hat{\mathbf{k}})$ becomes the renormalized Navier-Stokes equation with modified values for both the viscosity and the expansion parameter λ:

$$(-i\omega + \nu(\Lambda)k^2)\ v_l(\hat{\mathbf{k}}) = f_l(\hat{\mathbf{k}}) - \frac{i\lambda(\Lambda)}{2}\ P_{lmn}(\vec{\mathbf{k}}) \int \frac{d\hat{\mathbf{q}}}{(2\pi)^{d+1}}\ v_m(\hat{\mathbf{k}} - \hat{\mathbf{q}})\ v_n(\hat{\mathbf{q}}) \qquad (6)$$

This complete procedure is iterated a large number of times in order to eliminate a large amount of wavenumbers. Considering infinitesimal variations of the cutoff at each step of this iteration leads to similarly infinitesimal variations of both ν and λ. We then obtain two differential equations relating the viscosity and the expansion parameter to the parameter r.

$$\frac{d\nu}{dr} = \nu\big((z-2) + B_d^\alpha \overline{\lambda}^2\big) + O(\overline{\lambda}^4) \qquad (7)$$

$$\frac{d\lambda}{dr} = \lambda\Big(\frac{3-\alpha}{2}\big(z + \frac{y-d-2}{3}\big) + \frac{2\alpha c_1(\alpha)}{d^2+2d}\overline{\lambda}^2 + O(\overline{\lambda}^4)\Big) \qquad (8)$$

where $c_1(\alpha) = \dfrac{1}{\pi}\displaystyle\int_{-\infty}^{+\infty} dx\ \frac{|x|^\alpha}{(1+x^2)^2} = \frac{1-\alpha}{2\sin\big(\pi(\alpha+1)/2\big)}\,, \qquad -1 < \alpha < 3$

$$B_d^\alpha = c_1(\alpha)\ \frac{d^2 - y - 4 - (d-y+2)\alpha/3}{d(d+2)}$$

Since the first paper of Forster, Nelson and Stephen, it is known that the renormalization procedure leads to corrections to the force. These corrections are proportional to $k^2\omega^0$. They are irrelevant in the large scale and long time limit if $y > -2$ and $\alpha < 0$. To avoid any renormalization of the forcing, we limit our investigation to negative values of the parameter α. Moreover, to ensure the convergence of some integrals which appear in the elimination procedure, we have to require $\alpha > -1$. Taking into account all these conditions, we restrict the parameter α to the values $-1 < \alpha < 0$. In this case, the forcing amplitude D_0 is not modified by the small scales elimination. By assuming that D_0 is not altered by the rescaling either, one can relate the rescaling dimension γ to the other parameters:

$$\gamma = \frac{3-\alpha}{2}\ z + \frac{d+y}{2} - \frac{(y-d-2)\alpha}{6} \qquad (9)$$

We have already taken into account this relation in the equations (7) and (8).

The differential equations (7) and (8) give the variation of ν and λ as an expansion in $\overline{\lambda}$. The lowest order terms originate from the rescaling of the variable given by (5) and are not influenced by the $\overline{\lambda}$-expansion. The other terms are generated by the small scale elimination. It should be noted that, up to order two in $\overline{\lambda}$, the elimination of large wavenumbers does not affect the value of λ if the forcing is a white noise ($\alpha = 0$). This is a consequence of the Galilean invariance when $\alpha = 0$ and can be easily verified in equation (8).

From the definition (3) and the equations (7) and (8), we can deduce the asymptotic behaviour of the effective expansion parameter. It is easy to show that for large r, $\overline{\lambda}^2 \sim \epsilon = y + 4 - d$. The $\overline{\lambda}$−expansion is then equivalent to an ϵ−expansion. Let us stress that the value of ϵ is completely independent of the colour of the noise α. Therefore, the use of colored noises does not affect the convergent or divergent nature of the $\overline{\lambda}$−expansion.

3. The energy spectrum

The Kolmogorov-Obukhov spectrum[7] constitutes a very characteristic property of the hydrodynamic turbulence. For this reason, it often plays a major role when comparing the theory and the experimental results. It is thus important to evaluate the energy spectrum compatible with the renormalized Navier-Stokes equation. This spectrum is defined by:

$$E(k) = \frac{1}{2} S_d k^{d-1} \int \frac{d\omega}{2\pi} \int \frac{d\hat{\mathbf{q}}}{(2\pi)^{d+1}} < v_i(\hat{\mathbf{k}}) v_i(\hat{\mathbf{q}}) > \qquad (10)$$

To lowest order in $\overline{\lambda}$, one obtains:

$$\begin{aligned} E(k) &= \frac{1}{2} S_d k^{d-1} \int \frac{d\omega}{2\pi} \int \frac{d\hat{\mathbf{q}}}{(2\pi)^{d+1}} G(\hat{\mathbf{q}}) G(\hat{\mathbf{k}}) \; < f_i(\hat{\mathbf{k}}) f_i(\hat{\mathbf{q}}) > \\ &= \frac{(d-1)\, c_2(\alpha)}{2} \left(\frac{\epsilon}{3 B_d^{\alpha}} \right)^{\frac{1-\alpha}{3-\alpha}} D_0^{2/3} \, k^{1-2\epsilon/3} \end{aligned} \qquad (11)$$

where $\quad c_2(\alpha) = \frac{1}{\pi} \int_{-\infty}^{+\infty} dx \, \frac{|x|^{\alpha}}{(1+x^2)} = \frac{1}{2 \sin\left(\pi(\alpha+1)/2\right)} \qquad -1 < \alpha < 1$

This result is compatible with the Kolmogorov-Obukhov spectrum ($E(k) = C_K \, \overline{\varepsilon}^{2/3} k^{-5/3}$ where C_K is the Kolmogorov constant and $\overline{\varepsilon}$ the dissipation rate of energy) if $\epsilon = 4$. The use of $\epsilon = 4$ as a small parameter is the major difficulty of the RG approch of turbulence. Nevetheless, as we focus our attention on the influence of colored noises on the RG results, we do not discuss the problem of the convergence of the ϵ−expansion. Indeed, we noted at the end of the section 2 that the parameter α does not affect the convergence property of this expansion.

At this stage, the Kolmogorov constant remains an undetermined parameter. Indeed, as long as the forcing amplitude (D_0) is not related to the dissipation rate of energy, it is not possible to obtain a numerical value for C_K. This problem has been solved by Dannevik, Yakhot and Orszag.[8] Following their work, it is possible to derive a closed equation for the energy from the renormalized Navier-Stokes equation (6). This equation is exactly the same as the one deduced from the Eddy Damped Quasi Normal Markovian approximation (EDQNM):[9]

$$\frac{\partial E(k,t)}{\partial t} = T(k,t) - 2\nu k^2 E(k,t) + O(\lambda^3) \tag{12}$$

The term $T(k,t)$ represents the variation of energy which originates neither from an external source nor from the viscous effects. This term can then be associated to the transfer of energy from the large scales to the small ones. That is why it is named the transfer term. As the transfer effects are due to the non-linearity of the Navier-Stokes equation, it is not surprising that $T(k,t)$ is also non-linear:

$$T(k,t) = C_d \iint_{p-k<q<p+k} dq\, dp \left(\frac{\sin\zeta}{k}\right)^{d-3} \frac{k}{pq} \Big(a^d_{kpq}\, \theta^\alpha_{kpq}\, k^{d-1}\, E(p,t)\, E(q,t) \\ - b^d_{kpq}\, \theta^\alpha_{pkq}\, p^{d-1}\, E(q,t)\, E(k,t) \Big) \tag{13}$$

where $C_d = 4S_{d-1}/S_d(d-1)^2$ and ζ represents the angle between $\vec{\mathbf{q}}$ and $\vec{\mathbf{p}}$ in the triad $\{\vec{\mathbf{k}},\ \vec{\mathbf{q}},\ \vec{\mathbf{p}} = \vec{\mathbf{k}} - \vec{\mathbf{q}}\}$. The functions a and b are defined as follows:

$$a^d_{kpq} = \frac{1}{4k^2}\, P_{lmn}(\vec{\mathbf{k}})\, P_{mi}(\vec{\mathbf{p}})\, P_{nj}(\vec{\mathbf{q}})\, P_{lij}(\vec{\mathbf{k}})$$

$$b^d_{kpq} = \frac{1}{2k^2}\, P_{lmn}(\vec{\mathbf{k}})\, P_{nli}(\vec{\mathbf{p}})\, P_{mi}(\vec{\mathbf{q}})$$

and the time θ_{kpq} is given by:

$$\theta_{kpq} = \int_0^t ds\, e^{-\nu k^2(t-s)} \frac{U(p,t,s)U(q,t,s)}{U(p,t,t)U(q,t,t)} \tag{14}$$

The function $U(k,t,s)$ is related to the covariance of the velocity field:

$$< v_i(\vec{\mathbf{k}},t)\; v_j(\vec{\mathbf{q}},t') >= \frac{U(k,t,t')}{d-1}\, P_{ij}(\vec{\mathbf{k}})\, \delta(\vec{\mathbf{k}}+\vec{\mathbf{q}}) \tag{15}$$

To lowest order in the $\overline{\lambda}$-expansion, $U(k,t,s)$ and consequently θ_{kpq} can be directly deduced from the forcing correlations (2). In case of white noise forcing ($\alpha = 0$), the calculations can be analytically performed and lead to:

$$\theta_{kpq} = \frac{1}{\nu k^2 + \nu p^2 + \nu q^2} \tag{16}$$

We then obtain the same result as in the EDQNM approximation. Unfortunatly, if $\alpha \neq 0$, it is no longer possible to obtain the exact structure of θ_{kpq}. An approximation can however be deduced:

$$\theta^\alpha_{kpq} \approx \frac{1}{\nu k^2 + (1+\alpha)\nu p^2 + (1+\alpha)\nu q^2} \tag{17}$$

In the Kolmogorov-Richardson cascade, the transfer rate of energy is equal to the dissipation rate of energy. This leads to the relation:

$$\overline{\varepsilon} = -\frac{\partial}{\partial t} \int_0^k dk' \; E(k') \Bigg|_{\nu=0} = -\int_0^k dk' \; T(k') \tag{18}$$

By inserting (11), (17) and the asymptotic value of the renormalized viscosity into (18), we obtain a simple linear relation between the forcing amplitude and the dissipation rate of energy:

$$\overline{\varepsilon} = \frac{r_1(d,\alpha)\,(d-1)^2\; c_2(\alpha)^2}{4} \left(\frac{\epsilon}{3B_d^\alpha}\right)^{\dfrac{3-2\alpha}{3-\alpha}} D_0 \tag{19}$$

where $r_1(d) = C_d \int_0^K dk \iint_{p-k<q<p+k} dq\; dp \left(\frac{\sin\zeta}{k}\right)^{d-3} \frac{k}{pq}$

$$\times \left(a_{kpq}^d \frac{1}{k^{2/3} + (1-\alpha)p^{2/3} + (1-\alpha)q^{2/3}} k^{d-1}\; p^{-5/3}\; q^{-5/3} \right.$$

$$\left. -b_{kpq}^d \frac{1}{(1-\alpha)k^{2/3} + (1-\alpha)p^{2/3} + q^{2/3}} p^{d-1}\; q^{-5/3}\; k^{-5/3} \right)$$

From this relation, the Kolmogorov constant is completely determined. Its value as a function of the parameter α is plotted in the Fig.1.

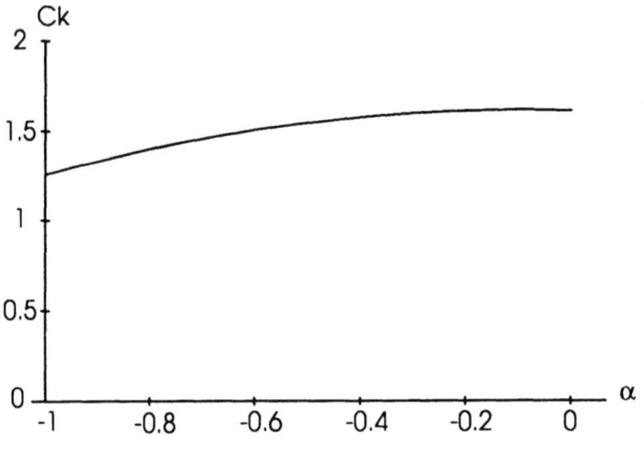

Figure 1: The Kolmogorov constant as a function of the parameter α. The experimentally observed values of C_K lie in the range $1.4 < C_K < 1.7$.

4. The temperature equation

The RG can also be applied to the evolution equation for a passive scalar like the temperature. The complete procedure of small scales elimination and rescaling of the variable is similar to that described in the section 2. In this case, a differential equation relating the thermal conductivity to the parameter r is obtained ($P = \nu/\kappa$ represents the Prandtl number):

$$\frac{d\kappa}{dr} = \kappa(z-2) + \nu\,\overline{\lambda}^{2}\,\frac{1-P^{1-\alpha}}{1-P^2}\,P\,C_d^\alpha + O(\overline{\lambda}^{4}) \tag{20}$$

where $C_d^\alpha = c_2(\alpha)\,d/(d-1)$

A differential equation for the Prandtl number can be constructed from (7) and (20).

$$\frac{dP}{db} = P\,\overline{\lambda}^{2}\left(B_d^\alpha - \frac{P^2(1-P^{1-\alpha})}{1-P^2}\,C_d^\alpha\right) \tag{21}$$

The renormalized value of P is then given by the fixed point of (21) and is represented in Fig. 2.

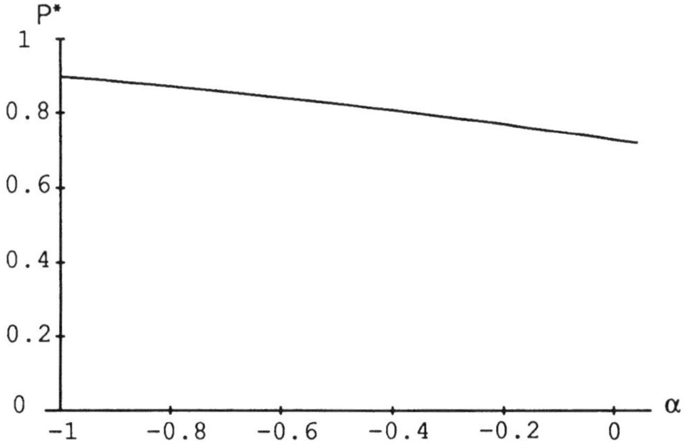

Figure 2: Renormalized Prandtl number (P^*) as a function of the parameter α. The value of P^* does not depend crucially of α and remains close to the experimental values of the turbulent Prandtl number ($0.7 < P_t < 0.9$).

5. Conclusion

We have shown in this work that the RG can be applied to the Navier-Stokes equation describing media forced by colored noises. The influence of the deviations from the white noise (characterized by the parameter α) is very small. Even if the numerical values of the Kolmogorov constant and the Prandtl number depend on α, this dependence is weak (see Fig.1 and 2). Moreover, due to the scattering of the experimental measures, it is not possible to select a particular value of α which would give the best fit to the data.

Other characteristic constants of fully developed turbulence (Batchelor constant, $K-\epsilon$ viscosity amplitude, or Smagorinsky constant) can also be obtained by the RG.[3] They are also very weakly dependent on α.[10]

Finally, let us stress again the fact that the colour of the noise does not affect the divergent or convergent nature of the ϵ-expansion. Indeed, the asymptotic value of the effective expansion parameter is always proportional to $\epsilon^{1/2}$ and ϵ is independent of α. Moreover, to recover the Kolmogorov spectrum it is necessary to assume $\epsilon = 4$ independently of the value of α.

References

1 D. Forster, D.R. Nelson, and M.J. Stephen, Phys. Rev. A **16**, 732 (1977)

2 J.D. Fournier and U. Frisch, Phys. Rev. A **28**, 1000 (1983); J.D. Fournier Phd. Thesis, Nice (1983)

3 V. Yakhot and S.A. Orszag, J. Sci. Comp. 1, 3 (1987)

4 V. Yakhot and S.A. Orszag, Phys. Rev. Lett. 57, 1722, (1986)

5 R. Panda, V. Sonnad, E. Clementi, S.A. Orszag and V. Yakhot, Phys. Fluids **1**, 1045 (1989)

6 S.K. Ma, Rev. Mod. Phys. 45, 589 (1973)

7 A.N. Kolmogorov, Dokl. AN USSR, **30**, 299 (1941); A. M. Obukhov, Izves. AN USSR, **4**, 453 (1941).

8 W.P. Dannevik, V. Yakhot and S.A. Orszag, Phys. Fluids **30**, 2021 (1987).

9 S.A. Orszag, J. Fluid Mech. **41**, 363 (1970).

10 D. Carati, Phd Thesis, Bruxelles (1991)

Structure of homogeneous turbulence observed in a direct numerical simulation

M. Meneguzzi and A. Vincent
CERFACS, 42 av. Gustave Coriolis, 31000 Toulouse, France

Abstract

Direct numerical resolution of the three-dimensional Navier-Stokes equation on a 240^3 *grid is used to obtain a three-dimensional homogeneous turbulent flow at Reynolds numbers* $R_L \approx 1000$ *and* $R_\lambda \approx 150$*, based respectively on integral scale* L *and Taylor microscale* λ*. An inertial subrange is present in the energy spectrum over more than one decade in wavenumbers. Visualization of the flow confirms that the vorticity is organized in very elongated thin tubes. These tubes seem to originate in shear instabilities or in merging of previously formed tubes.*

1. Introduction

Direct simulations of homogeneous turbulent flow at the highest presently reachable resolution are useful to gather some information on the statistical properties of such a flow and to explore its structure in space. Indeed, these "numerical experiments" allow measurement of many quantities unaccessible in the laboratory and visualization of the small scales structures of the flow. It becomes feasible, on computers of the last generation, to reach Reynolds numbers at which a genuine inertial subrange shows up in the energy spectrum (Kerr 1985, Yamamoto and Hosokawa 1988, She et al. 1988, 1990). The main goal of the present calculation was to obtain a more extended inertial subrange than in previous work, and to concentrate on the inertial domain properties.

2. The calculation

We solve the Navier-Stokes equation for incompressible fluids in rotational form

$$\frac{\partial \mathbf{v}}{\partial t} = \mathbf{v} \times \omega - \nabla(p + v^2/2) + \nu \nabla^2 \mathbf{v} + \mathbf{f} \tag{1}$$

with the continuity equation $\nabla \cdot \mathbf{v} = 0$. $\mathbf{v}$ is the velocity field, $\omega = \nabla \times \mathbf{v}$ the vorticity, p the pressure, ν the kinematic viscosity and $\mathbf{f}$ a force field. Since we are interested in (statistically) homogeneous turbulent flows, we take periodic boundary conditions in all directions. In Fourier space the two equations can be combined to give

$$\frac{\partial \mathbf{v_k}}{\partial t} = \mathbf{P}(\mathbf{k}) \cdot (\mathbf{v} \times \omega)_\mathbf{k} - \nu k^2 \mathbf{v_k} + \mathbf{f_k} \tag{2}$$

. Paper presented at the workshop on large-scale structures in nonlinear physics, Villefranche-sur-Mer, France, 14-18 Jan. 1991 where the tensor $\mathbf{P}$ is the projector on the space of solenoidal fields, defined as $P_{ij}(\mathbf{k}) = \delta_{ij} - \frac{k_i k_j}{k^2}$. A pseudospectral method is used to compute the right-hand side of this equation (see Gottlieb and Orszag 1977). The time marching is done using a second order finite-difference scheme. An Adams-Bashfort scheme is used for the non linear term while the dissipative term is integrated exactly. The resulting numerical scheme is

$$\frac{\mathbf{v_k}^{n+1} - \mathbf{v_k}^n e^{-\nu k^2 \delta t}}{\delta t} = \mathbf{P}(\mathbf{k}) \cdot \left[\frac{3}{2} (\mathbf{v} \times \omega)_{\mathbf{k}}{}^n e^{-\nu k^2 \delta t} - \frac{1}{2} (\mathbf{v} \times \omega)_{\mathbf{k}}{}^{n-1} e^{-2\nu k^2 \delta t} + \mathbf{f_k}{}^n e^{-\nu k^2 \delta t} \right].$$

To start (or restart) the calculation, we use a second order Runge-Kutta scheme.

We force the field at low wavenumbers in a deterministic way. All Fourier modes with $k < 1.5$ are forced with a constant amplitude $\mathbf{f}$ independent of $\mathbf{k}$ (the wavenumber $\mathbf{k}$ has integer components because the space period is 2π). This results in large fluctuations of the energy injection rate $\epsilon_i = < \mathbf{f} \cdot \mathbf{v} >$. One can only hope to reach a steady-state regime in the sense that ϵ_i fluctuates in time around a constant value.

The calculation presented here was done with 240^3 Fourier components, and a viscosity of 10^{-3}. One time-step takes 12 seconds on a Cray-2, using the 4 processors. Integration over one turnover time (defined below) takes of the order of 4 hours of Cray-2.

To estimate the degree of isotropy of the flow, we use the same method as Curry et al. (1984). For each wavenumber $\mathbf{k}$, we define two unit vectors $\mathbf{e_1}(\mathbf{k})$ and $\mathbf{e_2}(\mathbf{k})$ which form with $\mathbf{k}$ an orthogonal reference frame. Since $\mathbf{k} \cdot \mathbf{v} = 0$, each Fourier mode $\mathbf{v}(\mathbf{k})$ is defined by its two components in this frame $v_1(\mathbf{k})$ and $v_2(\mathbf{k})$. We define the isotropy I as $I = \left[\frac{<|v_1|^2>}{<|v_2|^2>} \right]^{1/2}$. In the calculations presented here, I fluctuates by a few percent around 0.95. Therefore, our flow is close to statistically isotropic.

Let us recall the definition of some characteristic quantities used in the following. Three characteristic lengths are used, the integral scale $L = \frac{\int_0^\infty k^{-1} E(k) dk}{\int_0^\infty E(k) dk}$, the Taylor microscale $\lambda = \left[\frac{\int_0^\infty E(k) dk}{\int_0^\infty k^2 E(k) dk} \right]^{\frac{1}{2}}$ and the Kolmogorov dissipation scale $l = \left(\frac{\nu^3}{\epsilon} \right)^{\frac{1}{4}}$, where ϵ is the mean energy dissipation rate per unit mass. The two characteristic time-scales of homogeneous turbulence are the eddy turnover time $\tau_0 = L/v_0$ where v_o is the root mean square velocity, and the dissipation time $\tau_\nu = L^2/\nu$. With these quantities one can define two Reynolds numbers, the integral scale Reynolds number $R_L = v_0 L/\nu$ and the Taylor microscale Reynolds number $R_\lambda = v_0 \lambda / \nu$.

3. The spectra of the flow

After integration over several tens of turnover times we obtain a statistically stationary regime in the above sense. The energy spectrum, shown in figure 1, displays a power law range for $k < 30$, with an exponent a little larger than -5/3. The fact that this is an inertial range is confirmed by inspection of the energy flux spectrum $\Phi(k) = \int_k^\infty T(k) dk$ where $T(k)$ is the energy transfer at wavenumber k. This function is found constant for k in the inertial range, as expected. After reaching this stationary regime, we integrate for 30 more turnover times in order to accumulate some statistics. The energy spectrum does not vary significantly during this period.

By fitting the spectrum with the Kolmogorov form $E(k) = C_K \epsilon^{2/3} k^{-5/3}$ we obtain a value of 2 for the Kolmogorov constant C_K, which does not vary by more than 3% in the statistically steady-state period. This value of C_K is a little larger than the experimental value 1.5 (see Monin and Yaglom 1975 for references). Our Reynolds numbers are $R_L \approx 1000$ and $R_\lambda \approx 150$.

4. The spatial structure of the flow

Figure 2 shows a 3D picture of the vorticity field. The vorticity at each grid point is represented by a vector, here so small that individual vectors can hardly be seen. Vectors are only plotted if their modulus is larger than a given threshold. By varying this threshold and rotating the figure on a graphic workstation screen we can explore the structure of the field in detail. One can see that the vorticity is organized in thin elongated tubes, as previously reported by Siggia (1981), Kerr (1985), She et al. (1990). Figure 3 shows the effect of lowering the threshold, and therefore letting smaller amplitude vorticity vectors appear. The length of these tubes can reach the integral scale L (the cube size is 2π). Their thickness is of the order of a few dissipation scales, here a few grid points. This is confirmed by a more detailed analysis. The dissipation scale l, in our simulation, is of the order of the mesh size, while the Taylor microscale is approximately ten mesh sizes. The characteristic tube thickness is a few dissipation lengths. Figure 4 shows a detailed view of a vorticity tube. It displays a sub-cube one sixth the size of the complete one, with 40 grid points on each side. Similar plots for the velocity field show mainly the forcing field if one uses a high threshold. But when the highest velocities are eliminated, the tubes are clearly visible.

From these visualizations, one is lead to the conclusion that the vorticity tubes, which seem to be the basic structure of three-dimensional homogeneous turbulence, involve all the scales of the flow. We have done the same kind of pictures of the vorticity field after removing all dissipation range scales. A smooth filter is applied in Fourier space in order to avoid spurious fluctuations. The large scales to which the forcing is applied were also removed. Figure 5 shows the same sub-cube as figure 4 when only inertial range scales are left. One can see the external regions of the tubes. Some helical structure is observed, as noted by She et al. (1990).

We have examined the shape of many of these tubes in order to see wether they are in fact rolled-up vorticity sheets, as suggested by Lundgren (1982), but this does not seem to be the case. Figure 6 gives an example of the projection of the velocity field on a plane perpendicular to a vorticity tube, while figure 7 shows the curves of constant vorticity on such a plane. As one can see on these figures, there is little evidence of spiral structure, or any other two dimensional structure.

In order to see how these tubes are produced, we have followed some of them back in time using a visualization sofware. What we find generally is that the largest vorticity tubes are the result of the merging of previous smaller tubes. We then follow the evolution of these smaller tubes back in time, we sometimes find two or even three tubes almost parallel, like in the example given in figure 8. Inspection of the associated velocity field shows that these tubes belong to one and the same shear zone (fig. 9). This strongly suggests that they have been produced by a shear instability. Unfortunately, this type of instability is difficult to observe in a forced turbulence calculation, because small vorticity regions fall below the amplitude threshold used to visualize the vorticity field, dominated by more intense features. The shear instabilities producing the vorticity tubes are easier to study in a decay calculation, with purely large-scale initial conditions. Indeed, Brachet et al. (1983) have seen the formation of vorticity tubes by what seems to be a Kelvin-Helmholtz type of instability. We are presently analysing a 256^3 decay calculaton with random initial conditions, in order to investigate this aspect of the dynamics.

6. Conclusion

We have obtained a turbulent homogeneous flow at $R_\lambda \approx 150$ and an inertial subrange more extended than in previous tridimensional simulations.

In physical space, we confirm the observation by Siggia (1981) that the vorticity is organized in thin tubes. We find that the thickness of these tubes is a few dissipation scales while their length is comparable to the integral scale of the flow. Therefore, the picture of intermittency that emerges in physical space from our simulations is that there exist phase relationships between all the scales present in the flow, due to these elongated structures. The study of the time history of these vorticity tubes reveals the frequent occurrence of vortex tube merging. Initial production of vorticity tubes seems to be due to shear instabilities. A more detailed study of these instabilities is under way.

REFERENCES

Ashurst W.T., Kerstein A.R., Kerr R.M., Gibson C.H. (1987), Phys. Fluids **30**, 2343.

Anselmet F., Gagne Y., Hopfinger E.J., Antonia R.A.(1984), J. Fluid Mech. **140**, 63.

Benzi R., Patarnello S., Santangelo(1987), ICE-0015 to appear in J.Phys. A.

Brachet M.E., Meiron D.I., Orszag S.A., Nickel B.G., Morf R.H., Frisch U (1983), J. Fluid Mech. **130**, 411. Small-scale structure of the Taylor-Green vortex.

Curry J.H., Herring J.R., Loncaric J., Orszag S.A. (1984), J. Fluid Mech. **147**,1.

Frisch U., Sulem P.L., Nelkin M (1978), J. Fluid Mech **87**, 719.

Gottlieb D., Orszag S.A. (1977), CBMS-NSF Regional Conference Series in Aplied Math., **26**. *"Numerical Analysis of Spectral Methods : Theory and Applications."*

Kerr R. (1985), J. Fluid Mech. **153**, 31.

Kolmogorov A.N. (1962), J. Fluid Mech. **13**, 82.

Lundgren T.S. (1982), Phys. Fluids **25**, 2193.

Monin A.S., Yaglom A.M. (1975), Statistical Fluid Mechanics, Vol 2, MIT Press.

Oboukhov A.M. (1962), J. Fluid Mech. **13**, 77.

She Z.S., Jackson E., Orszag S.A. (1988), J. Sci. Computing, **3**, 407.

She Z.S., Jackson E., Orszag S.A. (1990), Nature, **344**, 226.

Siggia E.D. (1981) J. Fluid Mech. **107**, 375.

Yamamoto K., Hosokawa I. (1988), J. Phys. Soc. Japan **57**, 1532.

Van Atta C.W., Chen W.Y. (1970), J. Fluid Mech. **44**, 145.

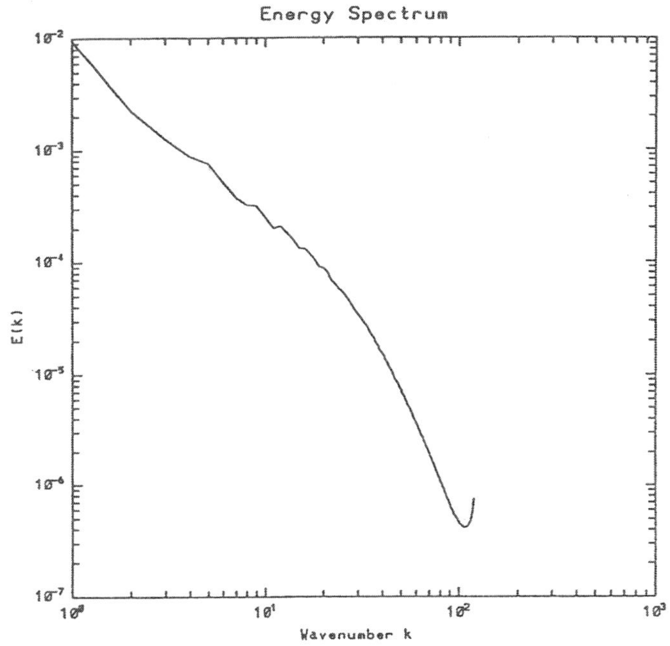

Figure 1: Energy spectrum averaged over 10 different times.

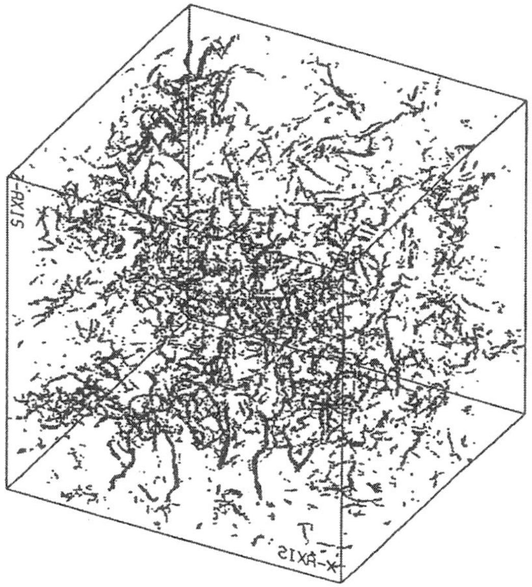

Figure 2: View of the vorticity field, represented by a vector of length proportional to the vorticity amplitude at each grid point. Only vectors larger than a given threshold value are shown.

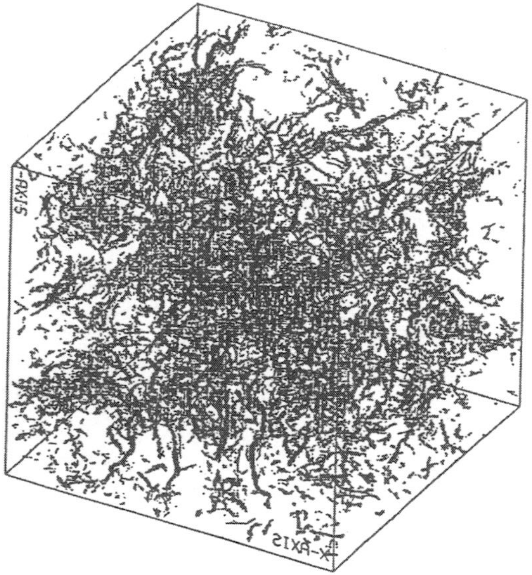

Figure 3: Same as figure 7, but with a lower threshold value, and therefore more vectors represented.

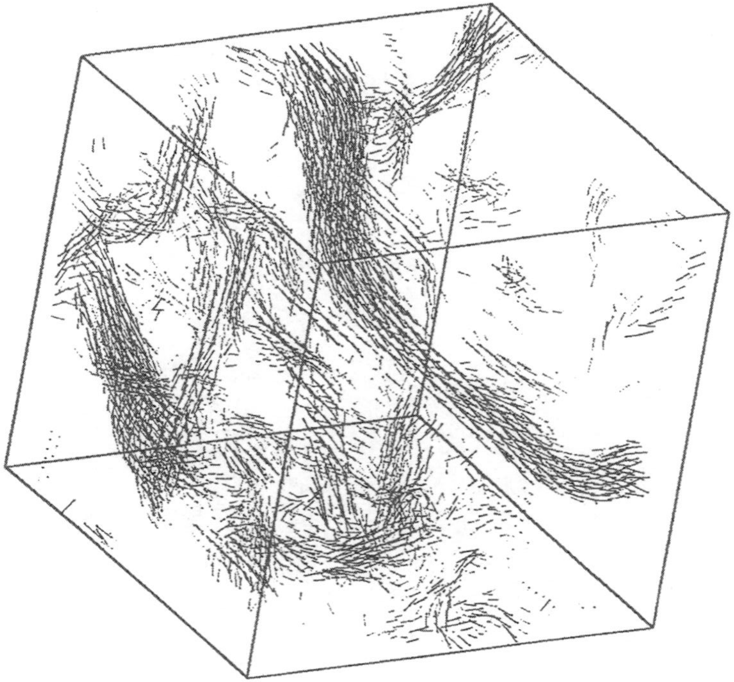

Figure 4: Detail of the vorticity field, showing a sub-cube of size one sixth of the complete cube, i.e. 40 grid points in each direction. The color are associated with vorticity amplitude. The red vectors are larger than the blue ones.

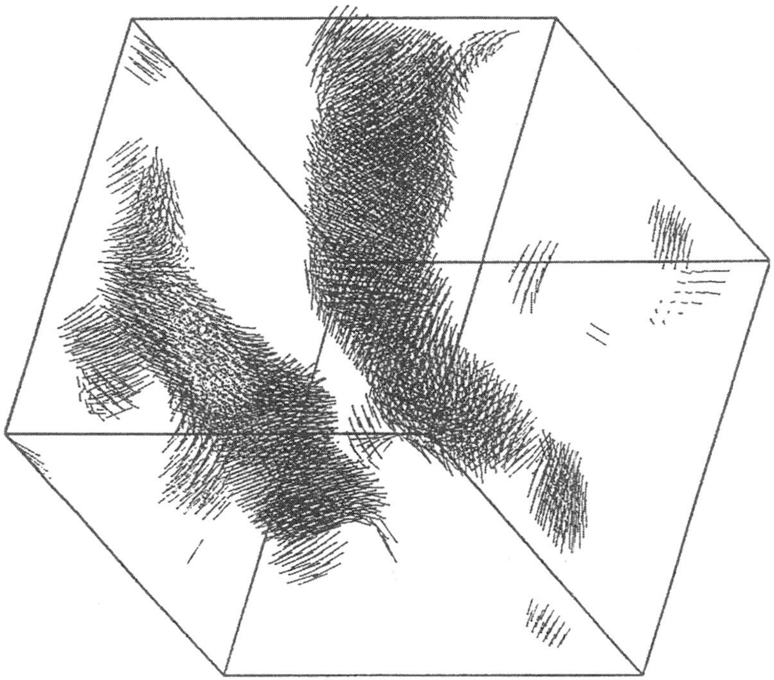

Figure 5: Same vorticity tube as in figure 9, but showing only inertial sub-range scales. Dissipation and energy injection scales have been removed by filtering in Fourier space.

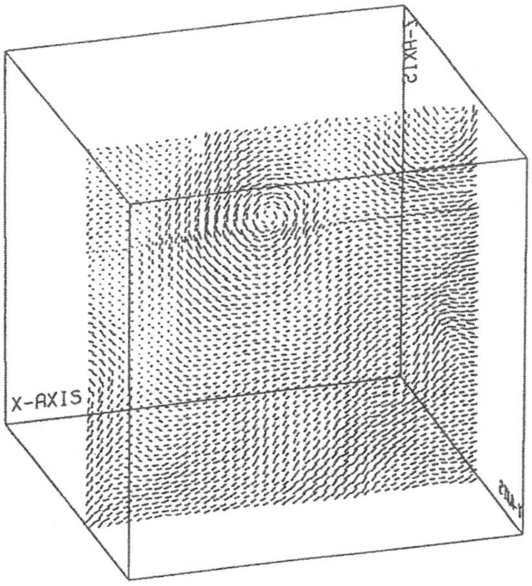

Figure 6: Projection of the velocity field on a plane perpendicular to a particular vorticity tube.

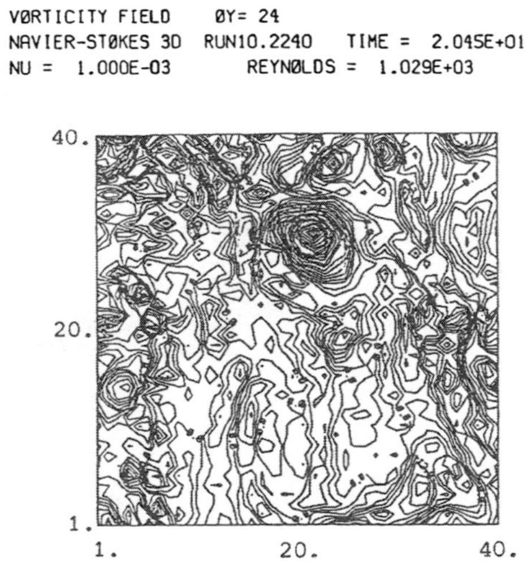

Figure 7: Curves of constant vorticity in a plane perpendicular to a selected vorticity tube.

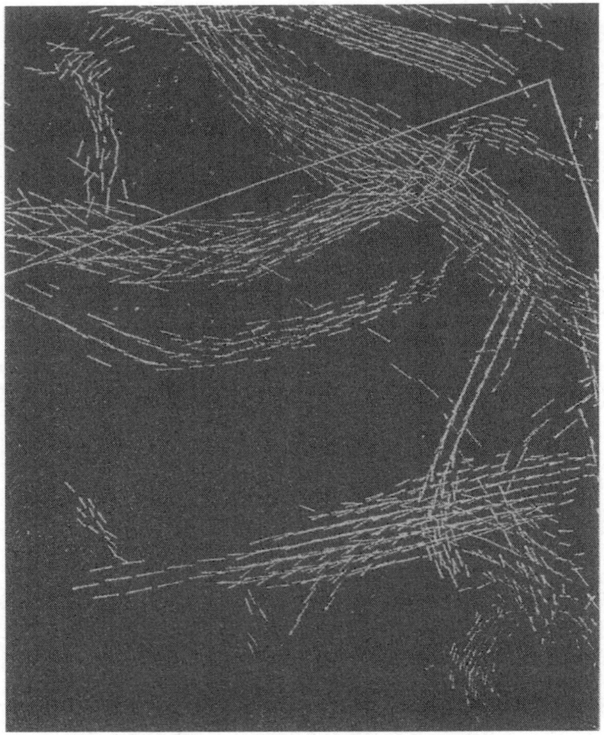

Figure 8: A detail of the vorticity field showing three approximately parallel tubes.

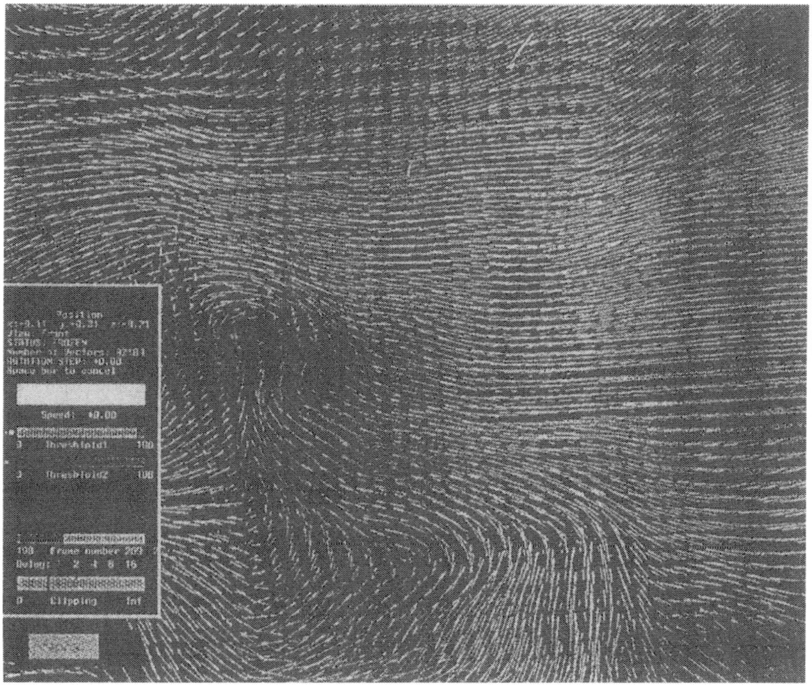

Figure 9: Velocity field associated with the vorticity tubes of figure 8, seen from a different position. Here, these tubes are approximately perpendicular to the page.

NEW RESULTS ON THE FINE SCALE STRUCTURE OF FULLY DEVELOPED TURBULENCE

M. VERGASSOLA[1] and U. FRISCH[1]

[1] CNRS, Observatoire de Nice, BP 139, 06003 Nice Cedex, France.

ABSTRACT

The multifractal model of fully developed turbulence predicts the existence of the *intermediate dissipation range*, where the singularity exponents of the multifractal spectrum manifest their existence because of the effect of viscosity [1]. The behaviour of the moments of the increments at a distance ℓ of the velocity field as a function of ℓ is of multiscaling type: a local power law, where the exponent is a slowly varying (logarithmically) function. The main consequence is the prediction of a new (other than the one predicted by the 1941 Kolmogorov theory) form of universality of the energy spectrum $E(k)$ with respect to the Reynolds number R: the function $\log E(k)/\log R$ is a universal function $F(\theta)$ of the variable $\theta = \log k/\log R$ (k is the wavenumber). From the curve $F(\theta)$ it should be possible to measure the multifractal spectrum of singularities. The prediction received a preliminar confirmation in convection [2] and in turbulence [3].

Similar arguments can be developed for the dynamical systems and, more generally, for the multifractal measures [4]. In this case, the viscosity is substituted by an artificial noise threshold introduced in the system. Multiscaling is observed in the behaviour of the moments of the measure with respect to the scale. The parameter entering in the rescaling is the value of the noise threshold.

References

[1] U. Frisch and M. Vergassola, *Europhys. Lett.*, **14**, 439, (1991).

[2] X.Z. Wu, L. Kadanoff, A. Libchaber and M. Sano, *Phys. Rev. Lett.*, **64**, 2140, (1990).

[3] Y. Gagne and B. Castaing, submitted to *C.R. Acad. Sci.*, (1991).

[4] M.H. Jensen, G. Paladin and A. Vulpiani, submitted to *Phys. Rev. Lett.*, (1991).

ON FAST DYNAMO ACTION IN STEADY CHAOTIC FLOWS

Andrew D. Gilbert.
D.A.M.T.P., Silver St., Cambridge, CB3 9EW, U.K.

The fast dynamo problem (Vainshtein & Zeldovich 1972) concerns the study of the induction equation:

$$\partial_t \mathbf{B} = \nabla \times (\mathbf{u} \times \mathbf{B}) + \eta \nabla^2 \mathbf{B}, \qquad \nabla \cdot \mathbf{B} = 0, \tag{1}$$

in the limit of very weak magnetic diffusion η. The equation governs the evolution of magnetic field in a flow $\mathbf{u}(\mathbf{x}, t)$ of a conducting fluid with conductivity $1/\eta$. We generally take the flow $\mathbf{u}(\mathbf{x}, t)$ to be prescribed and neglect the effects of the Lorentz force. This gives us a linear problem, the kinematic dynamo problem, which addresses the stability of the flow to weak magnetic fields. We call the flow $\mathbf{u}(\mathbf{x}, t)$ a dynamo if, for some value of η, the fastest growing solution of eq. (1) has a positive growth rate, $p_{\mathrm{max}}(\eta) > 0$; the flow is a fast dynamo if this growth rate remains positive and bounded away from zero in the limit of very weak diffusion η:

$$\lim_{\eta \to 0} p_{\mathrm{max}}(\eta) > 0.$$

In other words the growth rate remains on the convective time-scale (defined when the flow $\mathbf{u}(\mathbf{x}, t)$ is given) in the limit of very long diffusive time-scale. The fast dynamo problem is essentially to find flow fields that can be shown to be fast dynamos. The study of the kinematic dynamo problem and the fast dynamo problem is relevant to understanding the generation of magnetic field in the earth, sun, galaxy and other astrophysical bodies (see, for example, Moffatt 1978, Zeldovich *et al.* 1983).

Some of the difficulties of the fast dynamo problem may be seen by setting $\eta = 0$ in eq. (1); this gives the equation for the passive transport or Lie-dragging of the magnetic field in the flow $\mathbf{u}(\mathbf{x}, t)$. The equation is now first order and is solved by the Cauchy solution:

$$\mathbf{B}(\mathbf{x}, t) = \partial \mathbf{x} / \partial \mathbf{x}_0 \cdot \mathbf{B}(\mathbf{x}_0, 0). \tag{2}$$

Here $\mathbf{x}_0$ is the initial position of the fluid element which lies at $\mathbf{x}$ at time t. For a flow of any complexity the magnetic field becomes stretched and folded by the flow, and its scales decrease indefinitely. If the flow is chaotic field vectors will grow exponentially and different moments of the magnetic field will grow exponentially at different rates. Eigenfunctions of eq. (1) may only exist in a weak sense (Moffatt & Proctor 1985, Childress & Klapper 1991, Bayly 1991).

The introduction of very weak diffusion regularises the problem by smoothing out field on scales below some small-scale diffusive cut-off and leads to smooth eigenmodes of eq. (1); however these eigenmodes will still have complex structure above this cut-off. Furthermore diffusion can lead to complete decay of the field, for example if the flow field is two-dimensional (Zeldovich 1957). In two dimensions the flow stretches field and folds it in such a way that one obtains small-scale alternating field, which is then wiped out by diffusion. However in three dimensions it is possible to stretch and fold field in such a way as to align it (as in the "rope dynamo" of Vainshtein & Zeldovich 1972) and it seems likely that diffusion then merely smooths over small-scale variations and allows exponential growth of field and fast dynamo action. Thus, at least at a heuristic level, the question of fast dynamo action is connected with whether a chaotic flow folds field in a constructive way (so that the field tends to be aligned) or in a destructive way (alternating directions of field) (Finn & Ott 1988). This connection can be made rigorous in a number of idealised models (Bayly & Childress 1988, 1989, Finn *et al.* 1990, Childress & Klapper 1991, Klapper 1991).

Little progress has been made on proving mathematically that a smooth deterministic flow in a bounded or periodic domain of ordinary three-dimensional space can give fast dynamo action. It is known that exponential stretching is required (Vishik 1989) but it may be sufficient that this occur at isolated points or surfaces rather than in chaotic regions of positive measure. There is, however, strong numerical evidence of fast dynamo action in unsteady chaotic flows (Bayly & Childress 1988, 1989, Finn & Ott 1988, Otani 1989). Also fast dynamo action has been proved in a number of idealised models involving flows on manifolds of negative curvature (Arnold *et al.* 1981), maps with discontinuities (Bayly & Childress 1988, 1989, Finn & Ott 1988, Finn *et al.* 1989), laminar flows with singularities (Soward 1987, Gilbert 1988) and flows in unbounded space (Finn *et al.* 1990).

Our research concerns fast dynamo action in steady flows. It has been suggested that these are generically fast dynamos, on the basis of certain models (Finn *et al.* 1990). The flow fields that have attracted most interest as possible fast dynamos belong to the ABC family. This is a family of solutions of the Euler equation:

$$\mathbf{u}(\mathbf{x}) = (C\sin z + B\cos y, A\sin x + C\cos z, B\sin y + A\cos x), \tag{3}$$

given by three parameters, A, B and C. For general values of the parameters, these flows possess very complex streamline topology, with a mixture of chaotic regions and integrable vortices (Dombre *et al.* 1986). Dynamo action has been studied numerically in certain of these flows (Arnold & Korkina 1983, Galloway & Frisch 1986) but evidence for fast dynamo action is somewhat inconclusive.

In our work we have studied a model flow based on near-integrable ABC flows with parameter values $A = B = 1$, $C \ll 1$. If C is actually zero, the flow is integrable, having a cellular structure with a lattice of stagnation points joined by separatrices. As C is increased from zero these separatrices break up and form chaotic layers; the network of these layers, which extends periodically in space, is called a chaotic web (Beloshapkin *et al.* 1989). Within the web fluid elements may have chaotic or regular motion. We constructed a model based on such a near-integrable ABC flow; essentially we took leading order approximations to the ABC flow and patched these together (for details see Gilbert & Childress 1990). The resulting model flow has a number of qualitative features in comon with the near-integrable ABC flow; because of its construction the motion of particles and field vectors can be reduced to iterating maps, which allows relatively easy numerical exploration of magnetic field stretching and folding. Figure (1) shows a Poincaré section of the flow: points are plotted where fluid elements repeatedly intersect the section. There are both regular and chaotic orbits present.

Figure (1). Poincaré section of the model flow.

We have followed the evolution of various initial magnetic fields in this flow using the Cauchy solution (2). Figure (2) shows the vertical component of the magnetic field; the area depicted corresponds to the box marked in the Poincaré section in figure (1). White is plotted where the vertical field is negative, and grey where the field is positive, with black corresponding to the strongest positive field. The initial condition is:

$$\mathbf{B}(\mathbf{x},0) = (-\sin\sqrt{2}z, \cos\sqrt{2}z, 0). \tag{4}$$

There is clear visual evidence of constructive folding taking place within the flow: on the bottom left of both (2a) and (2b) there are bands of positive (grey) field accumulating; on the mid right of (2a) the field is largely negative (white) while in (2b) the field here is mostly positive (grey). We have confirmed this by computing the mean field in chaotic parts of the flow and have found clear growth with oscillations (Gilbert & Childress 1990), but at a rate less than the Liapunov exponent, because of cancellation. These calculations have been extended to real ABC flows in the case $A = B = C = 1$ and again we have found evidence for constructive folding and the exponential growth of mean fields in the absence of diffusion (Gilbert 1991) with growth rates similar to those obtained by Galloway & Frisch (1986).

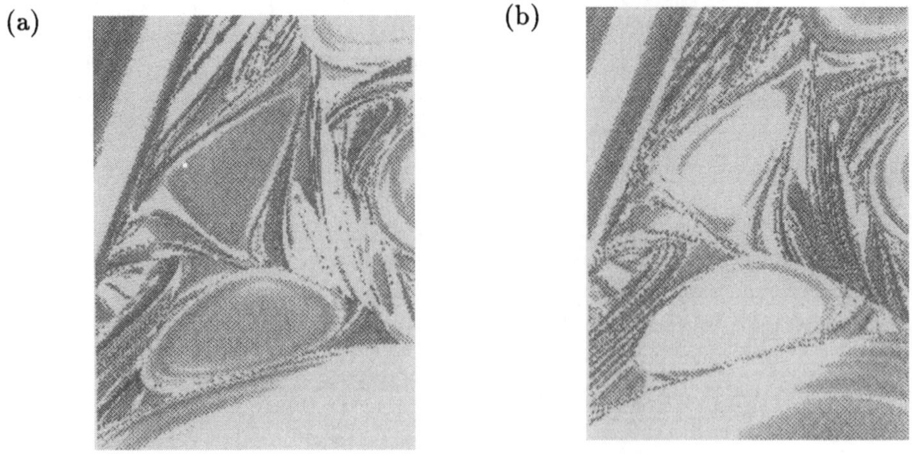

Figure (2). The vertical component B_z of magnetic field in the box shown in figure (1). White corresponds to negative B_z, and grey corresponds to positive B_z, the strongest field being shown in black. The initial condition (4) is evolved to (a) $t = 130$, (b) $t = 180$.

We have observed constructive folding in the case of ABC flow with $A = B = C = 1$, and a model flow, patched together and qualitatively similar to a near-integrable ABC flow. This is suggestive of fast dynamo action, since one would expect diffusion to smooth over the fine-scale structures while preserving the growing mean field that we observe. However it remains to be proved that constructive folding is sufficient for fast dynamo action in realistic flows (see Klapper 1991 for recent developments). Such a proof would reduce the fast dynamo problem to proving that constructive folding can occur in realistic chaotic flows, which is itself an open problem.

REFERENCES

Arnold, V.I. & Korkina, E.I. 1983 The growth of a magnetic field in a three-dimensional steady incompressible flow, *Vest. Mosk. Un. Ta. Ser. 1, Math. Mec.*, **3**, 43-46.

Arnold, V.I., Zeldovich, Ya. B., Ruzmaikin, A.A. & Sokoloff, D.D. 1981 A magnetic field in a stationary flow with stretching in Riemannian space, *Zh. Eksp. Teor. Fiz.*, **81**, 2052-2058 [*Sov. Phys. J.E.T.P.*, **54**, 1083-1086].

Bayly, B.J. 1991 Infinitely conducting dynamos and other horrible eigenproblems, Proc. Workshop on Nonlinear Phenomena in Atmospheric & Oceanic Sciences, I.M.A. (Minnesota) June 1990, to appear.

Bayly, B.J. & Childress, S. 1988 Construction of fast dynamos using unsteady flows and maps in three dimensions, *Geophys. Astrophys. Fluid Dyn.*, **44**, 211-240.

Bayly, B.J. & Childress, S. 1989 Unsteady dynamo effects at large magnetic Reynolds number, *Geophys. Astrophys. Fluid Dyn.*, **49**, 23-43.

Beloshapkin, V.V., Chernikov, A.A., Natenzon, M.Ya., Petrovichev, B.A., Sagdeev, R.Z. & Zaslavsky, G.M. 1989 Chaotic streamlines in pre-turbulent states, *Nature*, **337**, 133-137.

Childress, S. & Klapper, I. 1991 On some transport properties of baker's maps, *J. Stat. Phys.*, in press.

Dombre, T., Frisch, U., Greene, J.M., Hénon, M., Mehr, A. & Soward, A.M. 1986 Chaotic streamlines in the ABC flows, *J. Fluid Mech.*, **167**, 353-391.

Finn, J.M., Hanson, J.D., Kan, I. & Ott, E. 1989 Do steady fast dynamos exist? *Phys. Rev. Lett.*, **25**, 2965-2968.

Finn, J.M., Hanson, J.D., Kan, I. & Ott, E. 1990 Steady fast dynamo flows, Plasma Preprint UMLPR 90-015, University of Maryland.

Finn, J.M. & Ott, E. 1988 Chaotic flows and fast magnetic dynamos, *Phys. Fluids*, **31**, 2992-3011.

Finn, J.M. & Ott, E. 1990 The fast kinematic magnetic dynamo and the dissipationless limit, *Phys. Fluids*, **B2**, 916-926.

Galloway, D. & Frisch, U. 1986 Dynamo action in a family of flows with chaotic streamlines, *Geophys. Astrophys. Fluid Dyn.*, **36**, 58-83.

Gilbert, A.D. & Childress, S. 1990 Evidence for fast dynamo action in a chaotic web, *Phys. Rev. Lett.*, **65**, 2133-2136.

Gilbert, A.D. 1988 Fast dynamo action in the Ponomarenko dynamo, *Geophys. Astrophys. Fluid Dyn.*, **44**, 241-258.

Gilbert, A.D. 1991 Evidence for fast dynamo action in a chaotic ABC flow, *Nature*, **350**, 483-485.

Klapper, I. 1991 Shadowing and the role of small diffusivity in the chaotic advection of scalars, *Phys. Rev. Lett.*, submitted.

Moffatt, H.K. 1978 Magnetic field generation in electrically conducting fluids, Cambridge University Press.

Moffatt, H.K. & Proctor, M.R.E. 1985 Topological constraints associated with fast dynamo action, *J. Fluid Mech.*, **154**, 493-507.

Otani, N. 1989 E.O.S., Trans. Geophys. Union Vol. 69, No. 44, Nov. 1, Abstract No. HS51-15, p. 1366.

Soward, A.M. 1987 Fast dynamo action in a steady flow, *J. Fluid Mech.*, **180**, 267-295.

Vainshtein, S.I. & Zeldovich, Ya.B. 1972 Origin of magnetic fields in astrophysics, *Sov. Phys. Usp.*, **15**, 159-172.

Vishik, M.M. 1989 Magnetic field generation by the motion of a highly conducting fluid, *Geophys. Astrophys. Fluid Dyn.*, **48**, 151-167.

Zeldovich, Ya.B. 1957 The magnetic field in the two-dimensional motion of a conducting turbulent fluid, *Sov. Phys. J.E.T.P.*, **4**, 460-462.

Zeldovich, Ya.B., Ruzmaikin, A.A. & Sokoloff, D.D. 1983 Magnetic fields in astrophysics, Gordon and Breach.

MAGNETIC STRUCTURES IN FAST DYNAMO

Alexander A. Ruzmaikin
IZMIRAN, Troitsk Moscow Region, 142092, USSR

The fate of a smoke or other scalar in a turbulent air is well known. It disappears asymptotically in time in accordance with the maximal principle. The asymptotic behavior of the magnetic field vector can be completely different. The field in a given well conducting fluid will exponentially grow with a rate of growth independent on the conductivity for a class non–integrable or non–stationary flows called "fast dynamos"(see for instance Zeldovich et al.,1983).

Some examples of the fast dynamo are constructed numerically or analytically. Among them are a direct numerical simulation of the homogeneous isotropic turbulence of conducting fluid by Meneguzzi, Frisch and Pouquet (1981); the ABC flow $v(x) = (A\sin y + B\cos z, B\sin z + C\cos x, C\sin x + A\cos y)$ (Galloway and Frisch, 1986); a helical flow $v(x,t) = a\sin(qx + \psi) + b\cos(qx + \psi)$, $aq = 0$, $b \perp a$, where the amplitudes and phases of the wave are random values renewing after a time τ (Gilbert and Bayly, 1990); the short–correlated in time random flow for which $\tau\partial_j v_i$ is a Brownian process (Molchanov, Ruzmaikin and Sokoloff,1985). The other examples are the fast dynamo maps of the Baker's type (Finn and Ott, 1988). The dynamo theorem proved by Molchanov, Ruzmaikin and Sokoloff (1984) states that a three–dimensional random motion of general type renewing after a finite time τ can act as the fast dynamo.

Childress and Soward (1984) shows that the problem of evaluating the magnitude of so called α–effect important for the mean magnetic field generation (see Moffatt,1978) is closely connected with the fast dynamo problem in the limit of large magnetic Reynolds number.

A common feature of all fast dynamos is an intermittent structure of the self–excited magnetic field. At large magnetic Reynolds numbers (the high conductivity limit) the magnetic field is distributed in the form of ropes (or possibly also layers) occupying a small part of the volume but keeping almost all magnetic energy. As some examples demonstrates, in the limit of the infinite magnetic Reynolds number the magnetic field may be concentrated in a fractal set. The distribution of the intermittent magnetic field is essentially non–Gaussian, a probability of large deviations (the magnetic concentrations) is high. In that sense the intermittency has pure probabilistic origin in contrast to the non–linear nature of hydrodynamic and other known structures. The dynamo equations with a given velocity field are linear in magnetic field.

RIGOROUS WAVELET ANALYSIS FOR MULTIFRACTAL SETS

J.-M. Ghez[1] and **S. Vaienti**[2]

Centre de Physique Théorique
CNRS - Luminy - Case 907
13288 Marseille Cedex 9 (France)

Abstract

We present new rigorous results about the wavelet analysis of measures on a fractal set. The very definition of the wavelet transform allows us to introduce set functions exhibiting a transition point in correspondence with a particular fractal dimension of the set. We illustrate these results on some numerical examples.

In this article we present new rigorous results [1, 2, 3] about the wavelet analysis of fractal sets. These results can be used to compute numerically various fractal dimensions of a set or bounds to them and, in general, to investigate the multifractal structure of the sets generated in physical situations [4].

The wavelet transform of a finite measure μ on a compact set J with norm $||.||$, is defined as [5] :

$$T_p(\mu, a, x) = \frac{1}{a^p} \int_J g\left(\frac{||x-y||}{a}\right) d\mu(y) \tag{1}$$

We found the following assumption fundamental to establish rigorous results :
(W-1) g is a real function of class $C^1(\mathbb{R})$ and :

$$\lim_{a \to 0^+} \frac{1}{a^p} g\left(\frac{r}{a}\right) = 0 \text{ for } r > 0 \text{ and } p \geq 0. \tag{2}$$

In order to prove the theorems 1 and 2 below, we need a further assumption that, strictly speaking, goes beyond the usual properties of a wavelet [6], although numerical computations show that it can be weakened :

---

[1] PHYMAT, Département de Mathématiques, Université de Toulon et du Var, 83957 La Garde Cedex, France.

[2] Supported by Contrat CEE n. SC1*0281.

Does the intermittent structure conserve in the non–linear regime? The answer to this question is not evident because a back influence of the magnetic field on the motion first starts namely in the vicinities of the ropes. However one (not very realistic) example of a non–linear fast dynamo giving a stationary intermittent magnetic structure is constructed (Dittrich, Molchanov, Ruzmaikin and Sokoloff, 1988).

References

Childress,S., and Soward,A.D. 1984. On the rapid generation of magnetic fields, in 'Chaos in Astrophysics', NATO Advanced Research Workshop, Palm Cost, Florida, USA.

Dittrich,P., Molchanov,S.A., Ruzmaikin,A.A., and Sokoloff, D.D. 1988. The limiting distribution of magnetic field strength in a random flow. Magnitnaya Gidrodinamika, N 3, 9–12.

Finn,J., and Ott,E. 1988, Chaotic flows and fast magnetic dynamos. Phys.Fluids, 31, 2992–3011.

Galeeva,R.F.,Ruzmaikin,A.A., and Sokoloff,D.D. 1989. Typical realization of magnetic field in a random flow., Magnitnaya Gidrodinamika, N 4, 17–21.

Galloway,D., and Frisch, U.1986. Dynamo action in a family of flows with chaotic stream lines. Geophys.Astrohys.Fluid Dyn.,36, 58–83.

Gilbert,A., and Bayly,B.J. 1991. Magnetic field intermittency and random helical flows. Geophys.Astrophys.Fluid Dyn.(in press).

Moffatt,H.K. 1978. Magnetic Field Generation in Electrically Conducting Fluid, Cambridge Univ.Press, Cambridge.

Molchanov,S.A.,Ruzmaikin,A.A., and Sokoloff, D.D. 1984. A Dynamo Theorem. Geophys.Astrophys.Fluid Dyn.,30, 241–259.

Molchanov,S.A., Ruzmaikin,A.A., and Sokoloff, D.D. 1987. Short–correlated random flow as the fast dynamo. Soviet Doklady, 295, 576–579.

Meneguzzi,M.,Frisch,U., and Pouquet,A. 1981. Helical and non–helical dynamos. Phys.Rev.Lett.,47, 1060–1064.

Zeldovich,Ya.B.,Ruzmaikin,A.A., and Sokoloff,D.D. 1983. Magnetic Fields in Astrophysics Gordon and Breach New-York, Paris, London.

(W-2) $g(r)$ is monotone for $r \geq 0$.

This assumption is also explicity used in a recent paper by Falconer [7].

In [2] we introduced the integrated wavelet transform (IWT) defined as :

$$T_p(\mu, a) = \int_J T_p(\mu, a, x) d\mu(x) \tag{3}$$

and we showed that the asymptotic behavior of the IWT gives a fractal dimension of the set. To understand it, we now introduce the function:

$$V_p(\mu) = \limsup_{a \to 0^+} |T_p(\mu, a)| \tag{4}$$

By the very definition it follows the existence of a unique transition point p_μ such that : $V_p(\mu) = 0$ for $p < p_\mu$ and $V_p(\mu) = +\infty$ for $p > p_\mu$.

For a class of dynamic invariant sets, we found that p_μ coincides with the correlation dimension of the measure μ and so is independent of g. Those sets are the disconnected conformal mixing repellers [8], among which there are the disconnected hyperbolic Julia sets and the "cookie-cutter" Cantor sets. If J is such a set, we put on it the Gibbs measure $\mu_\beta, \beta \epsilon \mathbb{R}$, which is defined as the unique ergodic measure maximizing the expression :

$$P(\beta) = h(\mu_\beta) - \beta\gamma^+(\mu_\beta), \tag{5}$$

where $h(\mu)$ is the μ-Kolmogorov or metric entropy and $\gamma^+(\mu)$ the positive Lyapunov exponent of the measure μ [9]. $P(\beta)$ is called the topological pressure and is a real analytic convex function of β. In [2] we proved :

Theorem 1 *: If* $\limsup_{a \to 0^+} |T_p(\mu_\beta, a)|$ *is different from* 0 *and* $+\infty$*, then* $p = p_{\mu_\beta}$ *satisfies :* $P(2\beta - p) = 2P(\beta)$*. The same result holds if* $\liminf_{a \to 0^+} |T_p(\mu_\beta, a)| \neq (0, +\infty)$.

The value of p_{μ_β} coincides with the generalized dimension of order $2, D_{\mu_\beta}(2)$, defined by the usual partition function approach [10] : it is also called the correlation dimension and denoted with ν (without reference to the measure).

In [2] we indicated how to generalize the theorem by means of the following argument : the IWT can be written as :

$$|T_p(\mu, a)| \sim \frac{1}{a^{p+1}} \left| \int_0^\Delta C(r) g'\left(\frac{r}{a}\right) dr \right| \tag{6}$$

where Δ is the diameter of J and the symbol $\sim$ means that we neglect an additive term that is always zero in the limit $a \to 0^+$ (for non atomic measures). The function $C(r)$ is the correlation integral

$$C(r) = \int_J \int_J \theta(r - \|x - y\|) d\mu(x) d\mu(y) \tag{7}$$

and for several fractal sets it scales like [11, 12] :

$$C(r) \underset{r \to 0^+}{\sim} r^\nu \tag{8}$$

Assuming (8) up to the diameter Δ of the set and substituting in (6) we immediately get :

$$\log |T_p(\mu, a)| \underset{a\to 0^+}{\sim} (\nu - p)\log a + \text{oscillating terms} \tag{9}$$

We checked numerically the scaling (9) for linear Cantor sets with different scales and the Gibbs measure μ_0 (balanced or maximal entropy measure) and we recovered the correlation dimension with great accuracy [2]; moreover the asymptotic behavior of the IWT was affected by oscillations which is a phenomenon already known as *lacunarity* for the correlation integral [13]: in a few cases, the oscillations of the IWT can be completely explained [2]. For example for the ternary Cantor set, we get [2] :

$$T_0(\mu_0, a) = \frac{1}{2} T_0(\mu_0, 3a) + o(1), a \to 0^+$$

where the $o(1)$ term converges to zero faster than any power of a by (W-1). A general solution of the preceding relation without the $o(1)$ term is : $a^\alpha \psi(\log a)$, where ψ is a continuous periodic function of $\log a$ of period $(\log 3)/k, k \epsilon \mathbb{N}$ and $\alpha = \log 2/\log 3$. Figures 1a and 1b show the numerical results ; in particular, in figure 1a, we report $\log |T_o(a)|$ *vs.* $\log a$, but the amplitude of the oscillations is very small and does not appear. Nevertheless, it is particularly evident in figure 1b, where we plot ($\log |T_o(a)| - D_2 \log a$ *vs.* $\log a$), where D_2 is the true value of the correlation dimension. Apart a straightforward extension to the disconnected attractors of the hyperbolic iterated function systems [3], rigorous generalizations of the theorem above are not easy ; we have a partial result [3] :

- let us consider a system satisfying the condition :

$$\lim_{r\to 0^+} \frac{\log \mu(B(x,r))}{\log r} = \text{constant} = HD(\mu) \tag{10}$$

for x μ-almost everywhere and being $B(x,r)$ a ball of center x and radius r. If $\limsup_{a\to 0^+} |T_p(\mu, a)| < +\infty$, then $p \le HD(\mu) \le d_H(J)$, being $d_H(J)$ the Hausdorff dimension of J.

We conjecture that if $\limsup_{a\to 0^+} |T_p(\mu, a)| < +\infty$ whenever μ is an invariant measure on the class of conformal repellers quoted above, then $p \le d_H(J)$. Note that the Gibbs measures are dense in the space of the invariant measures.

The limit (10) often occurs for dynamical invariant sets endowed with ergodic measures [14] and its constant value is called the information dimension, or generalized dimension of order $1, D_\mu(1)$.

Our guess is that, in general, whenever $\limsup_{a\to 0^+} |T_p(\mu, a)| < +\infty$, then $p \le d_H(J)$. We can only prove a weaker version of this conjecture in terms of the wavelet transform $T_p(\mu, a, x)$. First of all we define the *p-wavelet capacity* of the set J as (from now on J will be a compact subset of $\mathbb{R}^n$):

$$D_p(J) = \sup \left\{ \frac{1}{W_p(\mu)}; \ J \text{ supports } \ \mu \ \textit{and} \ \mu(J) = 1 \right\} \tag{11}$$

where :

$$W_p(\mu) = \inf_A \left\{ || \limsup_{a\to 0^+} |T_p(\mu,a,x)| \, ||_{\infty,A} \right\} \tag{12}$$

being A any measurable set of positive μ-measure. We have the following :

Theorem 2 [3] *:*

$$D_p(J) = +\infty \; for\; p < d_H(J) \; and \; D_p(J) = 0 \; for \; p > d_H(J) \;, \tag{13}$$

that is, $D_p(J)$ behaves like the Hausdorff p-dimensional measure of J.

(See [15] for a definition of the Hausdorff measure).

This implies that whenever, for a finite measure μ, $|| \limsup_{a\to 0^+} |T_p(\mu,a,x)| \, ||_{\infty,A} < \infty$, with A any set of positive μ-measure, then $p \leq d_H(J)$. We guess that the same theorem holds substituting the function $W_p(\mu)$ with $V_p(\mu)$ defined in (4) in terms of the IWT ; in our opinion such results show the close relationship between the wavelet transform and the fractal geometry. The preceding theorem can be considered as the global version of a local analysis of the fractal measure, which holds for a class of wavelets larger than those satisfying (W-2) and containing in particular the mexican-hat $g(r) = (1-r^2)e^{-r^2/2}$, and that was qualitatively studied in [5]. We introduce, in analogy with (4), the function

$$V_p(\mu,x) = \limsup_{a\to 0^+} |T_p(\mu,a,x)| \tag{14}$$

and call $p_\mu(x)$ its transition point. Then, we have :

Theorem 3 [1] *: If $\beta(x) = \liminf_{r\to 0^+} \dfrac{\log \mu(B(x,r))}{\log r}$ is the local scaling exponent of the measure μ at the point $x \epsilon J$, then :*

$$p_\mu(x) \geq \beta(x) \tag{15}$$

Note that theorem 3 is not sufficient to conclude that, whenever $p > d_H(J)$, then $V_p(\mu,x) > 0$, but this follows directly from theorem 2, that is :

Corollary 4 *: If $p > d_H(J)$, then*

$$\limsup_{a\to 0^+} |T_p(\mu,a,x)| = +\infty \quad for \quad \mu - almost \; x\epsilon J.$$

Combining with Theorem 3, we finally get :

$$\beta(x) \leq p_\mu(x) \leq d_H(J) \quad \text{for} \quad \mu - \text{almost } x\epsilon J.$$

This result is important for the numerical pictures of the wavelet transform, because it guarantees us that, taking $p > d_H(J)$, the absolute value of the wavelet transform could in general oscillate, with relative maxima growing to infinity when $a \to 0^+$ for μ- almost all the points $x\epsilon J$, and the amplitude of these oscillations is just governed by the exponent $\beta(x)$.

We now illustrate this result by performing the numerical analysis of the wavelet transform for some dynamic invariant sets not previously investigated, namely :

i) The Lozi attractor generated by the mapping

$$\begin{cases} x' &= 1 + y - 1.7|x| \\ y' &= 0.5\ x \end{cases}$$

ii) The invariant set for the Baker transformation :

$$x' = \begin{cases} \gamma_a x & \text{for } y < \alpha \\ \frac{1}{2} + \gamma_b x & \text{for } y > \alpha \end{cases}$$

$$y' = \begin{cases} \frac{1}{\alpha} y & \text{for } y < \alpha \\ \frac{1}{1-\alpha}(y - \alpha) & \text{for } y > \alpha \end{cases}$$

We have chosen $\gamma_a = 0.2, \gamma_b = 0.3, \alpha = 0.4$.

iii) The connected Julia set for the mapping $z' = z^2 - \frac{1}{4}$.

We report in figures 2-3-4 the absolute value $|T_p(\mu, a, x)|$ of the wavelet transform of the measure μ, μ being the Sinai-Bowen-Ruelle measure for the Lozi and Baker maps and the balanced measure for the Julia set. We have chosen $p = 2$ according to the preceding discussion.

All the relations proposed in this paper could be summarized in the following approximative scheme, suitable for numerical investigations. Note that, apart (16ii) that is conjectured, all the others give the same transition point *independently of* g :

$$\begin{aligned} &|T_p(\mu, a)| \underset{a \to 0+}{\sim} a^{\nu - p} && (16i) \\ &\inf_{\mu} |T_p(\mu, a)| \underset{a \to 0+}{\sim} a^{p - d_H(J)} && (16ii) \\ &|T_p(\mu, a, x)| \underset{a \to 0+}{\sim} a^{\beta(x) - p} && (16iii) \\ &\inf_{\mu} \left\{ \sup_{x \epsilon A} |T_p(\mu, a, x)|, \mu(A) > 0 \right\} \underset{a \to 0+}{\sim} a^{p - d_H(J)} && (16iv) \end{aligned} \qquad (16)$$

The scalings (16) clearly show that the wavelet transform and the IWT are very useful tools to investigate both the local and global geometrical properties of a measure on a fractal set. These properties always arise as transition indices between two well defined (0 and ∞) asymptotic regimes and therefore they can be easily detected as the exponents of a power law decay in a $\log - \log$ plot. Moreover, thanks to (6) the choice of a wavelet decaying sufficiently fast at infinity could notably accelerate the convergence of the limit for $a \to 0^+$: in this respect the wavelet transform is more flexible than other integral methods to detect the fractal dimensions.

It is important to point out that the scalings (16) effectively occur in the numerical analysis of the fractal measures : see, for instance [2] and for (16iii) the extensive numerical study in [4]. Besides, in a few simple examples (smooth sets in $\mathbb{R}^n$ and the ternary Cantor sets), an analytic calculation of the IWT can be completely done [2]. In all these examples the choice of the wavelet is large : numerical computations indicates that the theorems quoted above are true for non-monotone wavelets, for example for the mexican-hat that is used throughout [4].

Our goal in this article was to show that, for the mixing repellers and, in full generality after having defined a suitable object (the wavelet-capacity), it is possible to put the wavelet analysis of fractal sets on a mathematical basis.

Acknowledgements

J.-M. G. would like to thank A. Arneodo who introduced him to the wavelet analysis of fractals. S. V. warmly thanks J-D. Fournier and P. Sulem for the kind invitation at the workshop in Villefranche.

References

[1] J.-M. Ghez, S. Vaienti, J. Stat. Phys. **57** (1989) 415.

[2] J.-M. Ghez, S. Vaienti, *Integrated wavelets on fractal sets*, Preprint CPT/2442.

[3] S. Vaienti, *A Frostman-like theorem for the wavelet transform on fractal sets*, Nonlinearity, to appear.

[4] F. Argoul, A. Arneodo, J. Elezgaray, G. Grasseau, in *Nonlinear dynamics*, G. Turchetti ed. (World Scientific, Singapore 1988) ;
F. Argoul, A. Arneodo, J. Elezgaray, G. Grasseau, R. Murenzi, *Wavelet Analysis of the Self-Similarity of Diffusion-Limited Aggregates and Electrodeposition Clusters*, Preprint CRPP 1990 ;
A. Arneodo, F. Argoul, G. Grasseau, *Transformation en ondelettes et renormalisation*, Preprint CRPP 1990.

[5] A. Arneodo , G. Grasseau, M. Holschneider, Phys. Rev. Lett. **61** (1988) 2281 ;
M. Holschneider, Journ. Stat. Phys. **50** (1988) 963 ;
F. Argoul, A. Arneodo, J. Elezgaray, G. Grasseau, R. Murenzi, Phys. Lett. **A135** (1989) 327 ;
M. Holschneider, Thesis, Marseille 1988.

[6] A. Grossmann, J. Morlet, Siam J. Math. Anal. **15** (1984) 723 ;
J.-M. Combes, A. Grossmann, P. Tchamitchian, eds. *Wavelets* (Springer-Verlag, Berlin 1989) and references therein ;
R.R. Coifman, Y. Meyer, *Ondelettes, Opérateurs et Analyse Nonlinéaire* (Hermann, Paris 1989) and references therein.

[7] K.J. Falconer, *Wavelets, fractals and order-two densities*, Lecture given at the I.M.A. meeting *Wavelets, fractals and Fourier transforms*, Cambridge, December 1990.

[8] D. Ruelle, Ergodic Th. and Dynam. Systems **2** (1982) 99 ;
D. Ruelle, in *Scaling and Self-Similarity in Physics*, J. Frölich ed. (Birkhaüser, Boston 1983) ;
S. Vaienti, J. Phys. A : Math. Gen. **21** (1988) 2023.

[9] D. Ruelle, *Thermodynamic Formalism* (Addison-Wesley, New York 1978);
R. Bowen , Lect. Not. Math. **470** (1975).

[10] H.G. Hentschel, I. Procaccia, Physica **D8** (1983) 435 ;
U. Frisch, G. Parisi, International School of Physics *Enrico Fermi*, course 88, p. 84, ed. M. Ghil (North-Holland, Amsterdam 1985) ;
P. Grassberger, Phys. Lett. **A107** (1985) 101 ;
T.C. Halsey, M.H. Jensen, L.P. Kadanoff, I. Procaccia and B.I. Shraiman, Phys. Rev. **A33** (1986) 1141 ;
D. Bessis, G. Paladin, G. Turchetti, S. Vaienti, J. Stat. Phys. **51** (1988) 109.

[11] P. Grassberger, I. Procaccia, Physica **D9** (1983) 189 ;
A. Cohen, I. Procaccia, Phys. Rev. **A31** (1985) 1872 ;
D. Bessis, J.D. Fournier, G. Servizi, G. Turchetti, S. Vaienti, Phys. Rev. **A36** (1987) 220.

[12] C.D. Cutler, J. Stat. Phys. **62**, (1991) 651.

[13] L.A. Smith, J.D. Fournier, E.A. Spiegel, Phys. Lett. **A114** (1986) 465 ;
J.D. Fournier, G. Turchetti, S. Vaienti, Phys. Lett. **A140** (1989) 331 ;
P. Badii, A. Politi, Phys. Lett. **A107** (1984) 303.

[14] L.S. Young, Ergodic Th. and Dyn. Systems **2** (1982) 109 ;
F. Ledrappier, L.S. Young, Ann. of Math. **122** (1985) 509 ;
A. Manning, Ann. of Math. **119** (1984) 425 ;
F. Ledrappier, C.R.A.S. Paris **299** (1984) 37 ;
F. Ledrappier, M. Misiurewicz, Ergodic Th. and Dyn. Systems, **5** (1985) 595 ;
P. Collet, Y. Lévy, Comm. Math. Phys. **93** (1984) 461 ;
R. Mañé, Lect. Not. Math. **1331** (1988).

[15] K.J. Falconer, *The Geometry of Fractal Sets* (Cambridge University Press, Cambridge 1985).

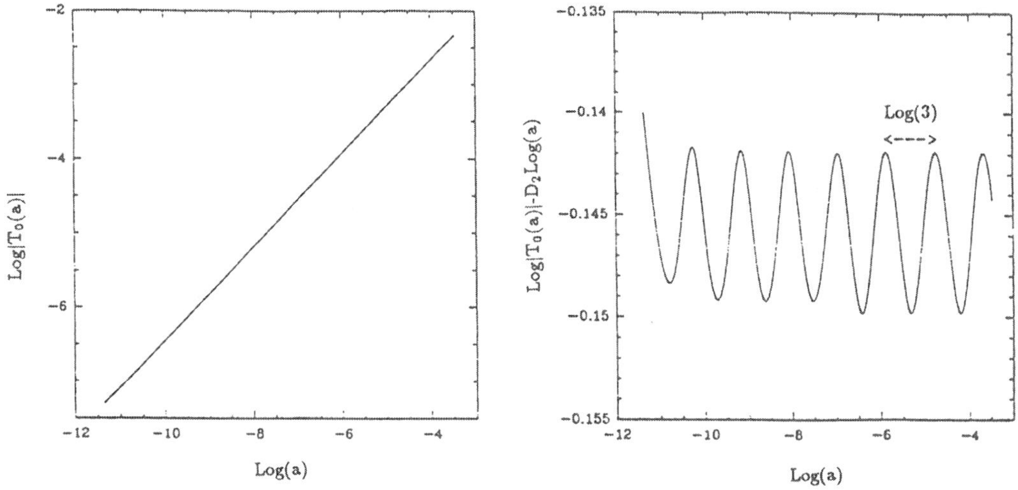

Figure 1

a : $\log|T_0(a)|$ *vs.* log a for the ternary Cantor set. The computed value of the correlation dimension is $\nu = 0.6392\ldots$, with $g(r) = e^{-r^2}$; **b** :$\left(\log|T_0(a)| - D_2 \log a\right)$*vs.* log a for the ternary Cantor set.

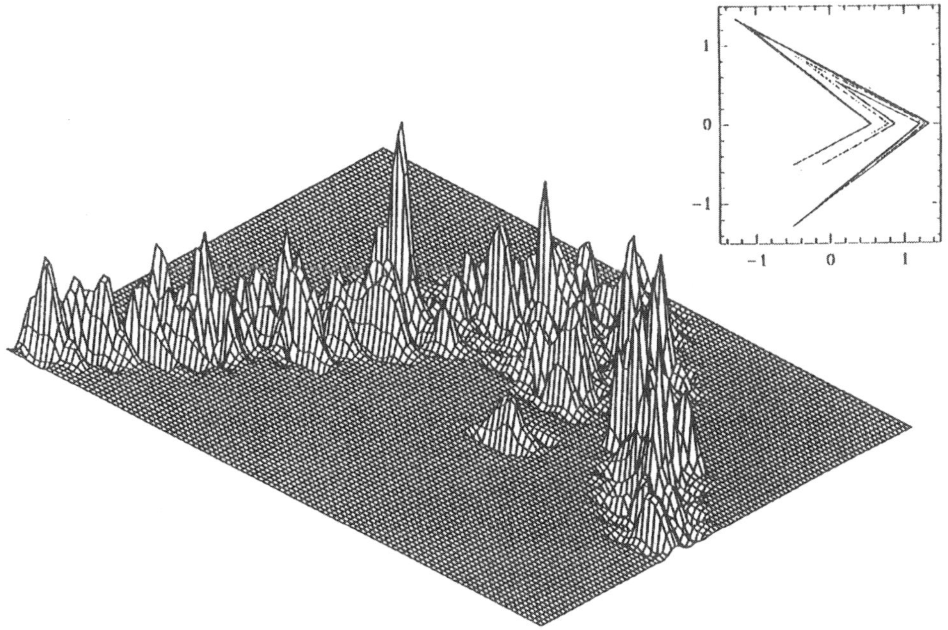

Figure 2

$|T_2(\mu, a, x)|$ (on the vertical axis) for $a = 10^{-3}$ and x on the Lozi attractor (in the horizontal plane) with $g(r) = (2-r^2)e^{-r^2/2}$. The Lozi attractor is reproduced in the upper right corner of the figure.

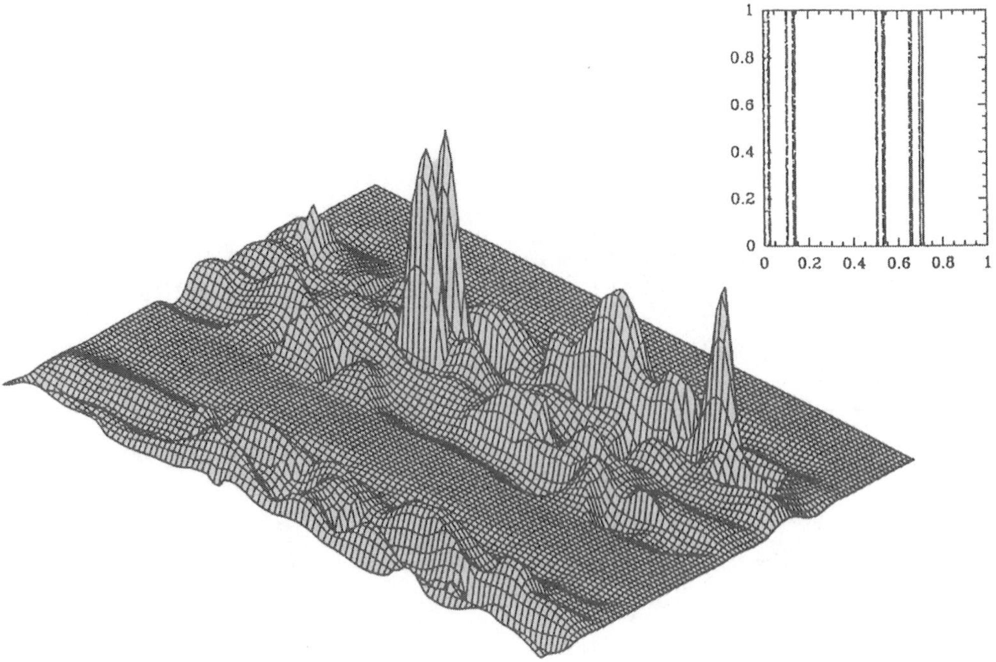

Figure 3

The same as fig. 2 for the Baker map.

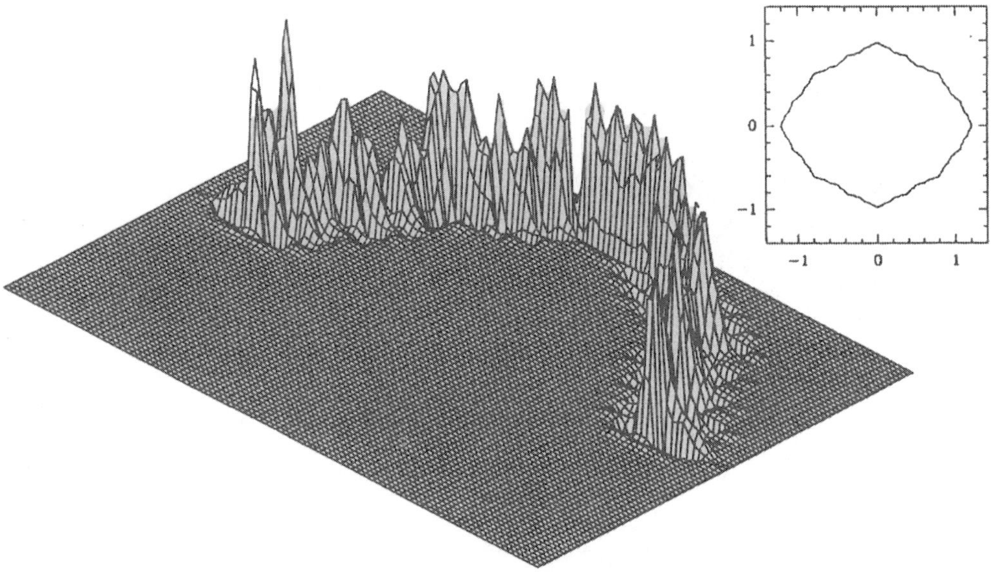

Figure 4

The same as fig. 2 for the Julia set.

Eigenfunction Analysis of Turbulent Mixing Phenomena

R.M. Everson†, L. Sirovich†, M. Winter‡ and T.J. Barber‡
† Center for Fluid Mechanics, Brown University, Providence, RI 02912, USA.
‡ United Technologies Research Center, East Hartford, CT 06109, USA.

An analysis of the inviscid mixing of a turbulent jet in crossflow is made. An experimental database is analyzed by means of a technique based on the Karhunen-Loéve procedure. It is shown that mixing which increases with downstream distance is characterized by the increasingly prominent role played by spatially complex eigenfunctions. A firm quantitative basis is presented which supports this visual perception of complexity.

Introduction

Mixing plays a vital role in a wide range of natural, engineering and technological processes. This paper is primarily concerned with the nature and description of these mixing processes, and provides the development of a framework, some tools and even a language by which to discuss mixing in quantitative terms. A longer range goal of this research is to develop methods and criteria for the management of mixing. While we focus on a problem arising from gas turbine combustor design, the methodology developed is equally applicable to virtually every aspect of technology and engineering where mixing processes are important.

The dilution zone of a gas turbine combustor is the region in which coolant air is injected and mixed with combustion gases to achieve the correct turbine inlet temperature. Current designs for the dilution zone resort to a series of round jets, injecting cold air at right angles to the primary gas flow to achieve the highest degree of mixing. The rate of mixing however, is strongly influenced by the hole shape, momentum ratio of the jet to primary flow, turbulence level, axial distance, etc. To date, mixing optimization has been largely an empirical process, while any assessment of induced mixing has been indirectly inferred. The present study will focus on a model problem of the dilution zone, viz. a subsonic jet in a crossflow. Such problems have been previously studied experimentally by Vranos.[1] In their experiment advanced data acquisition techniques using optically based diagnostics[2] furnish us with a wealth of detailed and highly resolved data.

While the data that will be analyzed below are from a physical experiment, the methodology presented is generally applicable and could just as easily have been applied to a computationally generated database. An additional feature of the investigation relates to the treatment of large databases in general. Our methods give a format for analyzing and compressing these in a physically relevant way.[3] As will be seen, all of this is accomplished by generating an intrinsically defined basis set of describing functions, which in a well defined sense are optimal.

Although it is not necessary to restrict attention to two-dimensions, all our deliberations will refer to concentration fields in two-dimensional slices of a flow. We will denote the fluctuation in a concentration at a time t by $c(\mathbf{x}, t)$ where $\mathbf{x} = (x, y)$ is a position in the plane. If we denote a time average by,

$$< c >= \frac{1}{T} \int_{-T/2}^{T/2} c(\mathbf{x}, t) dt, \tag{1}$$

then the fluctuation is defined such that

$$\int_A < c(\mathbf{x}) > d\mathbf{x} = 0, \tag{2}$$

where A denotes the area of the slice.

Perhaps the coarsest and most widespread measure of a passive scalar concentration field is it rms value:

$$\overline{c}(\mathbf{x}) =< c^2(\mathbf{x}) >^{1/2} \tag{3}$$

The smaller the average value of $\overline{c}$:

$$c_{rms} = \frac{1}{A} \int_A \overline{c}(\mathbf{x}) d\mathbf{x} \tag{4}$$

the better is the mixing. While (3) furnishes us with information that locates and measures the fluctuations, it does not address the issue of what are the scales of the fluctuations. This is of great importance since the *texture* of the concentration field determines how soon molecular diffusion plays its final role. When viewing fine grain versus coarse grain fluctuations of equal magnitude, it is clear that the former is more rapidly mixed than the latter.

This simple observation suggests that finer measures of mixedness should be employed. For example, if it makes sense to speak of the Fourier transform of $c(\mathbf{x}), \tilde{c}(k)$, then we can use the spectral entropy

$$S = - \int \epsilon(\mathbf{k}) \ln \epsilon(\mathbf{k}) d\mathbf{k}, \quad \epsilon = |\tilde{c}|^2 / c_{rms}^2, \tag{5}$$

where $\epsilon(\mathbf{k})$ is the energy at wavenumber $\mathbf{k}$. Two other measures which will be discussed later are,

$$D_1 = \frac{1}{A} \int \frac{| \nabla c | \, d\mathbf{x}}{c_{rms}} \tag{6}$$

and

$$D_2 = \frac{1}{A} \int \frac{| \nabla c |^2 \, d\mathbf{x}}{c_{rms}^2}. \tag{7}$$

The first of these (5) addresses the diversity of scales, while (6) and (7) bring out the importance of small scales in mixing. It should be noted that both (6) and (7) are dimensional and should be normalized with respect to the Kolmogorov scale. *More specifically, we are the addressing that part of the mixing process that is inviscid (large scale) and which precedes molecular diffusion (small scale).*

A shortcoming of the above discussion is the absence of *physics*, more specifically the isolation of the basic patterns by which the mixing process takes place. If the mixing process can itself be decomposed into fundamental patterns or modes of mixing then we can hope to proceed to the goal of managing the mixing process by manipulating these fundamental modes.

Karhunen-Loéve (K-L) procedure

The K-L procedure was proposed by Lorenz[4] and by Lumley[5] for the analysis of turbulence data. In the former instance to treat meteorological data, and in the latter case to isolate coherent structures. In each instance time series or their equivalent were considered. In recent years it has been shown that fully three dimensional phenomena can be treated by this method.[6,7] This has been accomplished by the method of *snapshots* which we now briefly develop within the framework of the jet in cross flow, which we analyze in detail in the Results section.

We consider the fluctuation field in a plane orthogonal to the jet and denote it, as earlier, by $c(\mathbf{x}, t)$. We assume that the mean value has been subtracted so that (2) is satisfied. The concentration fluctuation is captured *instantaneously* at a uniformly spaced sequence of instants and the resulting ensemble of such states is denoted by

$$\{c^{(n)}\} = \{c^{(n)}(\mathbf{x})\} = \{c(\mathbf{x}, t_n)\} \tag{8}$$

where as indicated the superscript denotes the instant of time. Typical concentration profiles are shown in Figure 1.

To develop the K-L procedure we pose the question: Is there some *concentration profile* which is most typical? Stated analytically, for what ϕ is

$$< (\phi, c)^2 >= \frac{1}{N} \sum_{n=1}^{N} \left(\int \phi(x,y) c^{(n)}(x,y) dx dy \right)^2 \tag{9}$$

a maximum? (Note that the time average is replaced by a discrete average.) For this to make sense there must be some constraint on ϕ. E.g., since we regard ϕ as a *concentration fluctuation* we require

$$\| \phi \|^2 = (\phi, \phi) =< (c,c) >= c_{rms}^2 = E, \tag{10}$$

although any such requirement will suffice. (In the following E will be termed the *energy*.) The result of the variational problem posed by (9) and (10) is that ϕ satisfy

$$\int K(\mathbf{x}, \mathbf{x}') \phi(\mathbf{x}) d\mathbf{x} = \lambda \phi(\mathbf{x}), \tag{11}$$

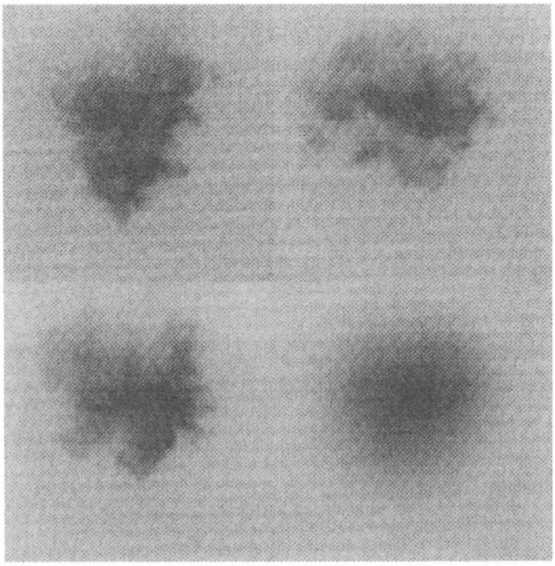

Figure 1: Three typical cross-sectional snapshots compared to the time-average concentration for a subsonic jet in crossflow. The jet orifice was at the top of the pictures.

where

$$K(\mathbf{x},\mathbf{x}') =< c(\mathbf{x})c(\mathbf{x}') >= \frac{1}{N}\sum_{j=1}^{N} c^{(j)}(\mathbf{x})c^{(j)}(\mathbf{x}'), \tag{12}$$

is the autocorrelation of the ensemble of concentrations. The eigenfunction equation (11) generates a complete set of functions each of which satisfies the extremal conditions (9) and (10), along with the side condition that it be orthogonal to all previously generated eigenfunctions. Without loss of generality we can take the set of eigenfunctions to be an orthonormal set.

A considerable reduction in the complexity of the problem is achieved by recognizing that an eigenfunction which satisfies (11) with K such that (10) holds must be an admixture of instantaneous *snapshots*, i.e., we can write

$$\phi(x,y) = \sum_{j=1}^{N} \alpha_j c^{(j)}(\mathbf{x}). \tag{13}$$

This reduces the integral equation, which in general may have support in many dimensions, to the relatively simple step of the diagonalization of matrix of order N. This version of the K-L procedure is known as the method of snapshots. As a result of the completeness of the eigenfunctions we can express the time dependent concentration in terms of the eigenfunctions $\{\phi_n(\mathbf{x})\}$ as follows

$$c(\mathbf{x},t) = \sum_{n=1}^{\infty} a_n(t)\phi_n(\mathbf{x}) \tag{14}$$

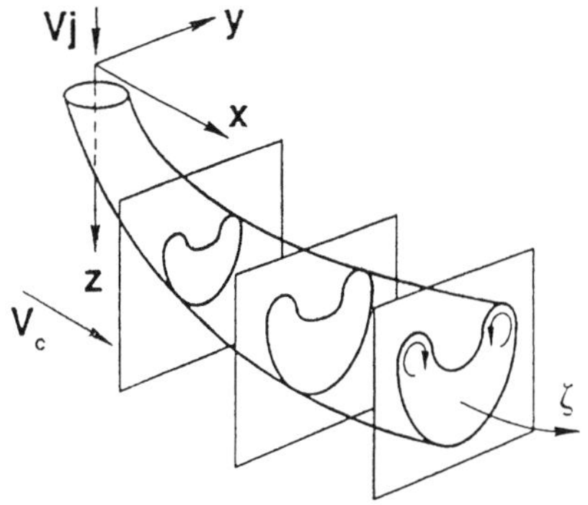

Figure 2: Schematic diagram of a subsonic jet in crossflow. The jet fluid is injected with velocity v_j into a crossflow having mean velocity v_c.

where

$$a_n = (\phi_n, c). \tag{15}$$

The set $\{\phi_n\}$ is termed the *empirical eigenfunctions* and they have the property that the variance

$$\sigma =<\| c - \sum_{j=1}^{N} a_n \phi_n \|^2>^{1/2} \tag{16}$$

is minimal (for any N) over the class of all admissible orthonormal basis sets. One additional characterization is that with the empirical eigenfunctions, and only with these, are the coefficients decorrelated

$$< a_n a_m >= 0, \quad m \neq n, \tag{17}$$

and from this the (information) entropy is minimal amongst all representations.[8] Using this it follows from the definition (12) that

$$< a_n^2 >= \lambda_n, \tag{18}$$

so the eigenvalue measures the mean square projection of the ensemble onto the associated eigenfunction.

Experimental procedure and database

A round jet injected at right angles to a subsonic primary stream induces the characteristic flow structure seen in Figure 2. Initially, the jet begins with an inviscid or potential core, persisting until viscous mixing effects predominate. The deflection of the jet by the primary stream induces a pair of counter-rotating vortices that result in the familiar kidney-shaped jet cross-section on average. An inviscid secondary flow mechanism[9] is responsible for their generation.

The character of the flow depends upon the momentum, $\rho_j V_j^2$, carried by the injected fluid gas to that of the crossflow gas, $\rho_c V_j^2$. This momentum ratio, $M = \rho_j V_j^2 / \rho_c V_j^2$, was varied by changing the velocity of the injected fluid at constant cross-flow velocity. The experiments were conducted over a Reynolds numbers in the range 50 000 – 80 000 based on jet diameter and crossflow conditions. The working fluid was air at room temperature and jet velocities were sufficiently low to eliminate compressibility effects.

Planar digital imaging of the light scattered from small aerosol particles was used to assay the flow. This technique, known as Lorenz-Mie scattering, is based on the principle that, for a gas seeded uniformly with aerosol particles, the intensity of elastically scattered light from the aerosols is proportional to the concentration of the seeded gas. Planar digital imaging has been demonstrated as an effective, quantitative diagnostic technique in small laboratory jets by Long *et al.*[2] Planar nephelometry measurements were made by marking the flowfield with small aerosol particles introduced into the nozzle gas and briefly illuminating transverse sections of the jet by reflecting a beam of unfocused argon-ion laser light with a rotating mirror. The elastically scattered light was imaged onto a low-light-level vidicon camera and digitized. The fluid motion was frozen by exposing the detector for only $10\mu s$ coincident with a single sweep of the laser beam through the flowfield. The rotation rate of the mirror was sufficiently fast so that all measurements within the frame are considered simultaneous. Each image consists of approximately 10 000 data points arranged in a 100 by 100 pixel format. The raw images were corrected for background response due to detector noise, scattered light and laser sheet non-uniformity.

The database we analyse here is comprised of three sets of images collected at 4, 6 and 8 jet diameters downstream from the jet orifice, together with two additional sets collected at the 8 diameter station; one with a higher momentum ratio and the other with a lower momentum ratio. Each set consists of approximately $N = 500$ snapshots.

Figure 1 shows three instantaneous realizations and the average concentration field of 500 realizations for the low momentum ratio. For each of the five databases described, we have doubled the ensemble size by adding to each ensemble the mirror images, in the vertical midline, of each ensemble member. In doing this we are exploiting the natural symmetry of the equations governing the flow, which imply that if $c_n(x,y)$ is a snapshot, then $c_n(-x,y)$ is also a valid snapshot. This addition to the ensemble produces eigenfunctions that are either odd or even in the vertical midline, a result that would not otherwise be realized unless the ensemble size were extremely large.

While the data represent the large scale flow structure well in the concentration profiles, a considerable degree of noise and drift were present in the images. To standardize this large database for comparison between different experimental conditions and data acquired over long periods of time, the data were normalized in the following manner. A reference value for each snapshot was determined from a 10 by 10 pixel area containing no seeded fluid in the corner of each snapshot. A systematic drift in the measured concentration was eliminated based on these values and the data were then rescaled so that the mean concentration flux through each picture was constant. In addition, filtering in Fourier space was used to remove some background noise. While this renormalization changes the concentration values and hence the dynamic range, the macroscopic flow structure remains preserved.

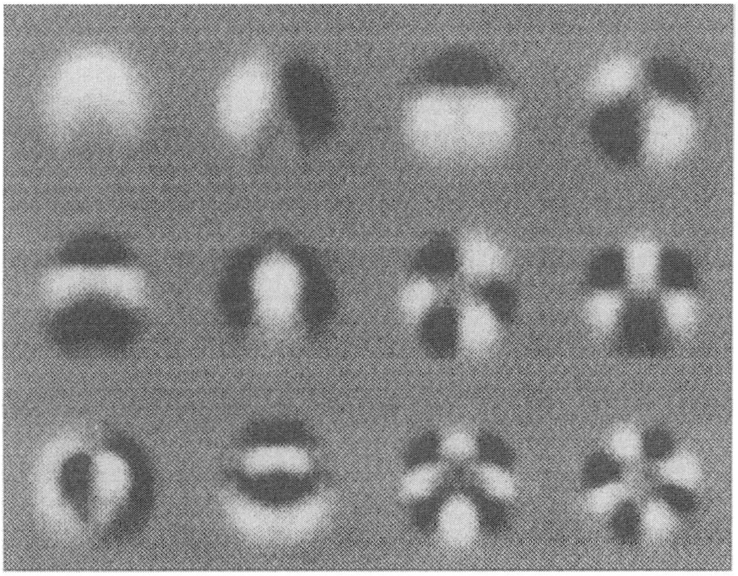

Figure 3: First twelve empirical eigenfunctions from the low momentum ratio ensemble. Light grey represents a postive fluctuation and dark grey a negative fluctuation, and the background grey represents zero

Results and discussion

The K-L procedure was applied to each of the five ensembles of concentration fluctuations to find the empirical eigenfunctions. In presenting the eigenfunctions we will generally arrange them in descending order of the magnitude of the eigenvalues, (18). Each eigenvalue is a measure of the degree to which the corresponding eigenfunction has been excited. If we divide the eigenvalues by the energy (10),

$$p_n = \frac{\lambda_n}{E}, \tag{19}$$

then p_n can be interpreted as being a probability. It tells us the probable degree to which the jet cross-section is in the state described by $\phi_n(\mathbf{x})$. Figure 3 contains the first twelve eigenfunctions for low momentum ratio ensemble. We have chosen this case to illustrate the nature of the eigenfunctions both because of the *lower degree of complexity* which results from its low momentum ratio, and the relatively high signal to noise ratio. As a result of the former, the seed concentration is still poorly mixed and the eigenfunctions then are naturally arranged in order of increasing *mixedness* by their corresponding eigenvalues. The first eigenfunction shown reflects the fact that although $< c >= 0$, at any instant the concentration need not have a zero spatial average since temporal pulsations are generally present. The second most probable eigenfunction allows for bilateral

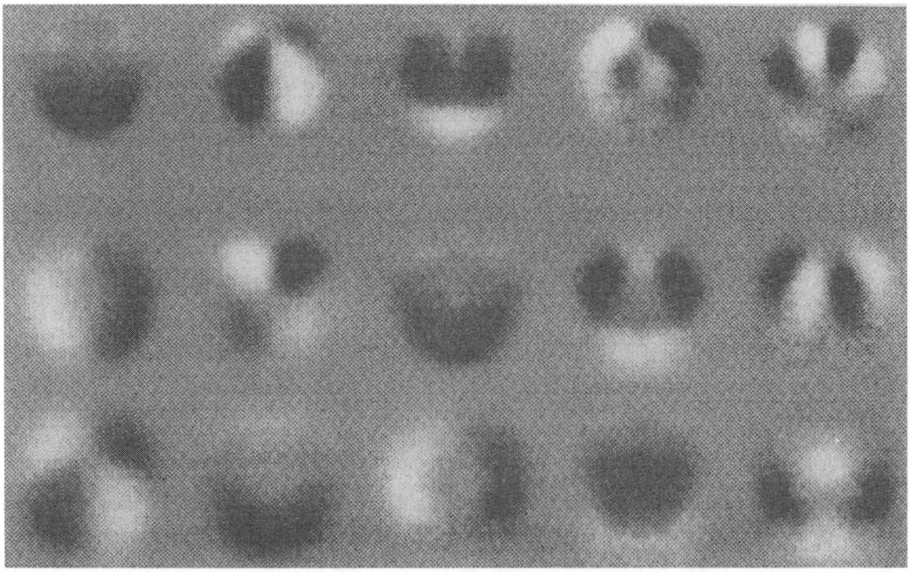

Figure 4: Comparison of the first five eigenfunctions at the 4 (top row), 6 and 8 (bottom row) diameter stations. The eigenfunctions in each row are ordered by the associated eigenvalues, with the most energetic at the left.

variation. It has two *cells*, the third has three cells and the fourth has four cells. The complexity and cell number is seen to increase as we proceed through the list of eigenfunctions. As we go down the list of eigenfunctions we see the possibility of increasing breakdown in the length scale by increasing complexity in the patterns. The notion of complexity is visual at this point and will be made more precise later in this section.

Figure 4 shows the first five eigenfunctions derived from data collected at the 4, 6 and 8 diameter stations. The first row corresponds to the section closest to the jet, where the mixing is poorest. There is, in fact, a close resemblance between the first row and Figure 3.

We adopt as a reasonable criterion for well mixedness, that more highly complex patterns become more probable. Such a trend is apparent from the rows of eigenpictures in Figure 4. It is apparent that mode crossing is occurring and that at 8 jet diameters downstream the dominant pattern is made up of four cells. Moving down the three rows, an increase in complexity is apparent, reflecting the progression in the mixing process as we move downstream.

Our use of the term *complexity* has been based, thus far, on visual appearance. We next make this notion more precise by defining several different measures of complexity. The first of these, termed the S-complexity, is based on the total length of the nodal lines of a function. In Figure 5 we show the nodal lines which correspond to the eigenfunctions of Figure 3. Simple functions, with few *cells*, have short nodal lines while convoluted functions have correspondingly longer nodal lines. In practice, the nodal line depends sensitively on the amount of experimental noise which contributes extraneous zero crossings

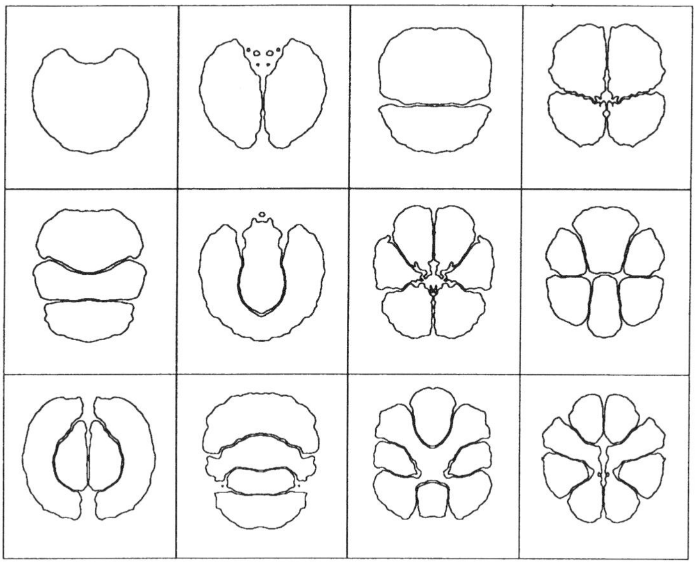

Figure 5: Nodal lines for eigenfunctions derived from the low momentum ratio data set (Figure 3) used to determine S-complexity.

and we prefer to use contours at heights $\pm z$:

$$S(\phi) = \frac{1}{2A^{1/2}} \left(\int_{\gamma(z)} dl + \int_{\gamma(-z)} dl \right), \tag{20}$$

where $\gamma(z)$ denotes the level curve of $\phi(x,y)$ at height z. The S-complexity is normalized by $A^{1/2}$, the square root of the area of the eigenfunction picture and is therefore dimensionless. The S-complexity fails to account for the characteristics of the eigenpictures that do not intersect the level curves $\gamma(\pm z)$. Two measures that may be regarded as the sum of the lengths of all the level curves are

$$D_1 = \frac{1}{A^{1/2}} \int \mid \nabla\phi \mid dA \tag{21}$$

and

$$D_2 = \int (\nabla\phi)^2 dA. \tag{22}$$

Both of these measures, which have been defined to be dimensional, are equivalent to (6) and (7).

The second of these is weighted by $\nabla\phi$ to emphasize level lines passing through regions where the function changes rapidly. These measures, too, are susceptible to experimental noise propagated through to the eigenpicture. However, an initial smoothing (by filtering in Fourier space) of the experimental snapshots yields broad agreement between D_1 and D_2.

The spectral entropy is given by

$$\sigma = - \int \epsilon(\mathbf{k}) \ln \epsilon(\mathbf{k}) d\mathbf{k}, \tag{23}$$

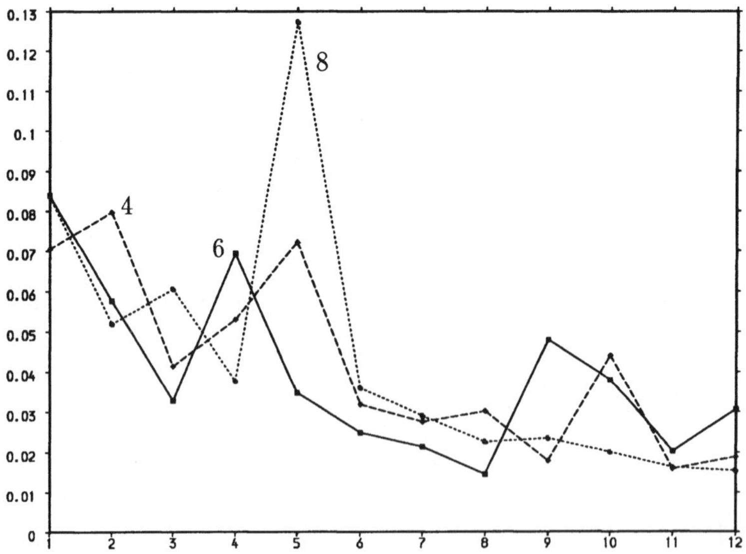

Figure 6: Eigenvalues at the 4, 6 and 8 jet diameter stations plotted in order of increasing D_2-complexity of the corresponding eigenfunction.

where $\epsilon(\mathbf{k})$, the energy at wavenumber $\mathbf{k}$, is the square magnitude of the Fourier transform of $\phi(\mathbf{x})$. It follows therefore that $\epsilon(\mathbf{k})$ is a probability, i.e.

$$\int \epsilon(\mathbf{k})d\mathbf{k} = 1. \tag{24}$$

The one-dimensional form of (23) has been used to characterize the complexity of certain patterns occurring in thermal convection.[10] If the eigenpicture is a single Fourier mode, then the spectral entropy is zero, whereas if the energy is distributed uniformly amongst all the modes, then $\sigma \to 1$. Note that the spectral entropy does not distinguish between high and low wavenumber modes, and for this reason is less useful than (21) or (22) for assessing mixedness.

There is broad agreement between the rankings of the eigenpictures according to the D_1, D_2 and σ complexities. Figure 6 shows the first few eigenvalues at the 4, 6 and 8 diameter stations plotted in order of increasing D_2-complexity of the corresponding eigenpicture. At the 4 diameter station the simplest, almost dc, eigenfunction contributes most to the flow. Increasingly complex eigenfunctions carry decreasing energies. At the 6 diameter station, there is no longer an almost dc eigenfunction and the mode carrying the most energy has two cells and odd symmetry about the centerline. Additionally, the second largest eigenvalue corresponds to the sixth least simple eigenpicture. This four-celled mode (also recognizable as the second eigenpicture at the 4 diameter station) carries most energy at the 8 diameter station. Here simpler eigenpictures contribute less to the flow's structure, indicating a more complex, better mixed flow.

Conclusions

This paper has focused on the problem of mixing and mixing assessment. The K-L procedure applied to data collected from a jet in crossflow provides optimal bases for the analysis of the flow. It is found that better mixed flows are characterized by the dominance of spatially complex eigenfunctions. Qualitative measures of the complexity are presented and it is anticipated that this technique will provide a sensitive *metric* of the mixedness of flows in a wide variety of applications.

Acknowledgements

The authors would like thank John Ringland, Andrew Webber and James Ferry of Brown University for their contributions to the K-L analysis. We are grateful to Al Vranos and David Liscinsky of UTRC for making their data available, and we would like to thank Mr. Harold Craig of Pratt & Whitney for his support and encouragement in this effort.

References

1. A. Vranos & D.S. Liscinsky, "Planar Imaging of Jet Mixing in Crossflow," *AIAA J.* 26 (1988), 1297–1298.
2. M.B. Long, B.T. Chu & R.K. Chang, "Instantaneous Two-Dimensional Gas Concentration Measuremnts by Light Scattering," *AIAA J.* 19 (1981), 1151–1157.
3. L. Sirovich & R.M. Everson, "Analysis and Management of Large Scientific Databases," *International Journal of Supercomputing Applications* (1991), (to appear).
4. E.N. Lorenz, "Empirical orthogonal functions and statistical weather prediction," Department of Metrology, MIT, 1, Statistical Forecasting Project, Cambridge, 1956.
5. J.L. Lumley, "The structure of inhomogeneous turbulent flows," in *Atmospheric Turbulence and Radio Wave Propagation*, A.M. Yaglom & V.I. Tatarski, eds., Nauka, Moscow, 1967, 166–178.
6. L. Sirovich, "Turbulence and the dynamics of coherent structures," *Quarterly of Applied Mathematics* XLV (1987), 561-590.
7. L. Sirovich, "Empirical Eigenfunctions and Low Dimensional Systems," in *New Perspectives in Turbulence*, L. Sirovich, ed., Springer-Verlag, New York–Heidelberg–Berlin, 1991.
8. L. Sirovich & C.H. Sirovich, "Low Dimensional Description of Complicated Phenomena," *Contemporary Mathematics* 99 (1989), 277–305.
9. W.R. Hawthorne, "Secondary Circulation in Fluid Flow," *Proc. Roy. Soc. London Ser. A* 206 (1951), 347-387.
10. C. Ciliberto, "Characterizing Space-Time Chaos in an Experiment of Thermal Convection," in *Measures of Complexity and Chaos*, N.B. Abraham, A.M. Albano, A. Passamante & P.E. Rapp, eds., Plenum Press, New York, NY, 1989, 445–456.

WAVELETS AND THE ANALYSIS OF ASTRONOMICAL OBJECTS

Albert BIJAOUI - URA CNRS 1361 A.Fresnel
Observatoire de la Côte d'Azur B.P. 139 F-06003 NICE CEDEX

I.The hierarchical structure of the Universe components.

1.The observations.

The analysis of the sky shows many kinds of non stellar components: planets and other solar system objects, planetary nebulae, neutral, ionised and molecular clouds, star clusters, galaxies, clusters of galaxies, large scale structural features, etc.. Each of these objects are irregular patterns, which cannot be fully described with analytical models.

The irregularity is generally associated to a hierarchical structure: a feature is often included in a larger one, which can also be a part of a larger structure, etc.. The analysis of quite all the listed components shows the existence of a such hierarchy. For example, the molecular clouds, in which generally stars born, are very complex structures, often described as fractals. Many galaxies show sets of non stellar components which seem hierarchically distributed.

Charlier [1] introduced the hierarchical structure of the Universe in years 20 taking into account the apparent distribution of the brighest galaxies. Neyman's statistical model [2] was based on a two-level hierarchy. This model was complex and was not developed to a larger number of levels. The philosophy about the matter distribution of the Universe evolved toward the use of the correlation function [3]. It is only recently that the existence of superclusters, filaments, sheets and voids was seriously considered [4] giving a new interest to hierarchical structures.

2.The hierarchical structures.

The first new approach was due to B.Mandelbrot [5] with the introduction of the fractal structures. Beyond the mathematical formalism, this work gives a new insight for natural structures, rejecting the idea of regular patterns, described with analytical models. For classical fractals the hierarchical structure is statistically uniform. At the opposite, multifractals can show an innovation at each scale, such as it can be generally observed. It is the reason why multifractal models fit quite well to the distribution of the galaxies [6].

For signal processing, the knowledge of the generating stochastical process is essential. For example, Wiener filter results from a Gaussian process, while Kalman one is derived from a Markovian process. Hierarchical structures are associated to other kinds of process, containing fractal and multifractal models. The identification of the underlying process can provide tools in order to generate similar structures with the minimum of parameters.

The existence of hierarchical structures are connected to the physical processes which are involved. Linear physical phenomena generate sets of independant structures, without coupling. Hierarchical structures are necessarly connected to non linear phenomena such those involved to describe turbulent flows.

3.The main questions.

Hierarchical structures can be analysed component by component. But, a such analysis is not generally sufficient to furnish a complete description. We consider that the irregularity are not a noisy-like feature but contains a part of the information on the basic stochastical process. The analysis of hierarchical structures need to give an adequate solution to some questions:

- In what conditions we can say that we detect some element of the hierarchy?
- How to describe the different components?
- What global statistical indicators we can furnish for the modeling?
- How to classify the structures?
- How to generate images which look like the observed objects?

We expected that the Wavelet Transform was the tool allowing us to build an analysis taking into account all the constraints.

II.The wavelet transform.

1.The continuous wavelet transform.

Morlet-Grossmann [7] definition for a $1D$ signal $f(x) \in L^2(R)$ is:

$$C(a,b) = \frac{1}{\sqrt{a}} \int_{-\infty}^{+\infty} f(x) g^*(\frac{x-b}{a}) dx$$

z^* designs the conjugate of z. $g^*(x)$ is the analysing wavelet. a (> 0) is the scale parameter. b is the position parameter.

It is a linear transformation which is essential for numerical algorithms, statistical computation and understanding of the results. The wavelet transform is invariant under translation. The analysis does not depend on the origin of the coordinate frame. It is the general property of convolution operators. It is also invariant under dilatation. This is the property which gives its originality to the wavelet transform. We get a mathematical microscope the properties of which do not change with the magnification.

2.The main restoration formula and the physical interpretation.

Consider now a function $C(a,b)$ which is the wavelet transform of a given function $f(x)$. It was shown [7] that $f(x)$ can be restored with the formula:

$$f(x) = \frac{1}{C_g} \int_0^{+\infty} \int_{-\infty}^{+\infty} \frac{1}{\sqrt{a}} C(a,b) g(\frac{x-b}{a}) \frac{da.db}{a^2}$$

where:

$$C_g = \int_0^{+\infty} \frac{|\hat{g}(\nu)|^2}{\nu} d\nu$$

The reconstruction is only available if C_g is defined (admissibility condition). This is generally true if $\hat{g}(0) = 0$, i.e. the mean of the wavelet function is 0.

This formula gives another insight on the transformation. The function is the sum of wavelets which are obtained by translation and dilatation of a given pattern. The amplitude of each wavelet is the correlation (scalar product) of the function with the wavelet. The value of the function at a given point is the sum of the weighted wavelet values at this point.

Other $L^2(R)$ representations provides such an insight, but the scalar products are done with different patterns leading to some difficulties to interprete the coefficients in physical terms. With the wavelet transform, we only work with one pattern.

The reconstruction formulae give a function $f(x)$ *even if the coefficients are not provided by a wavelet transform.* This situation is common in image processing. For example, we may threshold the wavelet transform in order to keep only the significant coefficients. The resulting function $C_0(a,b)$ does not belong to the subset S of $L^2(R^{*+} \times R)$ generate by all the wavelet transforms.

If we inverse $C_0(a,b)$ we get a function $f_0(x)$ which gives a wavelet transform $C_1(a,b)$. The operator linking $C_0(a,b)$ to $C_1(a,b)$ is called *the reproducing kernel.* This operator plays an important part in the case of restoration problems under constraints (thresholding, positivity, partiallity, etc.).

3.The two-dimensional continuous wavelet transform.

Many $2D$ extensions of the continuous wavelet transform are possible:
- From identical dilatations on coordinates, using an isotropic wavelet function;
- From independant dilatations;
- From identical dilatations, with rotation of the wavelet pattern in Fourier space, using an anisotropic wavelet function [8];
- From independant dilatations and rotation.

The dimension of the resulting transform depends of the choice: 3 for the first one, 4 for the two following ones, and 5 for the last one. That is one of the reasons which led us to choice to work with an isotropic wavelet. Another reason lies in the physical data interpretation. An isotropic wavelet transform provides an isotropic vision from easily understood values.

4.The direct discrete wavelet transform.

In classical image processing the discretisation is guided by the well-known Shannon's sampling theorem. If Fourier transform $\hat{f}(\nu)$ of a given function $f(x)$ is different of 0 only in the frequency band $|\nu \leq \nu_c|$, the function can be exactly computed from a set of samples $f(nh)$, where $n \in Z$. The sampling set h must be smaller than $\frac{1}{2\nu_c}$. ν_c is generally called the cut-off frequency.

The use of the wavelet transform with a computer can be foreseen through this theorem. We can process signal with a cut-off frequency. We have to work in Fourier space, computing the transform scale by scale. The number of elements for a scale can be reduced, if the frequency bandwith is also reduced. This is available only for wavelet having also a cut-off frequency. The Littlewood-Paley decomposition [9] provides a very nice illustration of the reduction of elements scale by scale. It is based on an iterative dichotomy of the frequency band.

Littlewood-Paley's decomposition provides a discrete wavelet transform leading to a perfect restoration with a pyramidal set of data. We can generalise the discretisation with a step proportional to the scale $a.b_0$, and with a logarithmic discretisation for the scale $a = a_0^i$ (figure 1). The dyadic wavelet transform corresponds to $a_0 = 2$.

The sampling theorem is easily extended to two dimensional functions. Generally the sample is done on a square grid.

5.Wavelets and the Multiresolution Analysis.

S.Mallat [10] introduced the concept of multiresolution analysis in order to bring a description of a signal or an image from a pyramidal set of details. This work gave a new insight on some previous coding techniques. This approach is an extension of Littlewood-Paley decomposition to a large class of wavelets.

The multiresolution analysis is based on an increasing sequence of closed linear subspace V_j, $j \in Z$ of $L^2(R)$. A function $f(x)$ is projected at each step j on the subset V_j. This projection is defined by the scalar product $c_j(k)$ of $f(x)$ with the function $\phi(x)$ which is dilated and translated:

$$c_j(k) = < f(x)|2^j\phi(2^jx - k) >$$

$\phi(x)$ is named the scaling function of the analysis. Its main property lies in the following relation:

$$\frac{1}{2}\phi(\frac{x}{2}) = \sum_n h(n)\phi(x-n)$$

Or, in Fourier space:

$$\hat{\phi}(2\nu) = \hat{h}(\nu).\hat{\phi}(\nu)$$

where $\hat{h}(\nu)$ is a 1-periodic function. This relation permits to compute the set $\{c_j(k)\}$ from $\{c_{j+1}(k)\}$:

$$c_j(k) = \sum_n h(n-2k)c_{j+1}(n)$$

At each step, the number of scalar products is divided by 2. An information is lost, and step by step the signal is smoothed. The remaining information can be restored using the complementary subspace W_j of V_j in V_{j+1}. This subspace can be generated by a suitable wavelet function $\psi(x)$ with translation and dilation. In Fourier space we have:

$$\hat{\psi}(2\nu) = \hat{g}(\nu)\hat{\phi}(\nu)$$

where $\hat{g}(\nu)$ is another 1-periodic function.

We compute the scalar products $< f(x)|2^j\psi(2^jx - k) >$ with:

$$w_j(k) = \sum_n g(n-2k)c_{j+1}(n)$$

With this analysis, we have built the first part of a filter bank [10]. The restoration is performed with (figure 1):

$$c_{j+1}(k) = 2\sum_l [c_j(l)\tilde{h}(k-2l) + w_j(l)\tilde{g}(k-2l)$$

The set of filters must satisfy the following relations:

$$\hat{h}(\nu + \frac{1}{2})\hat{\tilde{h}}(\nu) + \hat{g}(\nu + \frac{1}{2})\hat{\tilde{g}}(\nu) = 0$$

$$\hat{h}(\nu)\hat{\tilde{h}}(\nu) + \hat{g}(\nu)\hat{\tilde{g}}(\nu) = 1$$

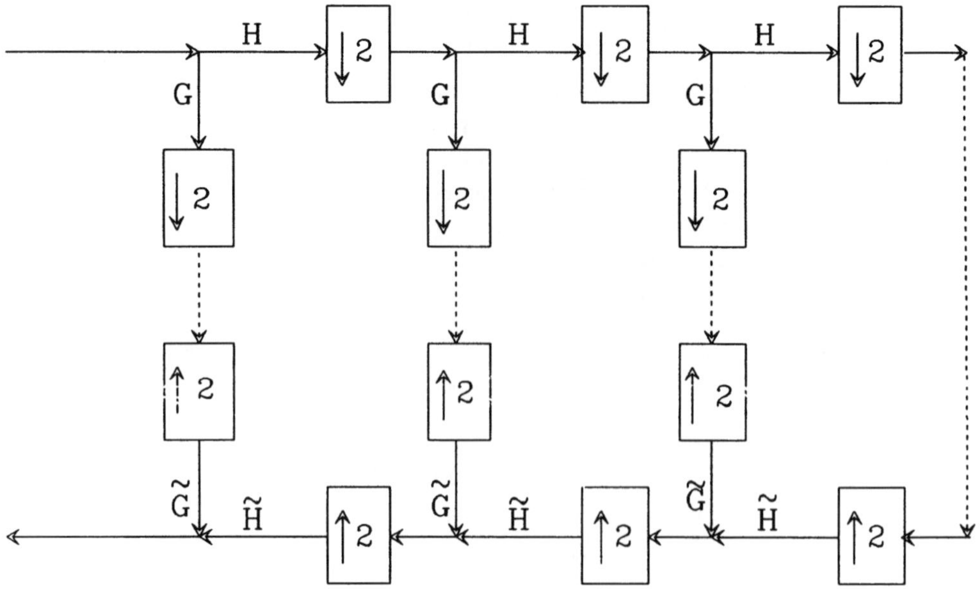

Figure 1: Decomposition and Restoration with a filter bank.

In the decomposition, the signal is successively smoothed with the two filters H (low frequencies) and G (high frequencies). Each resulting image is decimated by suppression of one sample on two. The high frequency signal is left, and we iterate with the low frequency signal. In the reconstruction, we restore the sampling by inserting a 0 between two samples, then we smoothed with the conjugate filters $\tilde{H}$ and $\tilde{G}$ and we add the resulting signals. We iterate up to smallest scales.

The multiresolution analysis furnishes a remarkable framework to code a signal, and more generally an image, with a pyramidal set of values.

III.Some astronomical Applications.

1.General applications.

The human vision is based on a pyramidal scheme of the information from the highest scales, giving the general feeling, to the smallest ones connected to the details [11]. The wavelet transform unfolds the information on the scale axis. The comparison between two images corresponds to the study of the intersection of two pyramids. Some technical applications of the use of the wavelet transform to astronomical imagery derived from this scheme:

• *Geometrical matching between two images:* We achieve a matching procedure [12] using the wavelet transform. The main idea lies in the pyramidal vision provided by the transform. We start to identify large scale structures, with a poor accuracy. Then the scale is decreased, and the matching is done on smaller structures, increasing the accuracy.

•*Optimal Image Addition:* The wavelet transform provides a very nice framework to obtain an optimal addition taking into account the variation of the resolution [12]. Each image is splitted into the wavelet space. For each scale, we weight wavelet images taking into account the amount of information given by each one.
•*Detection of different stuctural elements:* We have built a new procedure in order to detect and measure the objects on an astronomical image [13]. The detection of the statistically significant maxima in the Wavelet space is done.
•*Smoothing and Image Restoration:* The detection technique on the pyramid leads us to a new way in image smoothing and restoration based on the thresholding of the wavelet images [14].

2.Application to the study of the structure of the Universe.

The use of the wavelet transform is essentially guided by the study of hierarchically structured objects. The fractal or multifractal structure of natural objects was analysed by many authors [15]. As it was shown [16,17], the wavelet transform is a very nice tool for these studies. We start this analysis on the structure of the universe from the galaxy counts [18]. It is clear for us that this approach may be applied to the structure of a large set of astronomical objects like the HII emissive regions, the molecular clouds, the irregular galaxies, the planetary nebulae, etc..

The counts of galaxies are the first step allowing us to determine the distribution of matter in the Universe. We consider a galaxy count as an observation of a Poissonian process. We admit that the galaxies are distributed according an unknown density law, but their positions are independant. Consequently the catalogue of positions (x_i, y_i) is transform into an image $I(x, y)$ built as a sum of Dirac peaks:

$$I(x,y) = \sum_i \delta(x - x_i, y - y_i)$$

The computation of the wavelet transform of the image $I(x, y)$ becomes very easy:

$$C(a, b_x, b_y) = \frac{1}{a} \sum_i g^*(\frac{x_i - b_x}{a}, \frac{y_i - b_y}{a})$$

The wavelet function was choosen isotropic, and well localised in the direct and the frequency spaces. The *mexican hat* fits well to these conditions:

$$g(r) = (2 - r^2)e^{-\frac{r^2}{2}}$$

The analysis of the wavelet transform is done scale after scale. For each one we attribute to a structural element a probability to be real according to a theorical distribution law. Scanning along the scale axis, we are able to extract the hierarchical structure. We have processed, often in collaboration with other astronomers, many catalogues of galaxies: a owner count in the Coma supercluster [18], a catalogue of the Coma cluster [19], a part of the CfA slide in the Universe [20], the *Principal Catalogue of Galaxies* [21]. The hierarchy is always present. For example, on figure 2 we have plotted the used catalogue of Coma supercluster. We have superimposed the analysis

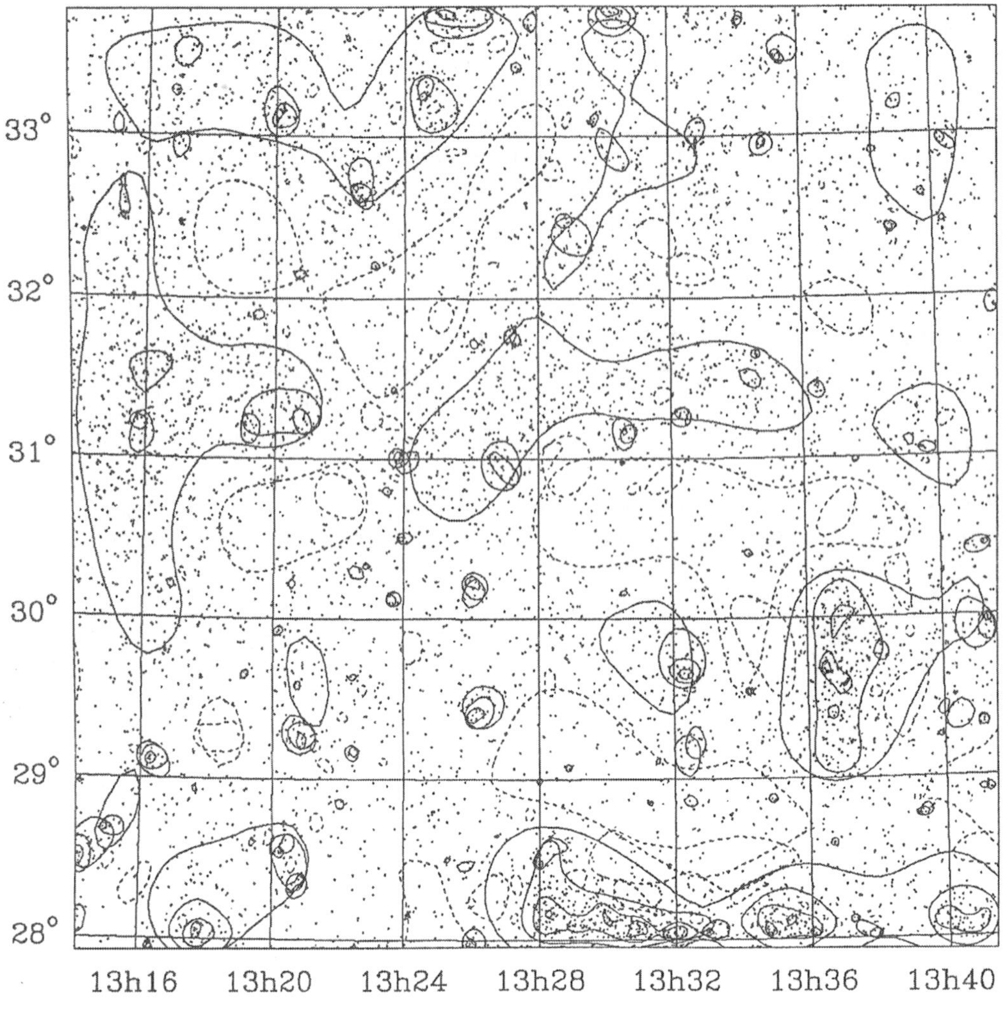

Figure 2: Groups and Voids detection in a field of the Coma supercluster.

We have plotted the catalogue resulting from an automated count of galaxies. An image built from this catalogue was analysed with the wavelet transform. The positions of the groups and the voids were extracted after a statistical analysis of the transform. A simple modeling of the results was done with an anisotropic Gaussian model. We have plotted the contour lines corresponding to the probability 99.5% that the wavelet coefficient values cannot be higher than the measured ones for a random distribution. Groups correspond to solid lines, while voids are plotted in dashed lines. The limits of this analysis can be forseen.

obtains from the detection of groups and voids with the wavelet transform at each scale. 4 scales was used for the analysis. A faint hierarchical structure can be detected.

References:

[1] C.V.L.Charlier: *Ark. Math. Astron. Fys.* **16** p.16 1922

[2] J.Neyman and E.L.Scott: *Ap. J.* **116** p.144-163 1952

[3] D.N.Limber: *Ap. J.* **110** p.134-144 1953

[4] M.J.Geller and J.P.Huchra: *Science* **246** p.897-903 1989.

[5] B.Mandelbrot: *Compt. Rend. Acad. Scien. A* **280** p.1551 1975.

[6] V.J.Martinez and B.J.T.Jones: *M.N.R.A.S.* **242** p.517-521.

[7] A.Grossmann and J.Morlet: *Mathematics and Physics, Lecture on recent results* World Scientific Publ. 1985.

[8] J.P.Antoine, P.Carrette, R.Murenzi and B.Piette: *Image Analysis with two-dimensional continuous wavelet transform* preprint UCL Louvain-la-Neuve.

[9] J.Littlewood and R.Paley: *Jour. London Math. Soc.* **6** p.230 1931.

[10] S.Mallat: *I.E.E.E. on P.A.M.I.* **11** p.674 1989.

[11] D.Marr: *Vision* Freeman New-York 1982.

[12] A.Bijaoui and M.Giudicelli: *Experimental Astronomy* to appear 1991.

[13] A.Bijaoui, E.Slezak and G.Mars: *12th GRETSI Conf.* p.785 1989.

[14] J.L.Starck and A.Bijaoui: *3rd ESO Data Analysis Workshop* 1991.

[15] B.Mandelbrot: *The fractal geometry of Nature* Freeman New-York 1983.

[16] M.Holdschneider: *J. Stat. Phys.* **50** p.953 1988.

[17] A.Arnéodo, G.Grasseau and M.Holdschneider: *Wavelets* p.182 Springer-Verlag 1989.

[18] E.Slezak, A.Bijaoui and G.Mars: *Astron. Astrophys.* **227** p.301 1990.

[19] E.Escalera, A.Mazure, E.Slezak and A.Bijaoui: *The matter distribution in the Universe* 2nd DAEC workshop ed. G.Mamon Observatoire de Paris 1991.

[20] E.Slezak, V.de Lapparent et A.Bijaoui: in preparation 1991.

[21] E.Slezak, G.Mars et A.Bijaoui: in preparation 1991.

VILLEFRANCHE 14-18 JANUARY 1991

LIST OF PARTICIPANTS

Postal Address
Phone, Fax
e-Mail.

A. ANTONI	Equipe Turbulence Plasma, IMT 13451 MARSEILLE Cédex 13, France. 91 05 45 31.
G. BAKER	Dept. of Mathematics, Ohio State University, 231 W. 18th. Ave., COLUMBUS, OH 43210, USA. 614-292-4010. GRB @SHAPE.MPS.OHIO.STATE.EDU
C. BENDER	Physics Dept., Campus Box 1105, Washington University, ST-LOUIS, MO 63130, USA. 314-889-6216. 314-889-6219. CMB@ARTHUR.WUSTL.EDU
A. BIJAOUI	Observatoire de Nice, BP 139, 06003 NICE Cédex, France. 92 00 30 27. 92 00 30 33. BIJAOUI@FRONI51.BITNET
D. CARATI	Plasmas, Phys. Statistique, CP321, ULB, Campus Plaine, Bd. du Triomphe,1050 BRUXELLES, Belgium. 2-650-58-13. 2-650-58-16. R09602@BBRBFU01.BITNET
H. CHATE	DPhG/PSRM, CEN-Saclay, 91191 GIF-SUR-YVETTE Cédex, France. (1) 69 08 73 46. (1) 69 08 87 86. CHATE @ POSEIDON.CEA.FR
P. COMTE	Equipe "Modélisation de la Turbulence", IMG, BP 53X, 38041 GRENOBLE, France. 76 82 51 21. 76 82 50 01. COMTE@FRGREN81
W. CRAIG	Math. Dept., Brown University, PROVIDENCE, RI 02912, USA. 401-863-1124. CRAIGW @ BROWNVM.BITNET

B. DUBRULLE — Observatoire Midi-Pyrénées,
14, av. E. Belin, 31400 TOULOUSE, France.
61 33 29 44.
DUBRULLE@FROMP51.BITNET

D. ESCANDE — Equipe Turbulence Plasma, IMT,
13451 MARSEILLE Cédex 13, France.
91 05 45 85. 91 05 43 43.
ESCANDE@FRMRS11.BITNET

R. EVERSON — Center for Fluid Mechanics, Brown University, Box 1966,
PROVIDENCE, RI 02915, USA.
401-863-2507.
RME@CFM.BROWN.EDU

J.-D. FOURNIER — Observatoire de Nice,
BP139, 06003 NICE Cédex, France.
92 00 30 19. 92 00 30 33.
FOURNIER@FRONI51.BITNET

H.M. FRIED — Physics Dept., Brown University,
PROVIDENCE, RI 02912, USA.
401-863-1467. 401-863-2024.
FRIED@BROWNVM.BITNET

U. FRISCH — Observatoire de Nice,
BP139, 06003 NICE Cédex, France.
92 00 30 35. 92 00 30 33.
URIEL@FRONI51.BITNET

B. GALANTI — Observatoire de Nice,
BP139, 06003 NICE Cédex, France.
92 00 30 44. 92 00 30 33.
GALANTI@FRONI51.BITNET

A. GILBERT — DAMTP, Silver St.,
CAMBRIDGE, CB3 9EW, UK.
223-337868.
ADG11@PHX.CAM.AC.UK

I. GOLDHIRSCH — Dept. Fluid Mechanics, Fac. Engineering, Tel Aviv Univ.,
Ramat-Aviv, TEL-AVIV 69 978, Israël.
972-3-545 09 29.
D73@TAUNOS.BITNET

A. GROSSMANN — CPT, CNRS-Luminy, Case 907,
13286 MARSEILLE Cédex 9, France.
91 26 95 00.

A. HEFF — Phys. Dept., Weizman Inst.,
76100 REHOVOT, Israël.
972-8-34 36 97.
FNHEFF@WEIZMANN

S. HOWISON	Oxford University, Mathematical Institute, 24-29 St. Giles, OXFORD OX1 3LB, UK. 44-865-27 05 05. HOWISON@VAX.OX.AC.UK
G. JONA-LASINIO	Dipartimento di Fisica, Università "La Sapienza" 00185 ROMA, Italy. 6-49914313. JONA@ROMA1.INFN.IT
M. LE BELLAC	INLN, Université de Nice-Sophia-Antipolis, Parc Valrose, 06034 NICE Cédex, France. 93 52 98 25. LEBELLAC@FRNICE51.BITNET
O. LEGRAND	Mat. Condensée, Université Nice-Sophia-Antipolis, Parc Valrose, 06034 NICE Cédex, France.
B. LEGRAS	Laboratoire de Météorologie Dynamique, ENS, 24, rue Lhomond, 75231 PARIS Cédex 05, France. (1)45 35 99 57. LEGRAS@ENS.ENS.FR
P. LOCHAK	Dept. de Mathématiques, ENS, 45, rue d'Ulm, 75005 PARIS Cédex 05, France. (1)43 29 12 25 ext. 3797. LOCHAK@FRULM63.BITNET
B. MALOMED	Starting Oct. 91: School of Math. Sciences, Tel-Aviv Univ., Ramat-Aviv, TEL-AVIV 69 978, Israël.
M. MENEGUZZI	CERFACS, 42, av Coriolis, 31057 TOULOUSE, France. 61 07 96 90. 61 07 96 13. MNGZ@ORION.CERFACS.FR
A. NEWELL	Dept. of Mathematics, University of Arizona, TUCSON, AZ 85721, USA. ANEWELL@MATH.ARIZONA.EDU
Th. PASSOT	Observatoire de Nice, BP139, 06003 NICE Cédex, France. 92 00 30 21. 92 00 30 33. PASSOT@FRONI51.BITNET
R. PESCHANSKI	CERN, Theory Division, 1211 GENEVE 23, Switzerland. 22-767 21 39. PESCH@CERNVM.BITNET
Y. POMEAU	Laboratoire de Physique Statistique, ENS, 24, rue Lhomond, 75231 PARIS Cédex 05, France.

A. POUQUET	Observatoire de Nice, BP 139, 06003 NICE Cédex, France. 92 00 30 57. 92 00 30 33. POUQUET@FRONI51.BITNET
V. ROM-KEDAR	Dept. of Mathematics, University of Chicago, CHICAGO, IL 60637, USA. 312-702-3064. 312-702-9787. VERED@ZAPHOD.UCHICAGO.EDU
A. RUZMAIKIN	IZMIRAN , IZMIRAN Troitzk, Moscow Region, 142092. USSR. 422-69-25. or 334-09-21. Télex 132 092 USSR
J. SOMMERIA	Lab. de Physique, ENS Lyon, 46, al. d'Italie, 69364 LYON Cédex 07, France. 72 72 84 39. SOMMERIA@FRENSL61.BITNET
D. SORNETTE	Mat. Condensée, Université Nice-Sophia-Antipolis, Parc Valrose, 06034 NICE Cédex, France. 93 52 33 73. SORNETTE@FRNICE51.BITNET
P.-L. SULEM	Observatoire de Nice BP139, 06003 NICE Cédex, France. 92 00 30 40. 92 00 30 33. SULEM@FRONI51.BITNET
S. VAIENTI	CPT, CNRS-Luminy, Case 907. 13 286 MARSEILLE Cédex 09, France. 91 26 95 44. VAIENTI@FRCPTM51.BITNET
L. VALDETTARO	CERFACS 42, av Coriolis, 31057 TOULOUSE, France. 61 07 96 29. VALDE@CERFACS.FR
A. VERGA	Equipe Turbulence Plasma, IMT, 13451 MARSEILLE Cédex13, France. 91 05 45 90. VERGA@FRMRS11.BITNET
M. VERGASSOLA	Università di Roma "La Sapienza", P. A.Moro, 2, 00187 ROMA, Italy.
&	Observatoire de Nice, BP 139, 06003 NICE Cédex, France. 92 00 31 19. 92 00 30 33. VERGASSOLA@FRONI51.BITNET

S. ZALESKI	Laboratoire de Physique Statistique, ENS, 24, rue Lhomond, 75231 PARIS Cédex 05, France. (1)45 87 02 84. ZALESKI@ASTERIX.ENS.FR
M. ZEKRI	Equipe Turbulence Plasma, IMT, 13457 MARSEILLE Cédex 13, France. 91 05 45 31.
V. ZHELIGOVSKY	Institute of Earthquake Prediction, MOSCOW WARSAWKOE av 79 k.2., USSR.
&	Observatoire de Nice, BP139, 06003 NICE Cédex, France. 92 00 31 32. 92 00 30 33. VLAD@FRONI51.BITNET
M. ZOLVER	Observatoire de Nice, BP 139, 06003 NICE Cédex, France. 92 00 31 17. 92 00 30 33. ZOLVER@FRONI51.BITNET

Lecture Notes in Physics

For information about Vols. 1–365
please contact your bookseller or Springer-Verlag

Vol. 366: M.-J. Goupil, J.-P Zahn (Eds.), Rotation and Mixing in Stellar Interiors. Proceedings, 1989. XIII, 183 pages. 1990.

Vol. 367: Y Osaki, H. Shibahashi (Eds.), Progress of Seismology of the Sun and Stars. Proceedings, 1989. XIII, 467 pages. 1990.

Vol. 368: L. Garrido (Ed.), Statistical Mechanics of Neural Networks. Proceedings. VI, 477 pages. 1990.

Vol. 369: A. Cassatella, R. Viotti (Eds.), Physics of Classical Novae. Proceedings, 1989. XII, 462 pages. 1990.

Vol. 370: H.-D. Doebner, J.-D. Hennig (Eds.), Quantum Groups. Proceedings, 1989. X, 434 pages. 1990.

Vol. 371: K.W. Morton (Ed.), Twelfth International Conference on Numerical Methods in Fluid Dynamics. Proceedings, 1990. XIV, 562 pages. 1990.

Vol. 372: F. Dobran, Theory of Structured Multiphase Mixtures. IX, 223 pages. 1991.

Vol. 373: C. B. de Loore (Ed.), Late Stages of Stellar Evolution. Computational Methods in Astrophysical Hydrodynamics. Proceedings, 1989. VIII, 390 pages. 1991.

Vol. 374: L. Ting, R. Klein, Viscous Vortical Flows. V, 222 pages. 1991.

Vol. 375: C. Bartocci, U. Bruzzo, R. Cianci (Eds.), Differential Geometric Methods in Theoretical Physics. Proceedings, 1990. XIX, 401 pages. 1991.

Vol. 376: D. Berényi, G. Hock (Eds.), High-Energy Ion-Atom Collisions. Proceedings, 1990. IX, 364 pages. 1991.

Vol. 377: W. J. Duschl, S.J. Wagner, M. Camenzind (Eds.), Variability of Active Galaxies. Proceedings, 1990. XII, 312 pages. 1991.

Vol. 378: C. Bendjaballah, 0. Hirota, S. Reynaud (Eds.), Quantum Aspects of Optical Communications. Proceedings 1990. VII, 389 pages. 1991.

Vol. 379: J. D. Hennig, W. Lücke, J. Tolar (Eds.), Differential Geometry, Group Representations, and Quantization. XI, 280 pages. 1991.

Vol. 380: I. Tuominen, D. Moss, G. Rüdiger (Eds.), The Sun and Cool Stars: activity, magnetism, dynamos. Proceedings, 1990. X, 530 pages. 1991.

Vol. 381: J. Casas-Vazquez, D. Jou (Eds.), Rheological Modelling: Thermodynamical and Statistical Approaches. Proceedings, 1990. VII, 378 pages. 1991.

Vol. 382: V.V. Dodonov, V. I. Man'ko (Eds.), Group Theoretical Methods in Physics. Proceedings, 1990. XVII, 601 pages. 1991.

Vol. 384: M. D. Smooke (Ed.), Reduced Kinetic Mechanisms and Asymptotic Approximations for Methane-Air Flames. V, 245 pages. 1991.

Vol. 385: A. Treves, G. C. Perola, L. Stella (Eds.), Iron Line Diagnostics in X-Ray Sources. Proceedings, Como, Italy 1990. IX, 312 pages. 1991.

Vol. 386: G. Pétré, A. Sanfeld (Eds.), Capillarity Today. Proceedings, Belgium 1990. XI, 384 pages. 1991.

Vol. 387: Y. Uchida, R. C. Canfield, T. Watanabe, E. Hiei (Eds.), Flare Physics in Solar Activity Maximum 22. Proceedings, 1990. X, 360 pages. 1991.

Vol. 388: D. Gough, J. Toomre (Eds.), Challenges to Theories of the Structure of Moderate-Mass Stars. Proceedings, 1990. VII, 414 pages. 1991.

Vol. 389: J. C. Miller, R. F. Haglund (Eds.), Laser Ablation-Mechanisms and Applications. Proceedings. IX, 362 pages, 1991.

Vol. 390: J. Heidmann, M. J. Klein (Eds.), Bioastronomy - The Search for Extraterrestrial Life. Proceedings, 1990. XVII, 413 pages. 1991.

Vol. 391: A. Zdziarski, M. Sikora (Eds.), Ralativistic Hadrons in Cosmic Compact Objects. Proceedings, 1990. XII, 182 pages. 1991.

Vol. 392: J.-D. Fournier, P.-L. Sulem (Eds.), Large-Scale Structures in Nonlinear Physics. Proceedings. VIII, 353 pages. 1991.

New Series m: Monographs

Vol. m 1: H. Hora, Plasmas at High Temperature and Density. VIII, 442 pages. 1991.

Vol. m 2: P. Busch, P. J. Lahti, P. Mittelstaedt, The Quantum Theory of Measurement. XIII, 165 pages. 1991.

Vol. m 3: A. Heck, J. M. Perdang (Eds.), Applying Fractals in Astronomy. IX, 210 pages. 1991.

Vol. m 4: R. K. Zeytounian, Mécanique des fluides fondamentale. XV, 615 pages, 1991.

Vol. m 5: R. K. Zeytounian, Meteorological Fluid Dynamics. XI, 346 pages. 1991.

A. Heck, J. M. Perdang (Eds.)

Applying Fractals in Astronomy

1991. IX, 210 pp. (Lecture Notes in Physics.
New Series m: Monographs. m3) Hardcover DM 46,–
ISBN 3-540-54353-8

W. D. D'haeseleer, W. N. G. Hitchon, J. D. Callen, J. L. Shohet

Flux Coordinates and Magnetic Field Structure

A Guide to a Fundamental Tool of Plasma Theory

1991. XII, 241 pp. 40 figs. (Springer Series in Computational Physics) Hardcover DM 158,–
ISBN 3-540-52419-3

A. V. Gaponov-Grekhov, M. I. Rabinovich (Eds.)

Nonlinear Physics: Oscillations – Chaos – Structures

1991. Approx. 120 pp.
180 figs. Hardcover DM 48,–
ISBN 3-540-51988-2